STUDENT'S SOLUTIONS MANUAL

SALVATORE SCIANDRA
Niagara County Community College

FINITE MATHEMATICS & ITS APPLICATIONS

TWELFTH EDITION

Larry J. Goldstein
Goldstein Educational Technologies

David I. Schneider
University of Maryland

Martha J. Siegel
Towson University

Steven M. Hair
Pennsylvania State University

The author and publisher of this book have used their best efforts in preparing this book. These efforts include the development, research, and testing of the theories and programs to determine their effectiveness. The author and publisher make no warranty of any kind, expressed or implied, with regard to these programs or the documentation contained in this book. The author and publisher shall not be liable in any event for incidental or consequential damages in connection with, or arising out of, the furnishing, performance, or use of these programs.

Reproduced by Pearson from electronic files supplied by the author.

Copyright © 2018, 2014, 2010 Pearson Education, Inc.
Publishing as Pearson, 330 Hudson Street, NY NY 10013

All rights reserved. No part of this publication may be reproduced, stored in a retrieval system, or transmitted, in any form or by any means, electronic, mechanical, photocopying, recording, or otherwise, without the prior written permission of the publisher. Printed in the United States of America.

1 17

ISBN-13: 978-0-13-446344-5
ISBN-10: 0-13-446344-7

Contents

Chapter 1

Exercises 1.1

1. Right 2, up 3

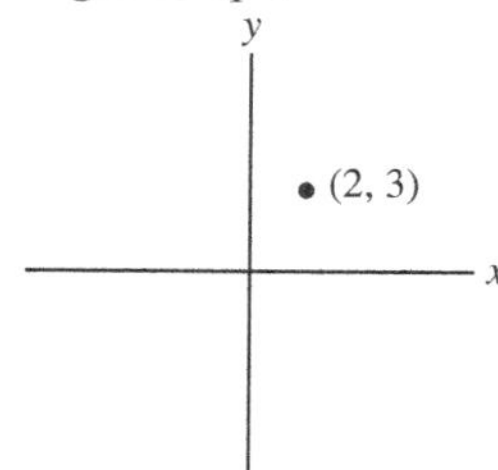

3. Down 2

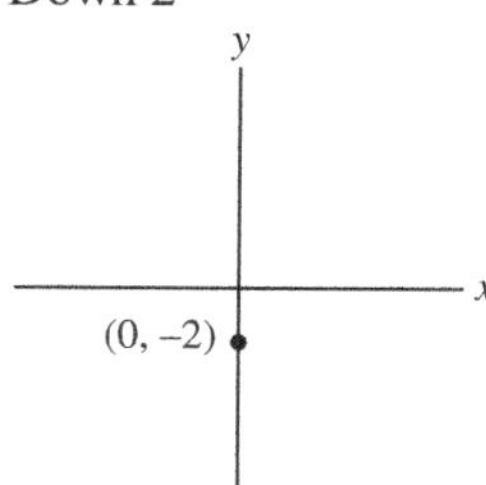

5. Left 2, up 1

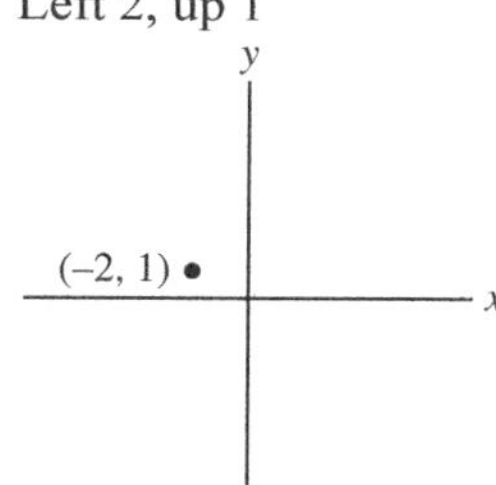

7. Left 20, up 40

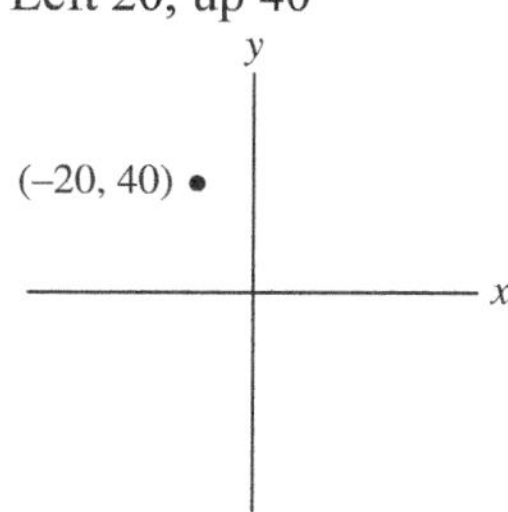

9. Point Q is 2 units to the left and 2 units up or $(-2, 2)$.

11. $-2(1)+\frac{1}{3}(3)=-2+1=-1$ so yes the point is on the line.

13. $-2x+\frac{1}{3}y=-1$ Substitute the x and y coordinates of the point into the equation: $\left(\frac{1}{2},3\right)\rightarrow -2\left(\frac{1}{2}\right)+\frac{1}{3}(3)=-1\rightarrow -1+1=-1$ is a false statement. So no the point is not on the line.

15. $m = 5, b = 8$

17. $y = 0x + 3; m = 0, b = 3$

19.
$$14x+7y=21$$
$$7y=-14x+21$$
$$y=-2x+3$$

21.
$$3x=5$$
$$x=\frac{5}{3}$$

23.
$$0=-4x+8$$
$$4x=8$$
$$x=2$$
x-intercept: (2, 0)
$y = -4(0) + 8$
$y = 8$
y-intercept: (0, 8)

25. When $y = 0, x = 7$
x-intercept: (7, 0)
$0 = 7$
no solution
y-intercept: none

27.
$$0=\frac{1}{3}x-1$$
$x = 3$
x-intercept: (3, 0)
$$y=\frac{1}{3}(0)-1$$
$y = -1$

Copyright © 2018 Pearson Education, Inc.

y-intercept: (0, –1)

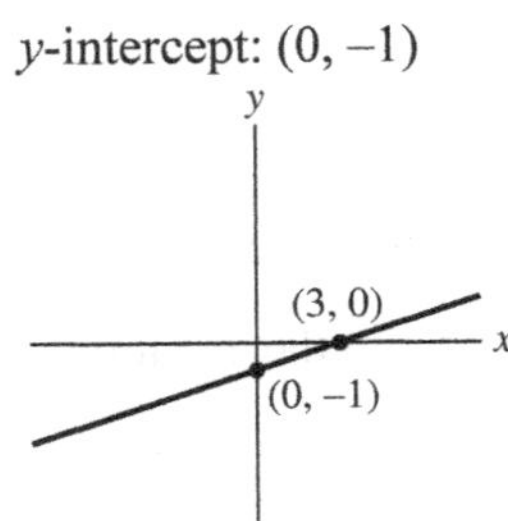

29. $0 = \frac{5}{2}$

no solution

x-intercept: none

When $x = 0$, $y = \frac{5}{2}$

y-intercept: $\left(0, \frac{5}{2}\right)$

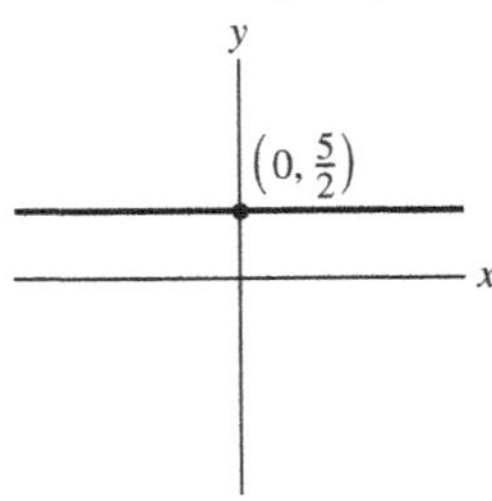

31. $3x + 4(0) = 24$

$x = 8$

x-intercept: (8, 0)

$3(0) + 4y = 24$

$y = 6$

y-intercept: (0, 6)

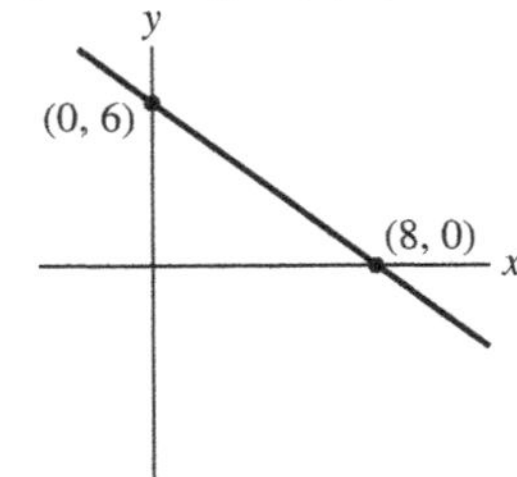

33. $x = -\frac{5}{2}$

35. $2x + 3y = 6$

$$3y = -2x + 6$$

$$y = -\frac{2}{3}x + 2$$

a. $4x + 6y = 12$

$$6y = -4x + 12$$

$$y = -\frac{2}{3}x + 2$$

Yes

b. Yes

c. $x = 3 - \frac{3}{2}y$

$$\frac{3}{2}y = -x + 3$$

$$y = -\frac{2}{3}x + 2$$

$$y = -\frac{2}{3}x + 2$$

Yes

d. $6 - 2x - y = 0$

$$y = 6 - 2x = -2x + 6$$

No

e. $y = 2 - \frac{2}{3}x = -\frac{2}{3}x + 2$

Yes

f. $x + y = 1$

$$y = -x + 1$$

No

 Copyright © 2018 Pearson Education, Inc.

37. a. $x+y=3$

$y=-x+3$

$m=-1,\ b=3$

L_3

b. $2x-y=-2$

$-y=-2x-2$

$y=2x+2$

$m=2,\ b=2$

L_1

c. $x=3y+3$

$3y=x-3$

$y=\frac{1}{3}x-1$

$m=\frac{1}{3},\ b=-1$

L_2

39. $y=30x+72$

a. When x = 0, y = 72. This is the temperature of the water at time = 0 before the kettle is turned on.

b. $y=30(3)+72$

$y=162^{\circ}F$

c. Water boils when y = 212 so we have $212=30x+72.$ Solving for x gives $x=4\frac{2}{3}$ minutes or 4 minutes 40 seconds.

41. a. x-intercept: $\left(-33\frac{1}{3},\ 0\right)$

y-intercept: (0, 2.5)

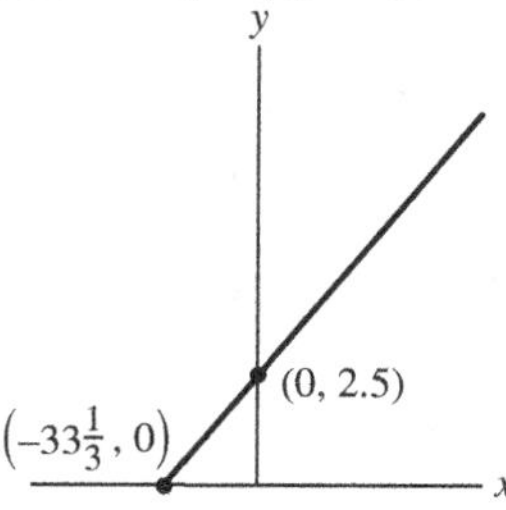

b. In 1960, 2.5 trillion cigarettes were sold.

c. $4=.075x+2.5$

$x=20$

$1960+20=1980$

d. $2024-1960=64$

$y=.075(64)+2.5$

$y=7.3$

7.3 trillion

43. a. x-intercept: (–32.9, 0)

y-intercept: (0, 756)

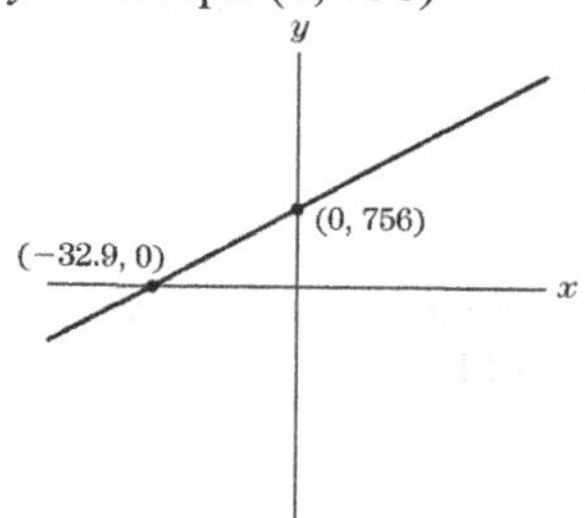

b. In 1999 the car insurance rate for a small car was $756.

c. $2007-1999=8$

$y=23(8)+756$

$y=940$

$940

d. $1308=23x+756$

$552=23x$

$x=24$

$1999+24=2023$

The yearly rate will be $1308 in 2023.

45. a. In 2000, 4.1% of entering college freshmen intended to major in biology.

b. $2014-2000=14$

$y=0.2(14)+4.1$

$y=6.9$

6.9% of college freshmen in 2014 intended to major in biology.

c. $5.5=0.2x+4.1$

$1.4=0.2x$

$x=7$

$2000+7=2007$

In 2007, the percent of college freshmen who intended to major in biology was 5.5.

Copyright © 2018 Pearson Education, Inc.

47. a. $2011-2004=7$
$y=461(7)+16800$
$y=20,027$
\$20,027 was the approximate average tuition in 2011.

b. $25000=461x+16800$
$8200=461x$
$x\approx 17.8$
$2004+17=2021$
In 2021, the approximate average cost of tuition will be more than \$25,000.

49. $y=mx+b$
$8=m(0)+b$
$b=8$
$0=m(16)+8$
$m=-\frac{1}{2}$
$y=-\frac{1}{2}x+8$

51. $y=mx+b$
$5=m(0)+b$
$b=5$
$0=m(4)+5$
$m=-\frac{5}{4}$
$y=-\frac{5}{4}x+5$

53. On the x-axis, $y=0$.

55. The equation of a line parallel to the y axis will be in the form $x=$ a.

57. $2x-y=-3$

59. $\frac{2}{3}x+y=-5$
$2x+3y=-15$

61. Since (a,0) and (0,b) are points on the line the slope of the line is (b-0)/(0-a) = -b/a. Since the y intercept is (0,b), the equation of the line is $y=-(b/a)x+b$ or $ay=-bx+ab$. In general form, the equation is bx + ay = ab.

63. One possible equation is $y=x-9$.

65. One possible equation is $y=x+7$.

67. One possible equation is $y=x+2$.

69. One possible equation is $y=x+9$.

71. The x – intercept has a y coordinate of 0, therefore the x coordinate of the first equation is:
$0=\frac{2}{3}x-2$
$2=\frac{2}{3}x$
$3=x$

Using this x coordinate in the second equation will find the value of c.
$0=-4(3)+c$
$0=-12+c$
$12=c$

73. a. y = –3x + 6

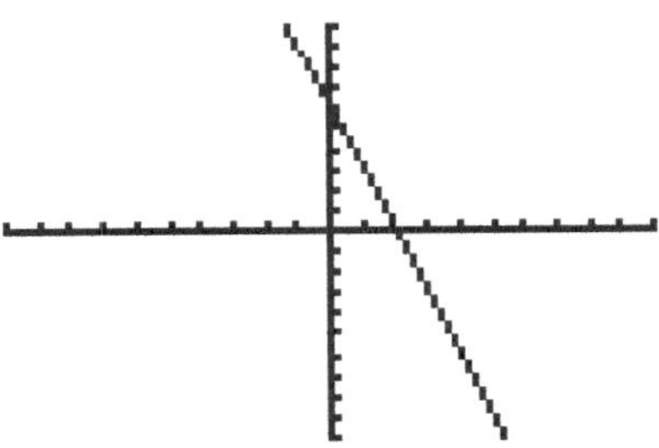

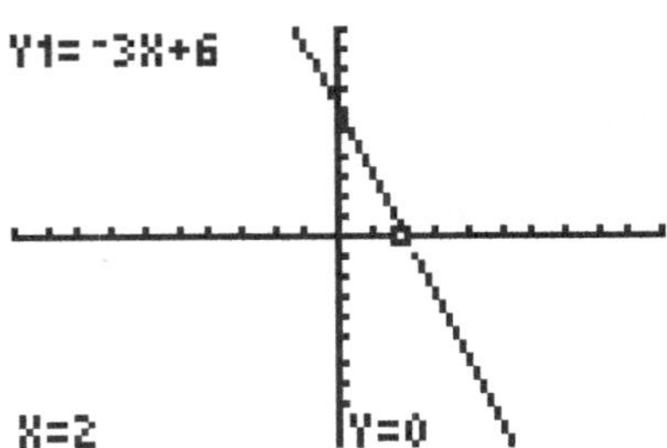

b. The intercepts are at the points (2, 0) and (0, 6)
c. When $x=2$, $y=0$

Copyright © 2018 Pearson Education, Inc.

75. a. $3y - 2x = 9$

$$3y = 2x + 9$$

$$y = \frac{2}{3}x + 3$$

Y1=(2/3)X+3

X=2 Y=4.3333333

b. The intercepts are at the points (–4.5, 0) and (0, 3).

c. When $x = 2$, $y = 4.33$ or $13 / 3$.

77. 2y + x = 100. When y = 0, x = 100, and when x = 0, y = 50. An appropriate window might be [-10, 110] and [-10,60]. Other answers are possible.

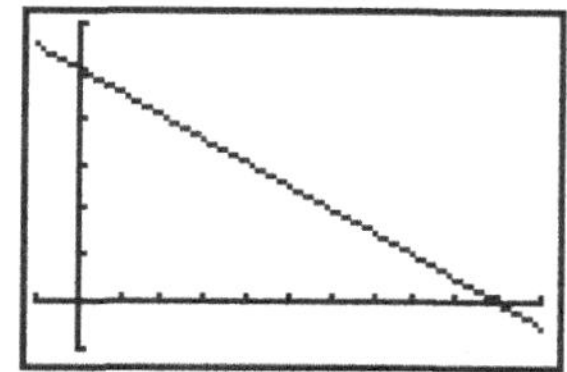

Exercises 1.2

1. $m = \frac{2}{3}$

3. $y - 3 = 5(x + 4)$

$y = 5x + 23$

$m = 5$

5. $\frac{x}{5} + \frac{y}{4} = 6$

$$\frac{4x}{5} + y = 24$$

$$y = -\frac{4}{5}x + 24$$

$$m = -\frac{4}{5}$$

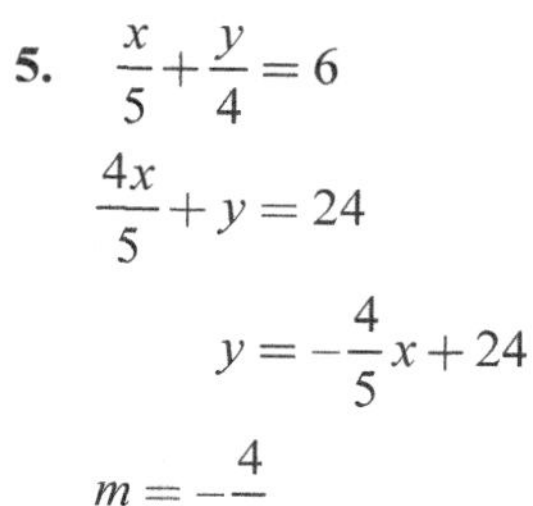

7.

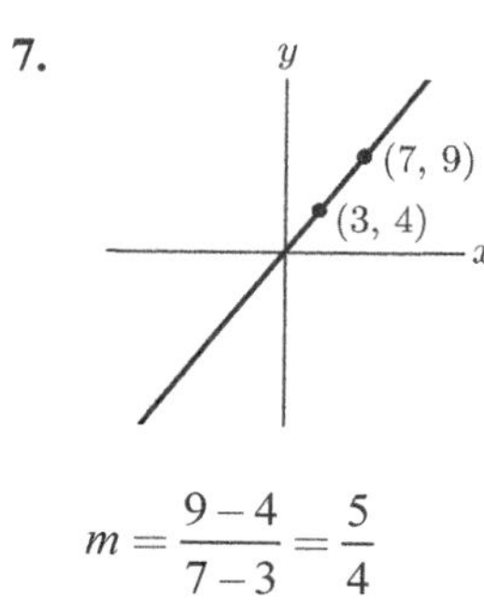

$$m = \frac{9-4}{7-3} = \frac{5}{4}$$

9.

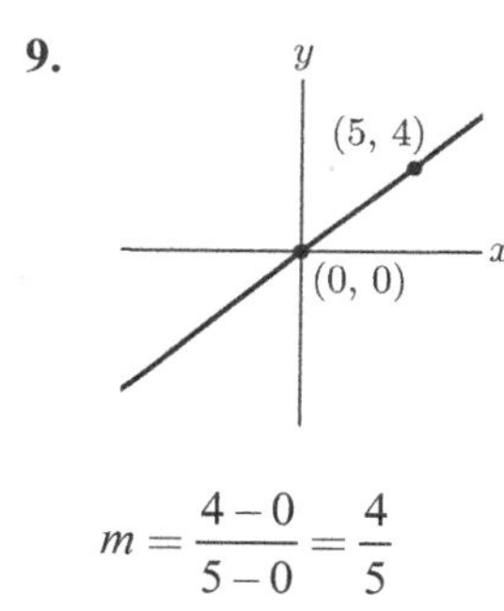

$$m = \frac{4-0}{5-0} = \frac{4}{5}$$

11. The slope of a vertical line is undefined.

13.

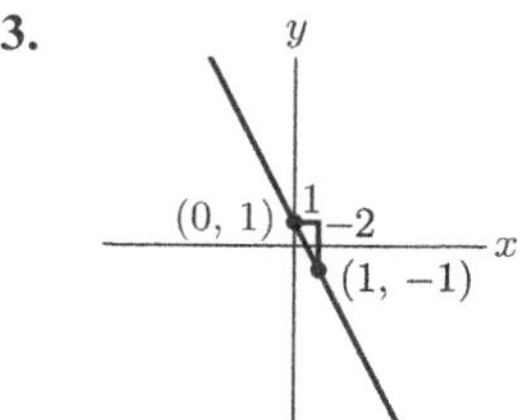

15.

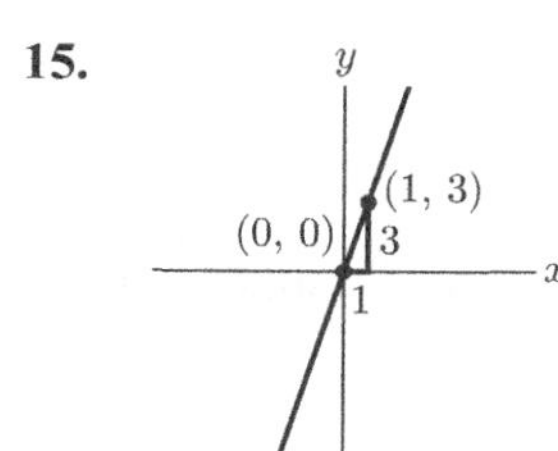

17. $m = \frac{-2}{1} = -2$
$y - 3 = -2(x - 2)$
$y = -2x + 7$

19. $m = \frac{0-2}{2-1} = -2$
$y - 0 = -2(x - 2)$
$y = -2x + 4$

21. $m = -\frac{1}{-4} = \frac{1}{4}$
$y - 2 = \frac{1}{4}(x - 2)$
$y = \frac{1}{4}x + \frac{3}{2}$

23. $m = -1$
$y - 0 = -1(x - 0)$
$y = -x$

25. $m = \frac{-1-1}{1-0} = \frac{-2}{1} = -2$
$y - (1) = -2(x - 0)$
$y = -2x + 1$

27. $m = \frac{0-2}{-4-0} = \frac{-2}{-4} = \frac{1}{2}$
$y - 0 = \frac{1}{2}(x + 4)$
$y = \frac{1}{2}x + 2$

29. $m = 0$
$y - 3 = 0(x - 2)$
$y = 3$

31. $y - 6 = \frac{3}{5}(x - 5)$
$y = \frac{3}{5}x + 3$
y-intercept: (0, 3)

33. m = undefined, therefore the equation is of the form $x = a$.
$x = 0$

35. Let y = cost in dollars.
$y = 4x + 2000$

37. a. Let x = altitude and y = boiling point.
$m = \frac{212 - 202.8}{0 - 5000} = -0.00184$
$y - 212 = -0.00184(x - 0)$
$y = -0.00184x + 212$

b. $y \approx -0.00184x + 212$
$y \approx -0.00184(29029) + 212$
$y \approx 158.6°\text{F}$

39. a. Let x = quantity and y = cost.
$m = \frac{9500 - 6800}{50 - 20} = 90$
$y - 6800 = 90(x - 20)$
$y = 90x + 5000$

b. \$5000

c. \$90

d.

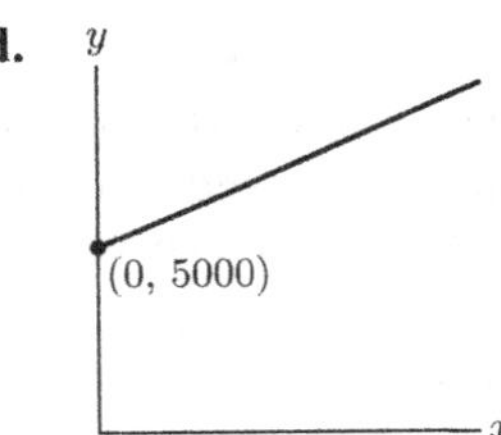

41. a. $100(300) = \$30,000$

b. $6000 = 100x$
$x = 60$ coats

c. $y = 100(0) = 0$
(0, 0); if no coats are sold, there is no revenue.

d. 100; each additional coat yields an additional \$100 in revenue.

 Copyright © 2018 Pearson Education, Inc.

43. **a.**

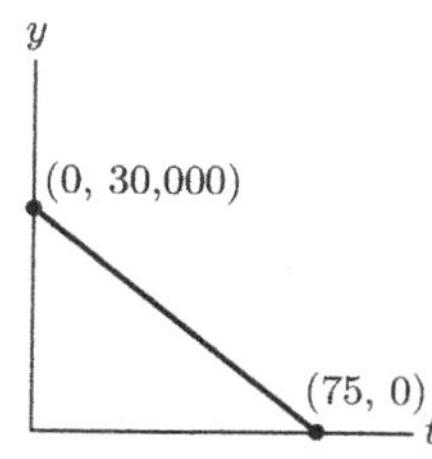

b. On February 1, 31 days have elapsed since January 1. The amount of oil y = 30,000 – 400(31) = 17,600 gallons.

c. On February 15, 45 days have elapsed since January 1. Therefore, the amount of oil would be y = 30,000- 400(45) = 12,000 gallons.

d. The significance of the y-intercept is that amount of oil present initially on January 1. This amount is 30,000 gallons.

e. The t-intercept is (75,0) and corresponds to the number of days at which the oil will be depleted.

45. **a.** $y = 0.10x + 220$

b. $y = 0.10(2000) + 220$

$y = 420$

c. $540 = 0.10x + 220$

$x = \$3200$

47. $m = -\frac{1}{2},\ b = 0$

$y = -\frac{1}{2}x$

49. $m = -\frac{1}{3}$

$y - (-2) = -\frac{1}{3}(x - 6)$

$y = -\frac{1}{3}x$

51. $m = \frac{1}{2}$

$y - (-3) = \frac{1}{2}(x - 2)$

$y = \frac{1}{2}x - 4$

53. $m = -\frac{2}{5}$

$y - 5 = -\frac{2}{5}(x - 0)$

$y = -\frac{2}{5}x + 5$

55. $m = \frac{3 - (-3)}{-1 - 5} = -1$

$y - 3 = -1[x - (-1)]$

$y = -x + 2$

57. $m = \frac{-1 - (-1)}{3 - 2} = 0$

$y - (-1) = 0(x - 2)$

$y = -1$

59. Changes in x-coordinate: 1, –1, –2
Changes in y-coordinate are m times that or 2, –2, –4: new y values are 5, 1, –1

61. The slope is $\frac{-1}{4}$ Changes in x coordinates are 1, 2, –1. Changes in y coordinates are m times the x coordinate changes. New y coordinates are $\frac{-5}{4}, \frac{-3}{2}, \frac{-3}{4}$

63. **a.** $x + y = 1$

$y = -x + 1$

(C)

b. $x - y = 1$

$y = x - 1$

(B)

c. $x + y = -1$

$y = -x - 1$

(D)

d. $x - y = -1$

$y = x + 1$

(A)

65. One possible equation is $y = x + 1$.

67. One possible equation is $y = 5$.

69. One possible equation is $y = -\frac{2}{3}x$.

Copyright © 2018 Pearson Education, Inc.

71. $m = \dfrac{212-32}{100-0} = \dfrac{9}{5}$

$$F - 32 = \frac{9}{5}(C - 0)$$
$$F = \frac{9}{5}C + 32$$

73. Let 2001 correspond to x = 0. So in 2013, x = 12. When x = 0, tuition is 3735. When x = 12, tuition is 8312. Using (0,3735) and (12,8312) as ordered pairs, find the slope of the line containing these points: $\dfrac{8312-3735}{12-0} = \dfrac{4577}{12}$. Since the y-intercept is 3735, the equation becomes $y = \dfrac{4577}{12}x + 3735$. Therefore, in 2009 when x = 8, the tuition should approximately be $y = \dfrac{4577}{12}(8) + 3735 = \$6786.33$.

75. Let x = number of pounds tires are under inflated. When x = 0, the miles per gallon (y) is 25. When x = 1, mpg decreases to 24.5. The equation is $y = -\dfrac{1}{2}x + 25$. Thus, when x = 8 pounds the miles per gallon will be $y = -\dfrac{1}{2}(8) + 25 = 21$ mpg.

77. Let 2001 correspond to x = 0 and 2013 correspond to x = 12. Then, the two ordered pairs are on the line: (0, 263515) and (12,360823). The slope of the line is $\dfrac{360,823-263,515}{12-0} = 8109$ The equation of the line is therefore $y = 8190x + 263,515$. In the year 2020, x = 19, so the number of Bachelor's degrees awarded can be estimated as $y = 8109(19) + 263,515 = 417,586$.

79. Let 2012 correspond to x = 0 and 2015 correspond to x = 3. Then, the two ordered pairs are on the line: (0, 3.5) and (3,4.5). The slope of the line is $\dfrac{4.5-3.5}{3-0} = \dfrac{1}{3}$. The equation of the line is therefore $y = \dfrac{1}{3}x + 3.5$. In the year 2014, x = 2, so the cost of a 30-second advertising slot (in millions) can be estimated as $y = \dfrac{1}{3}(2) + 3.5 \approx \$4.2$ million.

81. The slope is $\dfrac{3.4-3}{6-5} = 0.4$

$$p - p_1 = m(q - q_1)$$
$$p - 3 = 0.4(q - 5)$$
$$p - 3 = 0.4q - 2$$
$$p = 0.4q + 1$$

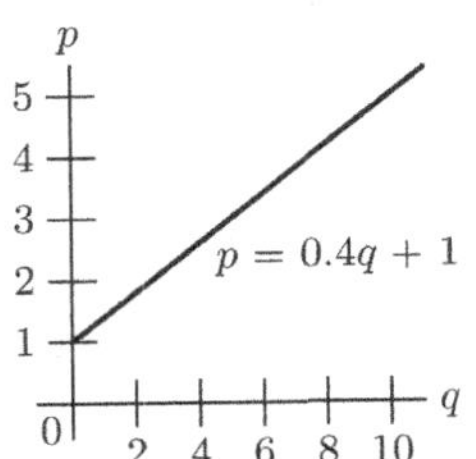

83. $m_1 = \dfrac{4-3}{2-1} = 1$

$m_2 = \dfrac{-1-4}{3-2} = -5$

$m_1 \neq m_2$

85. Set slopes equal:

$$\frac{-3.1-1}{2-a} = \frac{2.4-0}{3.8-(-1)}$$
$$\frac{-4.1}{2-a} = \frac{1}{2}$$
$$-8.2 = 2 - a$$
$$a = 10.2$$

87. Solve $mx + b = m'x + b'$

$$(m - m')x = b' - b$$
$$x = \frac{b' - b}{m - m'},$$

which is defined if and only if $m \neq m'$.

89. Let x = Centigrade temperature
y = Fahrenheit temperature

$$m = \frac{212-32}{100-0} = 1.8$$
$$y = 1.8x + 32$$
$$y = 1.8(30) + 32 = 86°\text{F}$$

Copyright © 2018 Pearson Education, Inc.

91. Let x = number of T-shirts
profit = revenue – cost
$$65,000 = 12.50x - (8x + 25,000)$$
$$90,000 = 4.50x$$
$$x = 20,000$$
So 20,000 T-shirts must be produced and sold.

93. $q = 800 - 4(150)$
$= 200$ bikes
revenue = 150(200) = \$30,000

95. Let x = variable costs
For 2015: profit = revenue – cost
$$400,000 = 100(50,000) - (50,000x + 600,000)$$
$$50,000x = 4,000,000$$
$$x = \$80 \text{ per unit}$$

For 2016:
Let y = 2016 price
profit = revenue – cost
$$400,000 = 50,000y - [80(50,000) + 600,000 + 200,000]$$
$$5,200,000 = 50,000y$$
$$y = \$104$$

97

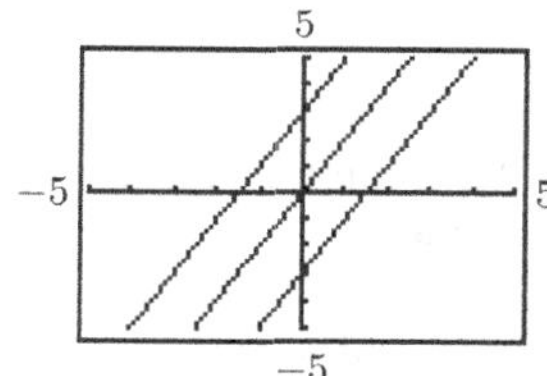

From left to right the lines are $y = 2x + 3$, $y = 2x$, and $y = 2x - 3$.
The lines are distinguished by their y-intercepts, which appear as b in the form $y = mx + b$.

99.

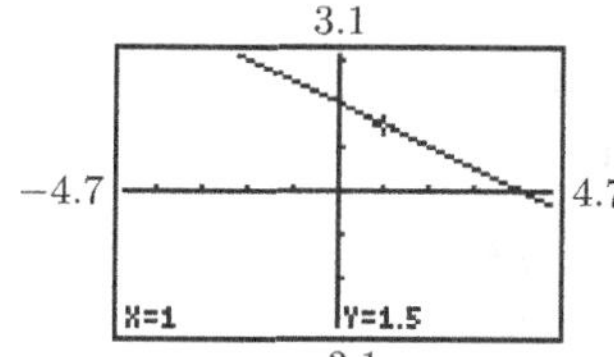

Since the slope equals $-\frac{1}{2}$, moving 2 units to the right requires moving $2 \cdot \left(-\frac{1}{2}\right) = -1$ unit up, or 1 unit down.

101.

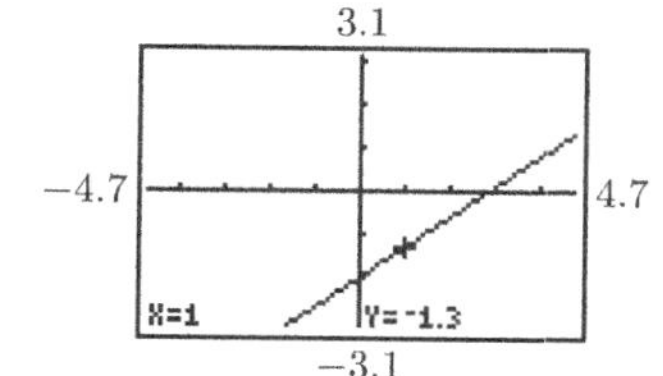

Since the slope equals 0.7, moving 2 units to the right requires moving $2 \cdot 0.7 = 1.4$ units up.

Exercises 1.3

1. $4x - 5 = -2x + 7$
$6x = 12$
$x = 2$
$y = 4(2) - 5 = 3$
(2, 3)

3. $x = 4y - 2$
$x = -2y + 4$
$4y - 2 = -2y + 4$
$6y = 6$
$y = 1$
$x = 4(1) - 2 = 2$
(2, 1)

5. $y = \frac{1}{3}(12) - 1 = 3$
(12, 3)

7. $\begin{cases} 6 - 3(4) = -6 \\ 3(6) - 2(4) = 10 \end{cases}$
$\begin{cases} -6 = -6 \\ 10 = 10 \end{cases}$
Yes

9. $\begin{cases} y = -2x + 7 \\ y = x - 3 \end{cases}$
$$-2x + 7 = x - 3$$
$$-3x = -10$$
$$x = \frac{10}{3}$$
$$y = \frac{10}{3} - 3 = \frac{1}{3}$$
$$x = \frac{10}{3},\ y = \frac{1}{3}$$

Copyright © 2018 Pearson Education, Inc.

11. $\begin{cases} y = \frac{5}{2}x - \frac{1}{2} \\ y = -2x - 4 \end{cases}$

$$\frac{5}{2}x - \frac{1}{2} = -2x - 4$$
$$\frac{9}{2}x = -\frac{7}{2}$$
$$x = -\frac{7}{9}$$
$$y = -2\left(-\frac{7}{9}\right) - 4 = -\frac{22}{9}$$
$$x = -\frac{7}{9},\ y = -\frac{22}{9}$$

13. $\begin{cases} x = 3 \\ 2x + 3y = 18 \end{cases}$

$$y = -\frac{2}{3}x + 6 = -\frac{2}{3}(3) + 6 = 4$$

$A = (3, 4)$

$\begin{cases} y = 2 \\ 2x + 3y = 18 \end{cases}$

$$x = -\frac{3}{2}y + 9 = -\frac{3}{2}(2) + 9 = 6$$

$B = (6, 2)$

15. $A = (0, 0)$

$\begin{cases} y = 2x \\ y = \frac{1}{2}x + 3 \end{cases}$

$$2x = \frac{1}{2}x + 3$$
$$x = 2$$
$$y = 2(2) = 4$$

$B = (2, 4)$

$\begin{cases} y = \frac{1}{2}x + 3 \\ x = 5 \end{cases}$

$$y = \frac{1}{2}(5) + 3 = \frac{11}{2}$$
$$C = \left(5, \frac{11}{2}\right)$$

$D = (5, 0)$

17. a. $p = 0.0001(19{,}500) + 0.05$
$= \$2.00$

b. $p = 0.0001(0) + 0.05$
$= \$0.05$

No units will be supplied for \$0.05 or less.

19. $\begin{cases} p = 0.0001q + 0.05 \\ p = -0.001q + 32.5 \end{cases}$

$$0.0001q + 0.05 = -0.001q + 32.5$$
$$0.0011q = 32.45$$
$$q = 29{,}500 \text{ units}$$
$$p = 0.0001(29{,}500) + 0.05$$
$$p = \$3.00$$

21. a.

$$p = -0.15q + 6.925$$
$$5.80 = -0.15q + 6.925$$
$$-1.125 = -0.15q$$
$$7.5 = q$$

$$p = 0.2q + 3.6$$
$$5.80 = 0.2q + 3.6$$
$$2.2 = 0.2q$$
$$11 = q$$

Demand will be 7.5 billion bushels and supply will be 11 billion bushels

b. The equilibrium point occurs when supply is the same as demand. Therefore,

$$-0.15q + 6.925 = 0.2q + 3.6$$
$$-0.35q = -3.325$$
$$q = 9.5$$

To find the equilibrium price, substitute the value into either equation.

$$p = -0.15(9.5) + 6.925$$
$$p = -1.425 + 6.925$$
$$p = 5.5$$

Equilibrium occurs when 9.5 billion bushels are produced and sold for \$5.50 per bushel.

Copyright © 2018 Pearson Education, Inc.

23. Let $C = F$, then

$$C = \frac{5}{9}(F-32)$$

$$F = \frac{5}{9}(F-32)$$

$$\frac{9F}{5} = F-32$$

$$\frac{4F}{5} = -32$$

$$F = -40$$

Therefore, when the temperature is -40°, it will be the same on both temperature scales.

25. Let x = numbers of shirts and y = cost of manufacture.

$$\begin{cases} y = 30x+1200 \\ y = 35x+500 \end{cases}$$

$$30x+1200 = 35x+500$$

$$-5x = -700$$

$$x = 140$$

$$y = 30x+1200$$

$$y = 30(140)+1200$$

$$y = 4200+1200$$

$$y = 5400$$

The manufactures will charge the same \$5400 if they produce 140 shirts.

27. Method A: $y = .45+.01x$

Method B: $y = .035x$

Intersection point:

$$.45+.01x = .035x$$

$$.45 = .025x$$

$$18 = x$$

For a call lasting 18 minutes, the costs for either method will be the same, $y = .035(18) = 63$. The cost will be 63cents.

29. $\begin{cases} 3x-y=3 \\ x+y=5 \\ y=0 \end{cases}$

$$\begin{cases} y = 3x-3 \\ y = -x+5 \\ y = 0 \end{cases}$$

$$\begin{cases} y = 3x-3 \\ y = -x+5 \end{cases} \Rightarrow (2,\ 3)$$

$$\begin{cases} y = -x+5 \\ y = 0 \end{cases} \Rightarrow (5,\ 0)$$

$$\begin{cases} y = 3x-3 \\ y = 0 \end{cases} \Rightarrow (1,\ 0)$$

Based on the above points of intersection, the base of the triangle is $5-1=4$ and the height is 3. Therefore the area of the triangle, in square units, is:

$$A = \frac{1}{2}bh$$

$$A = \frac{1}{2}(4)(3)$$

$$A = 6$$

31. Let x = weight of first contestant
y = weight of second contestant

$$\begin{cases} x+y = 700 \\ 2x = 275+y \end{cases}$$

$$\begin{cases} y = 700-x \\ y = 2x-275 \end{cases}$$

$$700-x = 2x-275$$

$$975 = 3x$$

$$x = 325 \text{ pounds}$$

33.

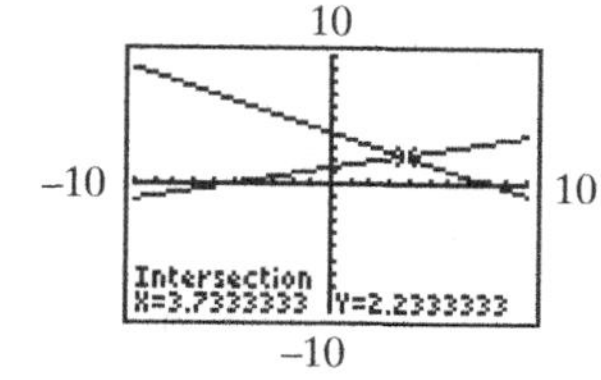

(3.73, 2.23)

35. $\begin{cases} x-4y=-5 \\ 3x-2y=4.2 \end{cases}$

$$\begin{cases} y = \frac{1}{4}x+\frac{5}{4} \\ y = \frac{3}{2}x-2.1 \end{cases}$$

Copyright © 2018 Pearson Education, Inc.

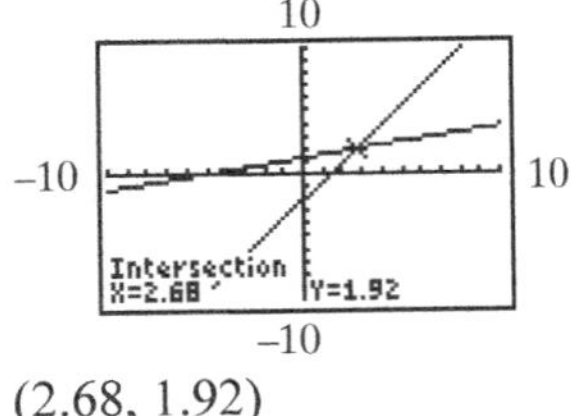

(2.68, 1.92)

Exercises 1.4

1.

Data Point	Point on Line	Vertical Distance
(1, 3)	(1, 4)	1
(2, 6)	(2, 7)	1
(3, 11)	(3, 10)	1
(4, 12)	(4, 13)	1

$1^2+1^2+1^2+1^2=4$

3. $E_1^2=[1.1(1)+3-3]^2=1.21$

$E_2^2=[1.1(2)+3-6]^2=.64$

$E_3^2=[1.1(3)+3-8]^2=2.89$

$E_4^2=[1.1(4)+3-6]^2=1.96$

$E=1.21+.64+2.89+1.96=6.70$

5.

x	y	xy	x^2
1	7	7	1
2	6	12	4
3	4	12	9
4	3	12	16
$\sum x=10$	$\sum y=20$	$\sum xy=43$	$\sum x^2=30$

$$m=\frac{4\cdot 43-10\cdot 20}{4\cdot 30-10^2}=-1.4$$

$$b=\frac{20-(-1.4)(10)}{4}=8.5$$

7. $\sum x=6, \sum y=18, \sum xy=45, \sum x^2=14$

$$m=\frac{3\cdot 45-6\cdot 18}{3\cdot 14-6^2}=4.5$$

$$b=\frac{18-(4.5)(6)}{3}=-3$$

$y=4.5x-3$

Copyright © 2018 Pearson Education, Inc.

9. $\sum x = 10, \sum y = 26, \sum xy = 55,$

$\sum x^2 = 30$

$$m = \frac{4\cdot 55 - 10\cdot 26}{4\cdot 30 - 10^2} = -2$$

$$b = \frac{26-(-2)(10)}{4} = 11.5$$

$y = -2x + 11.5$

11. a. $\sum x = 12, \sum y = 7, \sum xy = 41,$

$\sum x^2 = 74$

$$m = \frac{2\cdot 41 - 12\cdot 7}{2\cdot 74 - 12^2} = -0.5$$

$$b = \frac{7-(-.5)(12)}{2} = 6.5$$

$y = -0.5x + 6.5$

b. $m = \frac{4-3}{5-7} = -\frac{1}{2} = -0.5$

$y - 3 = -0.5(x-7)$

$y = -0.5x + 6.5$

c. The least–squares error for the line in (b) is E=0.

13. a. $\sum x = 7, \sum y = 20, \sum xy = 90,$

$\sum x^2 = 37$

$$m = \frac{2\cdot 90 - 7\cdot 20}{2\cdot 37 - 7^2} = 1.6 = \frac{8}{5}$$

$$b = \frac{20-(1.6)(7)}{2} = 4.4 = \frac{22}{5}$$

$$y = \frac{8}{5}x + \frac{22}{5}$$

b. $E = [1.6(4) + 4.4 - 5]^2 = 33.64$

15. a. Let x represent city and y represent highway, then $\sum x = 191, \sum y = 175, \sum xy = 8431,$

$\sum x^2 = 9201$

$$m = \frac{4\cdot 8431 - 191\cdot 175}{4\cdot 9201 - 191^2} \approx 0.9257$$

$$b = \frac{175-(0.9257)(191)}{4} \approx -0.452$$

$y = 0.9257x - 0.452$

b. $y = 0.9257(47) - 0.452$

$y = 43.06$ mpg

c. $47 = 0.9257x - 0.452$

$x = 51.26$ mpg

17. a. 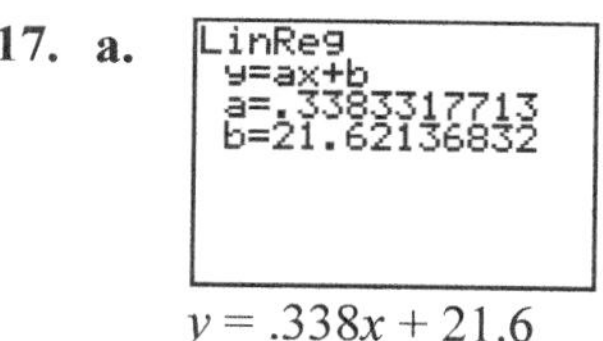

$y = .338x + 21.6$

b. $0.338(1100) + 21.6 = 393.4$
There will be about 393 deaths per million males

19. a. Let x be the number of years after 1989, then $y = 0.451x + 20.6$

b. $0.451(23) + 20.6 = 30.973$
About 31% completed four or more years of college

c. $35 = 0.451x + 20.6$

$x \approx 31.93$

Approximately 35% of persons 25 years and over will have completed 4 or more years of college in the year 1989 + 31 = 2020.

21. a. $y = 0.141x + 74.8$

b. $0.141(30) + 74.8 = 79.0$
A 30 – year old US male has a life expectancy of about 79.1 years

c. $0.141(50) + 74.8 = 81.9$
A 50 – year old US male has a life expectancy of about 81.9 years

d. $0.141(90) + 74.8 = 87.5$
Life expectancy will be about 87.5 years (This is an example of a fit that is not capable of extrapolating beyond the given data)

23. a. Let x be the number of years after 2000, then $y = 0.028x + 0.845$

b. $0.028(13) + 0.845 \approx 1.21$
In the year 2013, the price of a pound of spaghetti was about \$1.21.

c. $1.45 = 0.028x + 0.845$

$x \approx 21.6$

The price per pound of spaghetti will be \$1.45 in the year 2000 + 21 = 2021.

Copyright © 2018 Pearson Education, Inc.

Chapter 1 Review Exercises

1. $x = 0$

2.

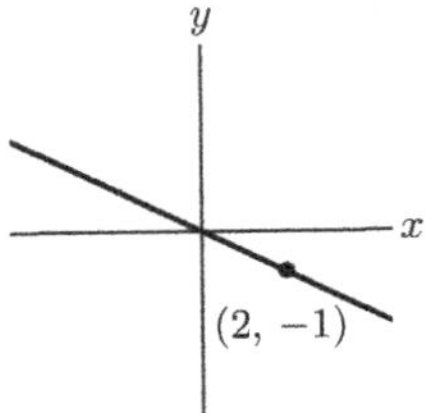

3. $\begin{cases} x-5y=6 \\ 3x=6 \end{cases}$

$\begin{cases} x=5y+6 \\ x=2 \end{cases}$

$5y+6=2$

$y=-\frac{4}{5}$

$\left(2, -\frac{4}{5}\right)$

4. $3x-4y=8$

$y=\frac{3}{4}x-2$

$m=\frac{3}{4}$

5. $m=\frac{0-5}{10-0}=-\frac{1}{2},\ b=5$

$y=-\frac{1}{2}x+5$

6. $\begin{cases} 2x-y=1 \\ x+2y=13 \end{cases}$

$\begin{cases} y=2x-1 \\ y=-\frac{1}{2}x+\frac{13}{2} \end{cases}$

$2x-1=-\frac{1}{2}x+\frac{13}{2}$

$\frac{5}{2}x=\frac{15}{2}$

$x=3$

$y=2(3)-1=5$

$(3, 5)$

7. $2x-10y=7$

$y=\frac{1}{5}x-\frac{7}{10}$

$m=\frac{1}{5}$

$y-16=\frac{1}{5}(x-15)$

$y=\frac{1}{5}x+13$

8. $y=3(1)+7=10$

9. $(5, 0)$

10. $\begin{cases} 3x-2y=1 \\ 2x+y=24 \end{cases}$

$\begin{cases} y=\frac{3}{2}x-\frac{1}{2} \\ y=-2x+24 \end{cases}$

$\frac{3}{2}x-\frac{1}{2}=-2x+24$

$\frac{7}{2}x=\frac{49}{2}$

$x=7$

$y=-2(7)+24=10$

$(7, 10)$

11. $y-9=\frac{1}{2}(x-4)$

$y=\frac{1}{2}x+7$

$b=7$

$(0, 7)$

12. The rate is \$35 per hour plus a flat fee of \$20.

13. $m_1=\frac{0-2}{2-1}=-2$

$m_2=\frac{1-0}{3-2}=1$

$m_1 \neq m_2$

No

14. $m=\frac{-2-0}{0-3}=\frac{2}{3},\ b=-2$

$y=\frac{2}{3}x-2$

 Copyright © 2018 Pearson Education, Inc.

15. $x+7y=30$
$-2y+7y=30$
$5y=30$
$y=6$

16. $\begin{cases}1.2x+2.4y=0.6\\4.8y-1.6x=2.4\end{cases}$
$\begin{cases}y=-0.5x+0.25\\y=\frac{1}{3}x+0.5\end{cases}$
$-0.5x+0.25=\frac{1}{3}x+0.5$
$-\frac{5}{6}x=0.25$
$x=-0.3$
$y=\frac{1}{3}(-0.3)+0.5=0.4$

17. $\begin{cases}y=-x+1\\y=2x+3\end{cases}$
$-x+1=2x+3$
$-3x=2$
$x=-\frac{2}{3}$
$y=-\left(-\frac{2}{3}\right)+1=\frac{5}{3}$
$\left(-\frac{2}{3},\frac{5}{3}\right)$
$m=\frac{\frac{5}{3}-1}{-\frac{2}{3}-1}=-\frac{2}{5}$
$y-1=-\frac{2}{5}(x-1)$
$y=-\frac{2}{5}x+\frac{7}{5}$

18. $\begin{cases}5x+2y=0\\x+y=1\end{cases}$
$\begin{cases}y=-\frac{5}{2}x\\y=-x+1\end{cases}$
$-\frac{5}{2}x=-x+1$
$-\frac{3}{2}x=1$
$x=-\frac{2}{3}$
$y=-\left(-\frac{2}{3}\right)+1=\frac{5}{3}$
Substitute $x=-\frac{2}{3}$ and $y=\frac{5}{3}$ in
$2x-3y=1$
$2\left(-\frac{2}{3}\right)-3\left(\frac{5}{3}\right)=1$
$-\frac{19}{3}=1$
No

19. $x+\frac{1}{2}y=4$
$y=-2x+8$
$m=-2$
y-intercept: (0, 8)
$0=-2x+8$
$x=4$
x-intercept: (4, 0)

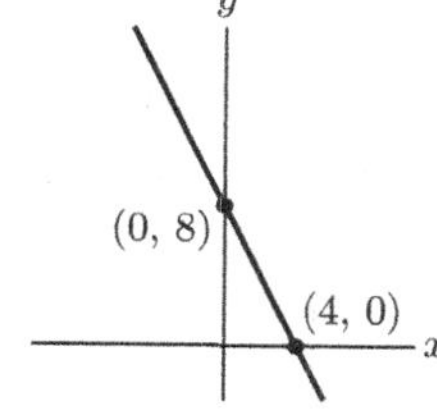

20. $\begin{cases}2x-3y=1\\3x+2y=4\end{cases}$
$\begin{cases}y=\frac{2}{3}x-\frac{1}{3}\\y=-\frac{3}{2}x+2\end{cases}$
$m_1=-\frac{1}{m_2}$

21. **a.** $4x+y=17$
$y=-4x+17$
L_3

b. $y=x+2$
L_1

c. $2x+3y=11$

$$y=-\frac{2}{3}x+\frac{11}{3}$$

L_2

22. Supply curve is $p = 0.005q + 0.5$
Demand curve is $p = -0.01q + 5$

$$\begin{cases} p=0.005q+0.5 \\ p=-0.01q+5 \end{cases}$$

$$0.005q+0.5=-0.01q+5$$
$$0.015q=4.5$$
$$q=300 \text{ units}$$
$$p = 0.005(300) + 0.5 = \$2$$

23. a. In 2004, approximately 28% of University of Alabama freshmen were from out of state.

b. $2009-2004=5$

$y=3.6(5)+28$

$y=46$

46% of the freshmen in 2009 were from out of state at the University of Alabama.

c. $82=3.6x+28$

$54=3.6x$

$x=15$

$2004+15=2019$

In 2019, the percent of college freshmen that are from out of state at the University of Alabama will be 82.

24. a. $m = 10$

$y-4000=10(x-1000)$

$y=10x-6000$

b. $0 = 10x - 6000$

$x = 600$

x-intercept: (600, 0)

y-intercept: (0, –6000)

c.

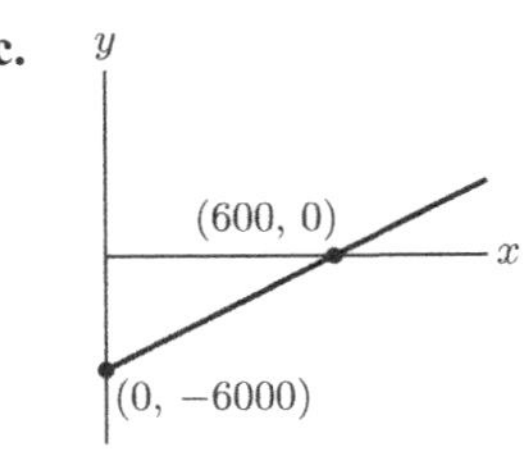

25. a. A: $y = 0.1x + 50$
B: $y = 0.2x + 40$

b. A: $0.1(80) + 50 = 58$
B: $0.2(80) + 40 = 56$
Company B

c. A: $0.1(160) + 50 = 66$
B: $0.2(160) + 40 = 72$
Company A

d. $0.1x + 50 = 0.2x + 40$

$-0.1x = -10$

$x = 100$ miles

26. a. $m=\dfrac{5.45-3.20}{12-0}=0.1875$

$y-3.20=0.1875(x-0)$

$y=0.1875x+3.20$

b. $4.60=0.1875x+3.20$

$1.40=0.1875x$

$x\approx 7.47$

Bacon was an average \$4.60 per pound in the year 2003 + 7 = 2010.

27. (0, 591300) ; in 2024: (10, 730200)

$$m=\frac{730,200-591,300}{10-0}=13,890$$

$y=13,890x+591,300$

For the year 2020, $x = 6$:

$y=13,890(6)+591,300=674,640.$

28. $0.03x+200=0.05x+100$

$-0.02x=-100$

$x=\$5000$

29. Let x = 0 correspond to year 2000. Then y = 20.4. When x = 10, y = 17.0. The rate of change (slope) = (17.0 – 20.4)/(10 – 0) = –.34. The equation of the line that predicts the percentage of market is y = –.34x + 20.4. When x = 8, y = 17.7%.

30. (0, 107238) ; in 2013: (7, 104647)

$$m=\frac{104,647-107,238}{7-0}=-370.1$$

$y=-370.1x+107,238$

For the year 2019, $x = 13$:

$y=-370.1(13)+107,238\approx 102,427.$

Copyright © 2018 Pearson Education, Inc.

31. **a.** $y = -0.65x + 76.1$

b. $-0.65(5) + 76.1 = 72.9$
About 72.9%

c. $67 = -0.65x + 76.1$
$-9.1 = -0.65$
$x = 14$
14 years after 2012 or 2026

32. **a.** $y = 1.53x - 36.8$

b. $1.53(77.8) - 36.8 = 82.2$
Life expectancy for women in Greece will be about 82.2 years

c. $85.0 = 1.53x - 36.8$
$x \approx 79.6$
Life expectancy for men in France will be about 79.6 years

33. **a.** $y = 0.152x - 3.063$

b. $0.152(160) - 3.063 = 21.257$
The breast cancer death rate in Denmark will be about 21.3 deaths per 100,000

c. $22 = 0.152x - 3.063$
$x \approx 164.888$
The daily fat intake by women in New Zealand is about 165 grams

34. Counterclockwise

35. Up; the value of b is the y-intercept

36. A line with undefined slope is a vertical line and a line with zero slope is a horizontal line.

37. The x – intercept and the y – intercept are the same at (0, 0).

38. Yes; since the data value is on the line, there will be no vertical distance added to the least squares line.

39. No; A line that is parallel to the x axis and is not the x axis will not have an x intercept.

No; A line that is parallel to the y axis and is not the y axis will not have a y intercept

40. **a.** Infinity many

b. Answers will vary.

Copyright © 2018 Pearson Education, Inc.

Chapter 2

Exercises 2.1

1. $\begin{cases} \frac{1}{2}x - 3y = 2 \\ 5x + 4y = 1 \end{cases}$

$\xrightarrow{2R_1} \begin{cases} x - 6y = 4 \\ 5x + 4y = 1 \end{cases}$

3. $\begin{array}{rr} 5\text{(first)} & 5x + 10y = 15 \\ +\text{(second)} & \underline{-5x + 4y = 1} \\ & 14y = 16 \end{array}$

$\begin{cases} x + 2y = 3 \\ -5x + 4y = 1 \end{cases}$

$\xrightarrow{R_2 + 5R_1} \begin{cases} x + 2y = 3 \\ 14y = 16 \end{cases}$

5. $\begin{array}{rr} -4\text{(first)} & -4x + 8y - 4z = 0 \\ +\text{(third)} & \underline{4x + y + 3z = 5} \\ & 9y - z = 5 \end{array}$

$\begin{cases} x - 2y + z = 0 \\ y - 2z = 4 \\ 4x + y + 3z = 5 \end{cases}$

$\xrightarrow{R_3 + (-4)R_1} \begin{cases} x - 2y + z = 0 \\ y - 2z = 4 \\ 9y - z = 5 \end{cases}$

7. $\left[\begin{array}{rr|r} 1 & -\frac{1}{2} & 3 \\ 0 & 1 & 4 \end{array}\right] \xrightarrow{R_1 + \frac{1}{2}R_2} \left[\begin{array}{rr|r} 1 & 0 & 5 \\ 0 & 1 & 4 \end{array}\right]$

9. $\left[\begin{array}{rr|r} -3 & 4 & -2 \\ 1 & -7 & 8 \end{array}\right]$

11. $\left[\begin{array}{rrr|r} 1 & 13 & -2 & 0 \\ 2 & 0 & -1 & 3 \\ 0 & 1 & 0 & 5 \end{array}\right]$

13. $\begin{cases} -2y = 3 \\ x + 7y = -4 \end{cases}$

15. $\begin{cases} 3x + 2y = -3 \\ y - 6z = 4 \\ -5x - y + 7z = 0 \end{cases}$

17. Multiply the second row of the matrix by $\frac{1}{3}$.

19. Change the first row of the matrix by adding to it 3 times the second row.

21. Interchange rows 2 and 3.

23. $\left[\begin{array}{rr|r} 1 & 2 & 0 \\ 0 & 10 & 5 \end{array}\right]$

25. $\left[\begin{array}{rr|r} 1 & 2 & 3 \\ 3 & -2 & 0 \end{array}\right]$

27. $\left[\begin{array}{rr|r} 1 & 3 & -5 \\ 0 & 1 & 7 \end{array}\right]$

29. Use $R_2 + 2R_1$ to change the –2 to a 0.

31. Use $R_1 + (-2) R_2$ to change the 2 to a 0.

33. Interchange rows 1 and 2 or rows 1 and 3 to make the first entry in row 1 nonzero.

35. Use $R_1 + (-3) R_3$ to change the 3 to a 0 or $R_2 + (-2) R_3$ to change the 2 to a 0.

37. $\left[\begin{array}{rrr|r} 1 & 1 & -1 & 6 \\ -3 & 7 & 5 & 0 \\ 2 & -4 & 3 & -1 \end{array}\right]$

$\xrightarrow[R_3 + (-2)R_1]{R_2 + 3R_1} \left[\begin{array}{rrr|r} 1 & 1 & -1 & 6 \\ 0 & 10 & \boxed{2} & \boxed{18} \\ 0 & -6 & \boxed{5} & \boxed{-13} \end{array}\right]$

39. $\begin{cases} x + y = 7 \\ x - y = 1 \end{cases}; x = 4, y = 3$

41. $\begin{cases} 3x - 4y = -27 \\ x + 2y = 11 \end{cases}; x = -1, y = 6$

Copyright © 2018 Pearson Education, Inc.

43. $\begin{cases} 2x+y+3z=31 \\ x+y-2z=3 \\ 4x-2y+5z=17 \end{cases}$; $x=3, y=10, z=5$

45. $\begin{cases} 3x+7y+2z=5 \\ 7x-6y-3z=4 \\ 10x+9y-7z=3 \end{cases}$; $x=1, y=0, z=1$

47. $\left[\begin{array}{cc|c} 1 & 9 & 8 \\ 2 & 8 & 6 \end{array}\right]$

$\xrightarrow{R_2+(-2)R_1} \left[\begin{array}{cc|c} 1 & 9 & 8 \\ 0 & -10 & -10 \end{array}\right]$

$\xrightarrow{-\frac{1}{10}R_2} \left[\begin{array}{cc|c} 1 & 9 & 8 \\ 0 & 1 & 1 \end{array}\right]$

$\xrightarrow{R_1+(-9)R_2} \left[\begin{array}{cc|c} 1 & 0 & -1 \\ 0 & 1 & 1 \end{array}\right]$

$x=-1, y=1$

49. $\left[\begin{array}{ccc|c} 1 & -3 & 4 & 1 \\ 4 & -10 & 10 & 4 \\ -3 & 9 & -5 & -6 \end{array}\right]$

$\xrightarrow{R_2+(-4)R_1} \left[\begin{array}{ccc|c} 1 & -3 & 4 & 1 \\ 0 & 2 & -6 & 0 \\ -3 & 9 & -5 & -6 \end{array}\right]$

$\xrightarrow{R_3+3R_1} \left[\begin{array}{ccc|c} 1 & -3 & 4 & 1 \\ 0 & 2 & -6 & 0 \\ 0 & 0 & 7 & -3 \end{array}\right]$

$\xrightarrow{\frac{1}{2}R_2} \left[\begin{array}{ccc|c} 1 & -3 & 4 & 1 \\ 0 & 1 & -3 & 0 \\ 0 & 0 & 7 & -3 \end{array}\right]$

$\xrightarrow{R_1+3R_2} \left[\begin{array}{ccc|c} 1 & 0 & -5 & 1 \\ 0 & 1 & -3 & 0 \\ 0 & 0 & 7 & -3 \end{array}\right]$

$\xrightarrow{\frac{1}{7}R_3} \left[\begin{array}{ccc|c} 1 & 0 & -5 & 1 \\ 0 & 1 & -3 & 0 \\ 0 & 0 & 1 & -\frac{3}{7} \end{array}\right]$

$\xrightarrow{R_1+5R_3} \left[\begin{array}{ccc|c} 1 & 0 & 0 & -\frac{8}{7} \\ 0 & 1 & -3 & 0 \\ 0 & 0 & 1 & -\frac{3}{7} \end{array}\right]$

$\xrightarrow{R_2+3R_3} \left[\begin{array}{ccc|c} 1 & 0 & 0 & -\frac{8}{7} \\ 0 & 1 & 0 & -\frac{9}{7} \\ 0 & 0 & 1 & -\frac{3}{7} \end{array}\right]$

$x=-\frac{8}{7}, y=-\frac{9}{7}, z=-\frac{3}{7}$

51. $\left[\begin{array}{cc|c} 2 & -2 & -4 \\ 3 & 4 & 1 \end{array}\right] \xrightarrow{\frac{1}{2}R_1} \left[\begin{array}{cc|c} 1 & -1 & -2 \\ 3 & 4 & 1 \end{array}\right]$

$\xrightarrow{R_2+(-3)R_1} \left[\begin{array}{cc|c} 1 & -1 & -2 \\ 0 & 7 & 7 \end{array}\right]$

$\xrightarrow{\frac{1}{7}R_2} \left[\begin{array}{cc|c} 1 & -1 & -2 \\ 0 & 1 & 1 \end{array}\right]$

$\xrightarrow{R_1+1R_2} \left[\begin{array}{cc|c} 1 & 0 & -1 \\ 0 & 1 & 1 \end{array}\right]$

$x=-1, y=1$

53. $\left[\begin{array}{ccc|c} 4 & -4 & 4 & -8 \\ 1 & -2 & -2 & -1 \\ 2 & 1 & 3 & 1 \end{array}\right]$

$\xrightarrow{\frac{1}{4}R_1} \left[\begin{array}{ccc|c} 1 & -1 & 1 & -2 \\ 1 & -2 & -2 & -1 \\ 2 & 1 & 3 & 1 \end{array}\right]$

$\xrightarrow{R_2+(-1)R_1} \left[\begin{array}{ccc|c} 1 & -1 & 1 & -2 \\ 0 & -1 & -3 & 1 \\ 2 & 1 & 3 & 1 \end{array}\right]$

$\xrightarrow{R_3+(-2)R_1} \left[\begin{array}{ccc|c} 1 & -1 & 1 & -2 \\ 0 & -1 & -3 & 1 \\ 0 & 3 & 1 & 5 \end{array}\right]$

$\xrightarrow{(-1)R_2} \left[\begin{array}{ccc|c} 1 & -1 & 1 & -2 \\ 0 & 1 & 3 & -1 \\ 0 & 3 & 1 & 5 \end{array}\right]$

$\xrightarrow{R_1+R_2} \left[\begin{array}{ccc|c} 1 & 0 & 4 & -3 \\ 0 & 1 & 3 & -1 \\ 0 & 3 & 1 & 5 \end{array}\right]$

$\xrightarrow{R_3+(-3)R_2} \left[\begin{array}{ccc|c} 1 & 0 & 4 & -3 \\ 0 & 1 & 3 & -1 \\ 0 & 0 & -8 & 8 \end{array}\right]$

$\xrightarrow{\left(-\frac{1}{8}\right)R_3} \left[\begin{array}{ccc|c} 1 & 0 & 4 & -3 \\ 0 & 1 & 3 & -1 \\ 0 & 0 & 1 & -1 \end{array}\right]$

Copyright © 2018 Pearson Education, Inc.

$$\xrightarrow{R_1+(-4)R_3}\left[\begin{array}{ccc|c}1&0&0&1\\0&1&3&-1\\0&0&1&-1\end{array}\right]$$

$$\xrightarrow{R_2+(-3)R_3}\left[\begin{array}{ccc|c}1&0&0&1\\0&1&0&2\\0&0&1&-1\end{array}\right]$$

$x = 1, y = 2, z = -1$

55. $\left[\begin{array}{cc|c}0.2&0.3&4\\0.6&1.1&15\end{array}\right]$

$$\xrightarrow{5R_1}\left[\begin{array}{cc|c}1&1.5&20\\0.6&1.1&15\end{array}\right]$$

$$\xrightarrow{R_2+(-.6)R_1}\left[\begin{array}{cc|c}1&1.5&20\\0&0.2&3\end{array}\right]$$

$$\xrightarrow{5R_2}\left[\begin{array}{cc|c}1&1.5&20\\0&1&15\end{array}\right]$$

$$\xrightarrow{R_1+(-1.5)R_2}\left[\begin{array}{cc|c}1&0&-2.5\\0&1&15\end{array}\right]$$

$x = -2.5, y = 15$

57. $\left[\begin{array}{ccc|c}1&1&4&3\\4&1&-2&-6\\-3&0&2&1\end{array}\right]$

$$\xrightarrow{R_2+(-4)R_1}\left[\begin{array}{ccc|c}1&1&4&3\\0&-3&-18&-18\\-3&0&2&1\end{array}\right]$$

$$\xrightarrow{R_3+3R_1}\left[\begin{array}{ccc|c}1&1&4&3\\0&-3&-18&-18\\0&3&14&10\end{array}\right]$$

$$\xrightarrow{\left(-\frac{1}{3}\right)R_2}\left[\begin{array}{ccc|c}1&1&4&3\\0&1&6&6\\0&3&14&10\end{array}\right]$$

$$\xrightarrow{R_1+(-1)R_2}\left[\begin{array}{ccc|c}1&0&-2&-3\\0&1&6&6\\0&3&14&10\end{array}\right]$$

$$\xrightarrow{R_3+(-3)R_2}\left[\begin{array}{ccc|c}1&0&-2&-3\\0&1&6&6\\0&0&-4&-8\end{array}\right]$$

$$\xrightarrow{\left(-\frac{1}{4}\right)R_3}\left[\begin{array}{ccc|c}1&0&-2&-3\\0&1&6&6\\0&0&1&2\end{array}\right]$$

$$\xrightarrow{R_1+2R_3}\left[\begin{array}{ccc|c}1&0&0&1\\0&1&6&6\\0&0&1&2\end{array}\right]$$

$$\xrightarrow{R_2+(-6)R_3}\left[\begin{array}{ccc|c}1&0&0&1\\0&1&0&-6\\0&0&1&2\end{array}\right]$$

$x = 1, y = -6, z = 2$

59. $\left[\begin{array}{ccc|c}-1&1&0&-1\\1&0&1&4\\6&-3&2&10\end{array}\right]$

$$\xrightarrow{(-1)R_1}\left[\begin{array}{ccc|c}1&-1&0&1\\1&0&1&4\\6&-3&2&10\end{array}\right]$$

$$\xrightarrow{R_2+(-1)R_1}\left[\begin{array}{ccc|c}1&-1&0&1\\0&1&1&3\\6&-3&2&10\end{array}\right]$$

$$\xrightarrow{R_3+(-6)R_1}\left[\begin{array}{ccc|c}1&-1&0&1\\0&1&1&3\\0&3&2&4\end{array}\right]$$

$$\xrightarrow{R_1+1R_2}\left[\begin{array}{ccc|c}1&0&1&4\\0&1&1&3\\0&3&2&4\end{array}\right]$$

$$\xrightarrow{R_3+(-3)R_2}\left[\begin{array}{ccc|c}1&0&1&4\\0&1&1&3\\0&0&-1&-5\end{array}\right]$$

$$\xrightarrow{(-1)R_3}\left[\begin{array}{ccc|c}1&0&1&4\\0&1&1&3\\0&0&1&5\end{array}\right]$$

$$\xrightarrow{R_1+(-1)R_3}\left[\begin{array}{ccc|c}1&0&0&-1\\0&1&1&3\\0&0&1&5\end{array}\right]$$

$$\xrightarrow{R_2+(-1)R_3}\left[\begin{array}{ccc|c}1&0&0&-1\\0&1&0&-2\\0&0&1&5\end{array}\right]$$

$x = -1, y = -2, z = 5$

Copyright © 2018 Pearson Education, Inc.

61. Let x = grams of cheddar cheese
y = grams of potato

$$\begin{cases} x+y=180 \\ 0.25x+0.02y=10.5 \end{cases}$$

$$\left[\begin{array}{cc|c} 1 & 1 & 180 \\ 0.25 & 0.02 & 10.5 \end{array}\right] \xrightarrow{R_2+(-0.25)R_1} \left[\begin{array}{cc|c} 1 & 1 & 180 \\ 0 & -0.23 & -34.5 \end{array}\right]$$

$$\xrightarrow{-\frac{1}{.23}R_2} \left[\begin{array}{cc|c} 1 & 1 & 180 \\ 0 & 1 & 150 \end{array}\right] \xrightarrow{R_1+(-1)R_2} \left[\begin{array}{cc|c} 1 & 0 & 30 \\ 0 & 1 & 150 \end{array}\right]$$

30 grams of cheddar cheese

63. Let x = cost of golf balls and y = cost of golf glove. Then $x + y = 20$.

Using **Statement I**: $y = 3x$

$$x + 3x = 20$$
$$4x = 20$$
$$x = 5.$$

Using **Statement II**: $y = 15$

$$x + 15 = 20$$
$$x = 5.$$

The box of balls costs \$5. Either statement is sufficient, so the answer is (d).

65. Let x = number of short sleeve shirts
y = number of long sleeve shirts

$$x + y = 350$$
$$10x + 14y = 4300$$

$$\left[\begin{array}{cc|c} 1 & 1 & 350 \\ 10 & 14 & 4300 \end{array}\right] \xrightarrow{R_2+(-10)R_1} \left[\begin{array}{cc|c} 1 & 1 & 350 \\ 0 & 4 & 800 \end{array}\right]$$

$$\xrightarrow{\frac{1}{4}R_2} \left[\begin{array}{cc|c} 1 & 1 & 350 \\ 0 & 1 & 200 \end{array}\right] \xrightarrow{R_1+(-1)R_2} \left[\begin{array}{cc|c} 1 & 0 & 150 \\ 0 & 1 & 200 \end{array}\right]$$

150 short sleeve, 200 long sleeve

67. Let x = adults, y = children

$$\begin{cases} x+y=275 \\ 11.25x+8.50y=2860 \end{cases}$$

$$\left[\begin{array}{cc|c} 1 & 1 & 275 \\ 11.25 & 8.50 & 2860 \end{array}\right] \xrightarrow{R_2+(-11.25)R_1} \left[\begin{array}{cc|c} 1 & 1 & 275 \\ 0 & -2.75 & -233.75 \end{array}\right]$$

$$\xrightarrow{-\frac{1}{2.75}R_2} \left[\begin{array}{cc|c} 1 & 1 & 275 \\ 0 & 1 & 85 \end{array}\right] \xrightarrow{R_1+(-1)R_2} \left[\begin{array}{cc|c} 1 & 0 & 190 \\ 0 & 1 & 85 \end{array}\right]$$

190 adults, 85 children

69.
$$\begin{cases} x+y+z=9.5 \\ -x+y=0.2 \\ x-0.5y=1.75 \end{cases}$$

$$\left[\begin{array}{ccc|c} 1 & 1 & 1 & 9.5 \\ -1 & 1 & 0 & 0.2 \\ 1 & -0.5 & 0 & 1.75 \end{array}\right] \xrightarrow[R_3+(-1)R_1]{R_2+R_1} \left[\begin{array}{ccc|c} 1 & 1 & 1 & 9.5 \\ 0 & 2 & 1 & 9.7 \\ 0 & -1.5 & -1 & -7.75 \end{array}\right]$$

$$\xrightarrow{\frac{1}{2}R_2} \left[\begin{array}{ccc|c} 1 & 1 & 1 & 9.5 \\ 0 & 1 & 0.5 & 4.85 \\ 0 & -1.5 & -1 & -7.75 \end{array}\right] \xrightarrow[R_3+1.5R_2]{R_1+(-1)R_2} \left[\begin{array}{ccc|c} 1 & 0 & 0.5 & 4.65 \\ 0 & 1 & 0.5 & 4.85 \\ 0 & 0 & -0.25 & -.475 \end{array}\right]$$

$$\xrightarrow{-4R_3} \left[\begin{array}{ccc|c} 1 & 0 & 0.5 & 4.65 \\ 0 & 1 & 0.5 & 4.85 \\ 0 & 0 & 1 & 1.9 \end{array}\right]$$

$$\xrightarrow[R_2+(-0.5)R_3]{R_1+(-0.5)R_3} \left[\begin{array}{ccc|c} 1 & 0 & 0 & 3.7 \\ 0 & 1 & 0 & 3.9 \\ 0 & 0 & 1 & 1.9 \end{array}\right]$$

United States is 3.7 million square miles, Canada is 3.9 million square miles and the other countries are 1.9 million square miles

Copyright © 2018 Pearson Education, Inc.

71. $\begin{cases} x+y+z=16 \\ 0.6x+0.5y+0.7z=9.70 \\ x-2y=0 \end{cases}$

$$\left[\begin{array}{ccc|c} 1 & 1 & 1 & 16 \\ 0.6 & 0.5 & 0.7 & 9.70 \\ 1 & -2 & 0 & 0 \end{array}\right]$$

$$\xrightarrow[R_3+(-1)R_1]{R_2+(-0.6)R_1} \left[\begin{array}{ccc|c} 1 & 1 & 1 & 16 \\ 0 & -0.1 & 0.1 & 0.1 \\ 0 & -3 & -1 & -16 \end{array}\right]$$

$$\xrightarrow{-10R_2} \left[\begin{array}{ccc|c} 1 & 1 & 1 & 16 \\ 0 & 1 & -1 & -1 \\ 0 & -3 & -1 & -16 \end{array}\right]$$

$$\xrightarrow[R_3+3R_2]{R_1+(-1)R_2} \left[\begin{array}{ccc|c} 1 & 0 & 2 & 17 \\ 0 & 1 & -1 & -1 \\ 0 & 0 & -4 & -19 \end{array}\right]$$

$$\xrightarrow{-\frac{1}{4}R_3} \left[\begin{array}{ccc|c} 1 & 0 & 2 & 17 \\ 0 & 1 & -1 & -1 \\ 0 & 0 & 1 & 4.75 \end{array}\right]$$

$$\xrightarrow[R_2+R_3]{R_1+(-2)R_3} \left[\begin{array}{ccc|c} 1 & 0 & 0 & 7.5 \\ 0 & 1 & 0 & 3.75 \\ 0 & 0 & 1 & 4.75 \end{array}\right]$$

7.5 ounces of the Brazilian, 3.75 ounces of the Columbian, and 4.75 ounces of the Peruvian

73. $\begin{cases} x+y+z=100{,}000 \\ 0.08x+0.07y+0.1z=8000 \\ x+y-3z=0 \end{cases}$

$$\left[\begin{array}{ccc|c} 1 & 1 & 1 & 100{,}000 \\ 0.08 & 0.07 & 0.1 & 8000 \\ 1 & 1 & -3 & 0 \end{array}\right]$$

$$\xrightarrow{R_2+(-.08)R_1} \left[\begin{array}{ccc|c} 1 & 1 & 1 & 100{,}000 \\ 0 & -0.01 & 0.02 & 0 \\ 1 & 1 & -3 & 0 \end{array}\right]$$

$$\xrightarrow{R_3+(-1)R_1} \left[\begin{array}{ccc|c} 1 & 1 & 1 & 100{,}000 \\ 0 & -0.01 & 0.02 & 0 \\ 0 & 0 & -4 & -100{,}000 \end{array}\right]$$

$$\xrightarrow{(-100)R_2} \left[\begin{array}{ccc|c} 1 & 1 & 1 & 100{,}000 \\ 0 & 1 & -2 & 0 \\ 0 & 0 & -4 & -100{,}000 \end{array}\right]$$

$$\xrightarrow{R_1+(-1)R_2} \left[\begin{array}{ccc|c} 1 & 0 & 3 & 100{,}000 \\ 0 & 1 & -2 & 0 \\ 0 & 0 & -4 & -100{,}000 \end{array}\right]$$

$$\xrightarrow{\left(-\frac{1}{4}\right)R_3} \left[\begin{array}{ccc|c} 1 & 0 & 3 & 100{,}000 \\ 0 & 1 & -2 & 0 \\ 0 & 0 & 1 & 25{,}000 \end{array}\right]$$

$$\xrightarrow{R_1+(-3)R_3} \left[\begin{array}{ccc|c} 1 & 0 & 0 & 25{,}000 \\ 0 & 1 & -2 & 0 \\ 0 & 0 & 1 & 25{,}000 \end{array}\right]$$

$$\xrightarrow{R_2+2R_3} \left[\begin{array}{ccc|c} 1 & 0 & 0 & 25{,}000 \\ 0 & 1 & 0 & 50{,}000 \\ 0 & 0 & 1 & 25{,}000 \end{array}\right]$$

$x = \$25{,}000$, $y = \$50{,}000$, $z = \$25{,}000$

75 Let x = pounds of first type
y = pounds of second type
z = pounds of third type.

$$\begin{aligned} 0.4x + \qquad 0.4z &= 90 \\ 0.6x+0.3y+0.3z &= 100 \\ 0.7y+0.3z &= 120 \end{aligned}$$

$$\left[\begin{array}{ccc|c} 0.4 & 0 & 0.4 & 90 \\ 0.6 & 0.3 & 0.3 & 100 \\ 0 & 0.7 & 0.3 & 120 \end{array}\right]$$

$$\xrightarrow{\frac{1}{0.4}R_1} \left[\begin{array}{ccc|c} 1 & 0 & 1 & 225 \\ 0.6 & 0.3 & 0.3 & 100 \\ 0 & 0.7 & 0.3 & 120 \end{array}\right]$$

$$\xrightarrow{R_2+(-0.6)R_1} \left[\begin{array}{ccc|c} 1 & 0 & 1 & 225 \\ 0 & 0.3 & -0.3 & -35 \\ 0 & 0.7 & 0.3 & 120 \end{array}\right]$$

$$\xrightarrow{\frac{1}{0.3}R_2} \left[\begin{array}{ccc|c} 1 & 0 & 1 & 225 \\ 0 & 1 & -1 & -\frac{350}{3} \\ 0 & 0.7 & 0.3 & 120 \end{array}\right]$$

$$\xrightarrow{R_3+(-.7)R_2} \left[\begin{array}{ccc|c} 1 & 0 & 1 & 225 \\ 0 & 1 & -1 & -\frac{350}{3} \\ 0 & 0 & 1 & \frac{605}{3} \end{array}\right]$$

$$\xrightarrow{R_1+(-1)R_3} \left[\begin{array}{ccc|c} 1 & 0 & 0 & \frac{70}{3} \\ 0 & 1 & -1 & \frac{-350}{3} \\ 0 & 0 & 1 & \frac{605}{3} \end{array}\right]$$

Copyright © 2018 Pearson Education, Inc.

$$\xrightarrow{R_2+R_3}\left[\begin{array}{ccc|c}1&0&0&\frac{70}{3}\\0&1&0&85\\0&0&1&\frac{605}{3}\end{array}\right]$$

$\frac{70}{3}=23\frac{1}{3}$ pounds of the first type, 85 pounds of the second type, and $\frac{605}{3}=201\frac{2}{3}$ pounds of the third type

77. $\left[\begin{array}{cc|c}1&0&-5\\0&1&4\end{array}\right]$

79. $\left[\begin{array}{ccc|c}1&0&0&\frac{175}{54}\\0&1&0&\frac{16}{9}\\0&0&1&\frac{26}{27}\end{array}\right]$

81.

	A	B
1	-2.5	4
2	15	15
3		
4	CELL CONTENT	
5	B1 =.2x+.3y	
6	B2 =.6x+1.1y	

83.

	A	B
1	1	3
2	-6	-6
3	2	1
4		
5	**Cell**	**Content**
6	B1	=x+y+4*z
7	B2	=4*x+y-2*z
8	B3	=-3*x+2*z

Exercises 2.2

1. $\begin{bmatrix}2&-4&6\\3&7&1\end{bmatrix}\xrightarrow[R_2+(-3)R_1]{\frac{1}{2}R_1}\begin{bmatrix}1&-2&3\\0&13&-8\end{bmatrix}$

3. $\begin{bmatrix}7&1&4&5\\-1&1&2&6\\4&0&2&3\end{bmatrix}$

$$\xrightarrow[R_3+(-2)R_2]{\substack{\frac{1}{2}R_2\\R_1+(-4)R_2}}\begin{bmatrix}9&-1&0&-7\\-\frac{1}{2}&\frac{1}{2}&1&3\\5&-1&0&-3\end{bmatrix}$$

5. $\begin{bmatrix}2&3\\6&0\\1&5\end{bmatrix}\xrightarrow[R_3+(-1)R_1]{\substack{\frac{1}{2}R_1\\R_2+(-6)R_1}}\begin{bmatrix}1&\frac{3}{2}\\0&-9\\0&\frac{7}{2}\end{bmatrix}$

7. $\begin{bmatrix}4&3&0\\\frac{2}{3}&0&-2\\1&3&6\end{bmatrix}\xrightarrow[R_2+2R_3]{\frac{1}{6}R_3}\begin{bmatrix}4&3&0\\1&1&0\\\frac{1}{6}&\frac{1}{2}&1\end{bmatrix}$

9. $\begin{cases}x+y+4z=6\\2x+y+z=10\end{cases}$

$z=$ any value, $y=2-7z,\ x=4+3z$

11. $\begin{cases}-5x+15y-10z=5\\x-3y+2z=0\end{cases}$

no solution

13. $\begin{cases}2x-y+5z=12\\-x-4y+2z=3\\8x+5y+11z=30\end{cases}$

$z=$ any value, $y=z-2,\ x=5-2z$

15. $\begin{cases}x+2y+3z-w=4\\2x+3y+w=-3\\4x+7y+6z-w=5\end{cases}$

$z=$ any value, $w=$ any value,
$y=11-6z+3w,\ x=9z-5w-18$

Copyright © 2018 Pearson Education, Inc.

17. $\left[\begin{array}{cc|c} \underline{2} & -4 & 6 \\ -1 & 2 & -3 \end{array}\right]$

$\left[\begin{array}{cc|c} 1 & -2 & 3 \\ 0 & 0 & 0 \end{array}\right]$

$\begin{cases} x-2y=3 \\ 0=0 \end{cases}$

y = any value, $x = 3 + 2y$

19. $\left[\begin{array}{cc|c} \underline{-1} & 3 & 11 \\ 3 & -9 & -30 \end{array}\right]$

$\left[\begin{array}{cc|c} 1 & -3 & 0 \\ 0 & 0 & 1 \end{array}\right]$

$\begin{cases} x-3y=0 \\ 0=1 \end{cases}$

No solution

21. $\left[\begin{array}{cc|c} \underline{1} & 2 & 5 \\ 3 & -1 & 1 \\ -1 & 3 & 5 \end{array}\right]$

$\left[\begin{array}{cc|c} 1 & 2 & 5 \\ 0 & \underline{-7} & -14 \\ 0 & 5 & 10 \end{array}\right]$

$\left[\begin{array}{cc|c} 1 & 0 & 1 \\ 0 & 1 & 2 \\ 0 & 0 & 0 \end{array}\right]$

$x = 1, y = 2$

23. $\left[\begin{array}{cc|c} \underline{4} & 5 & 3 \\ 3 & 6 & 1 \\ 2 & -3 & 7 \end{array}\right]$

$\left[\begin{array}{cc|c} 1 & \frac{5}{4} & \frac{3}{4} \\ 0 & \underline{\frac{9}{4}} & -\frac{5}{4} \\ 0 & -\frac{11}{2} & \frac{11}{2} \end{array}\right]$

$\left[\begin{array}{cc|c} 1 & 0 & 0 \\ 0 & 1 & 0 \\ 0 & 0 & 1 \end{array}\right]$

no solution

25. $\left[\begin{array}{ccc|c} \underline{1} & -1 & 3 & 3 \\ -2 & 3 & -11 & -4 \\ 1 & -2 & 8 & 6 \end{array}\right]$

$\left[\begin{array}{ccc|c} 1 & -1 & 3 & 3 \\ 0 & \underline{1} & -5 & 2 \\ 0 & -1 & 5 & 3 \end{array}\right]$

$\left[\begin{array}{ccc|c} 1 & 0 & -2 & 5 \\ 0 & 1 & -5 & 2 \\ 0 & 0 & 0 & 5 \end{array}\right]$

$\begin{cases} x-2z=5 \\ y-5z=2 \\ 0=5 \end{cases}$

No solution

27. $\left[\begin{array}{ccc|c} \underline{1} & 1 & 1 & -1 \\ 2 & 3 & 2 & 3 \\ 2 & 1 & 2 & -7 \end{array}\right]$

$\left[\begin{array}{ccc|c} 1 & 1 & 1 & -1 \\ 0 & \underline{1} & 0 & 5 \\ 0 & -1 & 0 & -5 \end{array}\right]$

$\left[\begin{array}{ccc|c} 1 & 0 & 1 & -6 \\ 0 & 1 & 0 & 5 \\ 0 & 0 & 0 & 0 \end{array}\right]$

$\begin{cases} x+z=-6 \\ y=5 \\ 0=0 \end{cases}$

z = any value, $x = -z - 6, y = 5$

29. $\left[\begin{array}{ccc|c} \underline{6} & -2 & 2 & 4 \\ 3 & -1 & 2 & 2 \\ -12 & 4 & -8 & 8 \end{array}\right]$

$\left[\begin{array}{ccc|c} 1 & -\frac{1}{3} & 0 & 0 \\ 0 & 0 & 1 & 0 \\ 0 & 0 & 0 & 1 \end{array}\right]$

$\begin{cases} x-\frac{1}{3}y=0 \\ z=0 \\ 0=1 \end{cases}$

No solution

Copyright © 2018 Pearson Education, Inc.

31. $\left[\begin{array}{ccc|c} 1 & 2 & 8 & 1 \\ 3 & -1 & 4 & 10 \\ -1 & 5 & 10 & -8 \\ 1 & 1 & 1 & 3 \end{array}\right]$

$\left[\begin{array}{ccc|c} 1 & 0 & 0 & 0 \\ 0 & 1 & 0 & 0 \\ 0 & 0 & 1 & 0 \\ 0 & 0 & 0 & 1 \end{array}\right]$

$\begin{cases} x = 0 \\ y = 0 \\ z = 0 \\ 0 = 1 \end{cases}$

no solution

33. $\left[\begin{array}{cccc|c} \underline{1} & 1 & -2 & 2 & 5 \\ 2 & 1 & -4 & 1 & 5 \\ 3 & 4 & -6 & 9 & 20 \\ 4 & 4 & -8 & 8 & 20 \end{array}\right]$

$\left[\begin{array}{cccc|c} 1 & 1 & -2 & 2 & 5 \\ 0 & \underline{-1} & 0 & -3 & -5 \\ 0 & 1 & 0 & 3 & 5 \\ 0 & 0 & 0 & 0 & 0 \end{array}\right]$

$\left[\begin{array}{cccc|c} 1 & 0 & -2 & -1 & 0 \\ 0 & 1 & 0 & 3 & 5 \\ 0 & 0 & 0 & 0 & 0 \\ 0 & 0 & 0 & 0 & 0 \end{array}\right]$

$\begin{cases} x - 2z - w = 0 \\ y + 3w = 5 \\ 0 = 0 \\ 0 = 0 \end{cases}$

z = any value, w = any value, $x = 2z + w$, $y = 5 - 3w$

35. $\left[\begin{array}{cccc|c} \underline{1} & -1 & 1 & 1 & 1 \\ 0 & 1 & 3 & 2 & -7 \\ 0 & 1 & -1 & -3 & 1 \\ 1 & 0 & 4 & 3 & 0 \end{array}\right]$

$\left[\begin{array}{cccc|c} 1 & 0 & 0 & -2 & 0 \\ 0 & 1 & 0 & -1.75 & 0 \\ 0 & 0 & 1 & 1.25 & 0 \\ 0 & 0 & 0 & 0 & 1 \end{array}\right]$

$\begin{cases} x - 2w = 0 \\ y - \frac{7}{4}w = 0 \\ z + \frac{5}{4}w = 0 \\ 0 = 1 \end{cases}$

no solution

37. $\left[\begin{array}{ccc|c} \underline{1} & 2 & 1 & 5 \\ 0 & 1 & 3 & 9 \end{array}\right]$

$\left[\begin{array}{ccc|c} 1 & 2 & 1 & 5 \\ 0 & \underline{1} & 3 & 9 \end{array}\right]$

$\left[\begin{array}{ccc|c} 1 & 0 & -5 & -13 \\ 0 & 1 & 3 & 9 \end{array}\right]$

$\begin{cases} x - 5z = -13 \\ y + 3z = 9 \end{cases}$

z = any value, $x = 5z - 13$, $y = -3z + 9$
Possible answers: $z = 0, x = -13, y = 9$; $z = 1, x = -8, y = 6$; $z = 2, x = -3, y = 3$

39. $\left[\begin{array}{ccc|c} \underline{1} & 7 & -3 & 8 \\ 0 & 0 & 1 & 5 \end{array}\right]$

$\left[\begin{array}{ccc|c} 1 & 7 & -3 & 8 \\ 0 & 0 & \underline{1} & 5 \end{array}\right]$

$\left[\begin{array}{ccc|c} 1 & 7 & 0 & 23 \\ 0 & 0 & 1 & 5 \end{array}\right]$

$\begin{cases} x + 7y = 23 \\ z = 5 \end{cases}$

y = any value, $x = -7y + 23$, $z = 5$
Possible answers: $y = 0, x = 23, z = 5$; $y = 1, x = 16, z = 5$; $y = 2, x = 9, z = 5$

41. $\left[\begin{array}{ccc|c} 2 & 4 & 6 & 1000 \\ 3 & 7 & 10 & 1600 \\ 5 & 9 & 14 & 2400 \end{array}\right]$

$\left[\begin{array}{ccc|c} 1 & 0 & 1 & 300 \\ 0 & 1 & 1 & 100 \\ 0 & 0 & 0 & 0 \end{array}\right]$

$\begin{cases} x + z = 300 \\ y + z = 100 \\ 0 = 0 \end{cases}$

Food 3: z = any value between 0 and 100, Food 2: $y = 100 - z$, Food 1: $x = 300 - z$

Copyright © 2018 Pearson Education, Inc.

43. $\left[\begin{array}{ccc|c} 3 & 10 & 15 & 72 \\ 4 & 12 & 8 & 68 \\ 5 & 14 & 1 & 60 \end{array}\right]$

$\left[\begin{array}{ccc|c} 1 & 0 & -25 & 0 \\ 0 & 1 & 9 & 0 \\ 0 & 0 & 0 & 1 \end{array}\right]$

$\begin{cases} x-25z=0 \\ y+9z=0 \\ 0=1 \end{cases}$

There would be no solution

45. $\left[\begin{array}{ccc|c} \underline{1} & 3 & 6 & 380 \\ 3 & 6 & 3 & 450 \end{array}\right]$

$\left[\begin{array}{ccc|c} 1 & 0 & -9 & -310 \\ 0 & 1 & 5 & 230 \end{array}\right]$

$\begin{cases} x-9z=-310 \\ y+5z=230 \end{cases}$

Possible answers: 50 ottomans, 30 sofas, 40 chairs; 5 ottomans, 55 sofas, 35 chairs; 95 ottomans, 5 sofas, 45 chairs

47. $\begin{cases} g+b+f=8\times 12 \\ g+b-15f=0 \\ 3g+3b+5f=300 \end{cases}$

$\left[\begin{array}{ccc|c} \underline{1} & 1 & 1 & 96 \\ 1 & 1 & -15 & 0 \\ 3 & 3 & 5 & 300 \end{array}\right]$

$\left[\begin{array}{ccc|c} 1 & 1 & 1 & 96 \\ 0 & 0 & \underline{-16} & -96 \\ 0 & 0 & 2 & 12 \end{array}\right]$

$\left[\begin{array}{ccc|c} 1 & 1 & 0 & 90 \\ 0 & 0 & 1 & 6 \\ 0 & 0 & 0 & 0 \end{array}\right]$

$\begin{cases} g+b=90 \\ f=6 \\ 0=0 \end{cases}$

6 floral squares, the other 90 any mix of solid green and blue

49. $\left[\begin{array}{cc|c} 2 & -3 & 4 \\ -6 & 9 & k \end{array}\right]$

$\left[\begin{array}{cc|c} 1 & -\frac{3}{2} & 2 \\ 0 & 0 & 12+k \end{array}\right]$

No solution if $0 \neq 12 + k$, which happens when $k \neq -12$.
Infinitely many if $0 = 12 + k$, which happens when $k = -12$.

51. None; There is no point that satisfies all three equations at the same time.

53. 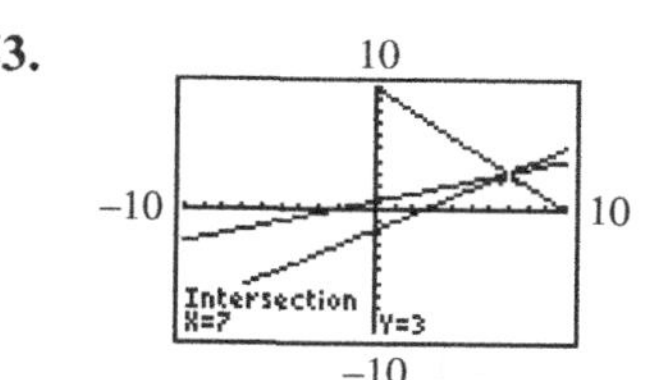

One solution when $x = 7$ and $y = 3$

55. 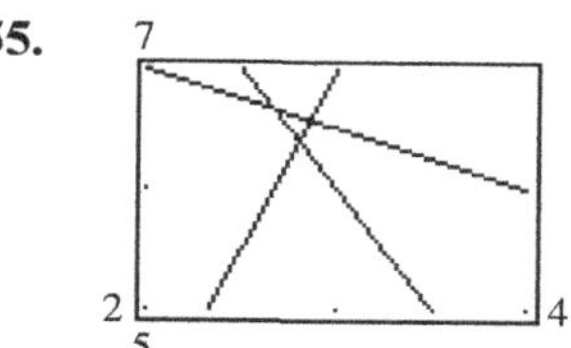

No Solution

57. There has been a pivot about the bottom right entry.

Exercises 2.3

1. 2 by 3

3. 1 by 3, row matrix

5. 2 by 2, square matrix

7. $a_{12} = -4$; $a_{21} = 0$

9. $i = 1$; $j = 3$

11. $\begin{bmatrix} 4+5 & -2+5 \\ 3+4 & 0+(-1) \end{bmatrix} = \begin{bmatrix} 9 & 3 \\ 7 & -1 \end{bmatrix}$

Copyright © 2018 Pearson Education, Inc.

13. $\begin{bmatrix} 1.3+0.7 & 5-1 & 2.3+0.2 \\ -6+0.5 & 0+1 & 0.7+0.5 \end{bmatrix} = \begin{bmatrix} 2 & 4 & 2.5 \\ -5.5 & 1 & 1.2 \end{bmatrix}$

15. $\begin{bmatrix} 2-1 & 8-5 \\ \frac{4}{3}-\frac{1}{3} & 4-2 \\ 1-(-3) & -2-0 \end{bmatrix} = \begin{bmatrix} 1 & 3 \\ 1 & 2 \\ 4 & -2 \end{bmatrix}$

17. $\begin{bmatrix} -5-2 \\ \frac{1}{2}-\frac{1}{3} \end{bmatrix} = \begin{bmatrix} -7 \\ \frac{1}{6} \end{bmatrix}$

19. $[5\cdot 1+3\cdot 2]=[11]$

21. $\left[6\cdot\frac{1}{2}+1(-3)+5\cdot 2\right]=[10]$

23. $\begin{bmatrix} \frac{2}{3}\cdot 6 & \frac{2}{3}\cdot 0 & \frac{2}{3}\cdot -1 \\ \frac{2}{3}\cdot -9 & \frac{2}{3}\cdot\frac{3}{4} & \frac{2}{3}\cdot\frac{1}{2} \end{bmatrix} = \begin{bmatrix} 4 & 0 & -\frac{2}{3} \\ -6 & \frac{1}{2} & \frac{1}{3} \end{bmatrix}$

25. $\begin{bmatrix} 2\cdot .5 & 2\cdot -1 \\ 2\cdot 4 & 2\cdot 0 \end{bmatrix} = \begin{bmatrix} 1 & -2 \\ 8 & 0 \end{bmatrix}$

$\begin{bmatrix} 3\cdot\frac{2}{3} & 3\cdot 7 \\ 3\cdot 5 & 3\cdot 1 \end{bmatrix} = \begin{bmatrix} 2 & 21 \\ 15 & 3 \end{bmatrix}$

$\begin{bmatrix} 1+2 & -2+21 \\ 8+15 & 0+3 \end{bmatrix} = \begin{bmatrix} 3 & 19 \\ 23 & 3 \end{bmatrix}$

27. Yes, columns of A = rows of B; 4, therefore the product will be size 3×5

29. No, columns of $A \neq$ rows of B

31. Yes, columns of A = rows of B; 3, therefore the product will be size 3×1

33. $\begin{bmatrix} 3\cdot 1+1\cdot 3 & 3\cdot 4+1\cdot 5 \\ 0\cdot 1+2\cdot 3 & 0\cdot 4+2\cdot 5 \end{bmatrix} = \begin{bmatrix} 6 & 17 \\ 6 & 10 \end{bmatrix}$

35. $\begin{bmatrix} 4\cdot 5+1\cdot 1+0\cdot 2 \\ -2\cdot 5+0\cdot 1+3\cdot 2 \\ 1\cdot 5+5\cdot 1+(-1)2 \end{bmatrix} = \begin{bmatrix} 21 \\ -4 \\ 8 \end{bmatrix}$

37. Multiplication by identity matrix: $\begin{bmatrix} 5 & 6 \\ 7 & 8 \end{bmatrix}$

39. $\begin{bmatrix} (0.6)(0.6)+(0.3)(0.4) & (0.6)(0.3)+(0.3)(0.7) \\ (0.4)(0.6)+(0.7)(0.4) & (0.4)(0.3)+(0.7)(0.7) \end{bmatrix}$

$= \begin{bmatrix} 0.48 & 0.39 \\ 0.52 & 0.61 \end{bmatrix}$

41. $\begin{bmatrix} 2\cdot 4+(-1)3+4\cdot 5 & 2\cdot 8+(-1)(-1)+4\cdot 0 & 2\cdot 0+(-1)2+4\cdot 1 \\ 0\cdot 4+1\cdot 3+0\cdot 5 & 0\cdot 8+1(-1)+0\cdot 0 & 0\cdot 0+1\cdot 2+0\cdot 1 \\ \frac{1}{2}\cdot 4+3\cdot 3+(-2)5 & \frac{1}{2}\cdot 8+3(-1)+(-2)0 & \frac{1}{2}\cdot 0+3\cdot 2+(-2)\cdot 1 \end{bmatrix} = \begin{bmatrix} 25 & 17 & 2 \\ 3 & -1 & 2 \\ 1 & 1 & 4 \end{bmatrix}$

43. $\begin{bmatrix} (\frac{1}{3})(\frac{1}{3})+(\frac{2}{3})(\frac{1}{3}) & (\frac{1}{3})(\frac{2}{3})+(\frac{2}{3})(\frac{2}{3}) \\ (\frac{1}{3})(\frac{1}{3})+(\frac{2}{3})(\frac{1}{3}) & (\frac{1}{3})(\frac{2}{3})+(\frac{2}{3})(\frac{2}{3}) \end{bmatrix} = \begin{bmatrix} \frac{1}{3} & \frac{2}{3} \\ \frac{1}{3} & \frac{2}{3} \end{bmatrix}$

45. $[2(0)+5(6) \quad 2(3)+5(7)]=[30 \quad 41]$

47. $\begin{bmatrix} 2(5)+0(0) & 2(0)+0(5) \\ 0(5)+3(0) & 0(0)+3(5) \end{bmatrix} = \begin{bmatrix} 10 & 0 \\ 0 & 15 \end{bmatrix}$

49. $\begin{bmatrix} 0(2.34)+0(-3.7) & 0(5.6)+0(0.08) \\ 0(2.34)+0(-3.7) & 0(5.6)+0(0.08) \end{bmatrix} = \begin{bmatrix} 0 & 0 \\ 0 & 0 \end{bmatrix}$

51. $\begin{bmatrix} 23(1)+24(0) & 23(0)+24(1) \\ 25(1)+26(0) & 25(0)+26(1) \end{bmatrix} = \begin{bmatrix} 23 & 24 \\ 25 & 26 \end{bmatrix}$

Copyright © 2018 Pearson Education, Inc.

53. $\begin{cases} 2x+3y=6 \\ 4x+5y=7 \end{cases}$

55. $\begin{cases} x+2y+3z=10 \\ 4x+5y+6z=11 \\ 7x+8y+9z=12 \end{cases}$

57. $\begin{bmatrix} 3 & 2 \\ 7 & -1 \end{bmatrix}\begin{bmatrix} x \\ y \end{bmatrix}=\begin{bmatrix} -1 \\ 2 \end{bmatrix}$

59. $\begin{bmatrix} 1 & -2 & 3 \\ 0 & 1 & 1 \\ 0 & 0 & 1 \end{bmatrix}\begin{bmatrix} x \\ y \\ z \end{bmatrix}=\begin{bmatrix} 5 \\ 6 \\ 2 \end{bmatrix}$

61. $\left(\begin{bmatrix} 1 & 2 \\ 0 & 3 \end{bmatrix}+\begin{bmatrix} 3 & -2 \\ 4 & 5 \end{bmatrix}\right)\begin{bmatrix} 1 & 6 \\ 2 & 0 \end{bmatrix}=\begin{bmatrix} 4 & 0 \\ 4 & 8 \end{bmatrix}\begin{bmatrix} 1 & 6 \\ 2 & 0 \end{bmatrix}=\begin{bmatrix} 4 & 24 \\ 20 & 24 \end{bmatrix}$

$\begin{bmatrix} 1 & 2 \\ 0 & 3 \end{bmatrix}\begin{bmatrix} 1 & 6 \\ 2 & 0 \end{bmatrix}+\begin{bmatrix} 3 & -2 \\ 4 & 5 \end{bmatrix}\begin{bmatrix} 1 & 6 \\ 2 & 0 \end{bmatrix}=\begin{bmatrix} 5 & 6 \\ 6 & 0 \end{bmatrix}+\begin{bmatrix} -1 & 18 \\ 14 & 24 \end{bmatrix}=\begin{bmatrix} 4 & 24 \\ 20 & 24 \end{bmatrix}$

63. $\begin{bmatrix} 3\cdot 1+(-1)2 & 3\cdot 2+(-1)6 \\ -1\cdot 1+\frac{1}{2}\cdot 2 & -1\cdot 2+\frac{1}{2}\cdot 6 \end{bmatrix}=\begin{bmatrix} 1 & 0 \\ 0 & 1 \end{bmatrix}$

$\begin{bmatrix} 1\cdot 3+2(-1) & 1(-1)+(2)\frac{1}{2} \\ 2\cdot 3+6(-1) & 2(-1)+(6)\frac{1}{2} \end{bmatrix}=\begin{bmatrix} 1 & 0 \\ 0 & 1 \end{bmatrix}$

65. a. $\begin{bmatrix} 6 & 8 & 2 \\ 2 & 5 & 3 \end{bmatrix}\begin{bmatrix} 20 \\ 15 \\ 50 \end{bmatrix}=\begin{bmatrix} 340 \\ 265 \end{bmatrix}$

b. Mike's clothes are worth $340; Don's clothes are worth $265.

c. $\begin{bmatrix} 1.25\cdot 20 \\ 1.25\cdot 15 \\ 1.25\cdot 50 \end{bmatrix}=\begin{bmatrix} 25 \\ 18.75 \\ 62.50 \end{bmatrix}$

d. The matrix represents the costs of the three items of clothing after a 25% increase.

67. a. $\begin{bmatrix} 210 & 175 & 135 \end{bmatrix}\begin{bmatrix} 3 & 3 & 5.8 \\ 2.5 & 3.5 & 6 \\ 9 & 8 & 9.5 \end{bmatrix}$

$=\begin{bmatrix} 2282.50 & 2322.50 & 3550.50 \end{bmatrix}$

The total value of the store's plain items was $2282.50, of the milk chocolate-covered items was $2322.50, and of the dark chocolate covered items was $3550.50.

b. $\begin{bmatrix} 3 & 3 & 5.8 \\ 2.5 & 3.5 & 6 \\ 9 & 8 & 9.5 \end{bmatrix}\begin{bmatrix} 105 \\ 390 \\ 285 \end{bmatrix}=\begin{bmatrix} 3138.00 \\ 3337.50 \\ 6772.50 \end{bmatrix}$

The store's weekly sales of peanuts was $3138.00, of raisins was $3337.50, and of espresso beans was $6772.50.

Copyright © 2018 Pearson Education, Inc.

c. $\begin{bmatrix} .9\cdot 105 \\ .9\cdot 390 \\ .9\cdot 285 \end{bmatrix} = \begin{bmatrix} 94.50 \\ 351 \\ 256.50 \end{bmatrix}$

The store's weekly sales if there were a 10% reduction in the number of pound sold.

69. a. $\begin{bmatrix} 0.25 & 0.35 & 0.30 & 0.10 & 0 \\ 0.10 & 0.20 & 0.40 & 0.20 & 0.10 \\ 0.05 & 0.10 & 0.20 & 0.40 & 0.25 \end{bmatrix} \begin{bmatrix} 4 \\ 3 \\ 2 \\ 1 \\ 0 \end{bmatrix} = \begin{bmatrix} 2.75 \\ 2.00 \\ 1.30 \end{bmatrix}$

I: 2.75; II: 2, III: 1.3

b. $[240\ 120\ 40] \begin{bmatrix} 0.25 & 0.35 & 0.30 & .10 & 0 \\ 0.10 & 0.20 & 0.40 & .20 & .10 \\ 0.05 & 0.10 & 0.20 & .40 & .25 \end{bmatrix}$

$= [74\ 112\ 128\ 64\ 22]$

A: 74, B: 112, C: 128, D: 64, F: 22

71. $[6000\ 8000\ 4000] \begin{bmatrix} 0.65 & 0.35 \\ 0.55 & 0.45 \\ 0.45 & 0.55 \end{bmatrix} = [10{,}100\ 7900]$

10,100 voting Democratic, 7900 voting Republican

73. $\begin{bmatrix} 50 & 20 & 10 \\ 30 & 30 & 15 \\ 20 & 20 & 5 \end{bmatrix} \begin{bmatrix} 20 \\ 30 \\ 40 \end{bmatrix} = \begin{bmatrix} 2000 \\ 2100 \\ 1200 \end{bmatrix}$

Carpenters: $2000, Bricklayers: $2100, Plumbers: $1200

75. a. $BN = [162\quad 150\quad 143]$, number of units of each nutrient consumed at breakfast

b. $LN = [186\quad 200\quad 239]$, number of units of each nutrient consumed at lunch

c. $DN = [288\quad 300\quad 344]$, number of units of each nutrient consumed at dinner

d. $B + L + D = [5\quad 8]$, total number of ounces of each food that Mikey eats during a day

e. $(B + L + D)N = [636\quad 650\quad 726]$, number of units of each nutrient consumed per day

77. a. $AP = \begin{bmatrix} 720 \\ 646 \end{bmatrix}$, the average amount taken in daily by the pool and the weight room

b. $720

79. a.

	Boston Cream Pie	Carrot Cake	
$T =$	30	45	Preparation
	30	50	Baking
	15	10	Finishing

Copyright © 2018 Pearson Education, Inc.

b. $S=\begin{bmatrix}20\\8\end{bmatrix}\begin{matrix}\text{Boston Cream Pie}\\\text{Carrot Cake}\end{matrix}$

$TS=\begin{bmatrix}960\\1000\\380\end{bmatrix}\begin{matrix}\text{Preparation}\\\text{Baking}\\\text{Finishing}\end{matrix}$

c. Total baking time is 1000 minutes or $16\frac{2}{3}$ hours. Total finishing time is 380 minutes or $6\frac{1}{3}$ hours.

81. a.
$$T=\begin{matrix} & \text{Cutting} & \text{Sewing} & \text{Finishing} & \\ & \left[\begin{matrix}2\\1.5\end{matrix}\right. & \begin{matrix}3\\2\end{matrix} & \left.\begin{matrix}2\\1\end{matrix}\right] & \begin{matrix}\text{Huge One}\\\text{Regular Joe}\end{matrix}\end{matrix}$$

b. $S=\begin{bmatrix}32\\24\end{bmatrix}\begin{matrix}\text{Huge One}\\\text{Regular Joe}\end{matrix}$

c.
$$A=\begin{matrix}\text{Huge One} & \text{Regular Joe}\\ [\ 27 & 56\ ]\end{matrix}$$

$$AT=\begin{matrix}\text{Cutting} & \text{Sewing} & \text{Finishing}\\ [\ 138 & 193 & 110\ ]\end{matrix}$$

$AS=[2208]$

d. 193 hours are needed for sewing.

e. The total revenue would be \$2208.

83. answers will vary.

85. $(A+B)-A=\begin{bmatrix}3&-2&1\\-5&6&7\end{bmatrix}$

87. 4×4

89. The matrix product:

$[\ 9257\quad 57{,}718\quad 89{,}389]\begin{bmatrix}13.9\\14.9\\14.2\end{bmatrix}$ gives the total number of pupils in the three states.

91. $A+B=\begin{bmatrix}6.4&-2&-2.7\\20.5&22.5&-2.4\\-14&17.6&16\end{bmatrix}$

93. $BA=\begin{bmatrix}-171.3&40.8&-31.8\\454.6&-22.5&22.7\\-2.6&122.3&53.56\end{bmatrix}$

95. $3A=\begin{bmatrix}1.2&21&-9\\57&1.5&4.8\\-27&33&6\end{bmatrix}$

97. Answer may vary. One possibility is with the message ERR:DIM MISMATCH

Exercises 2.4

1. $\begin{bmatrix}1&-2\\-\frac{1}{2}&2\end{bmatrix}\begin{bmatrix}4\\1\end{bmatrix}=\begin{bmatrix}2\\0\end{bmatrix}$

$x=2, y=0$

Copyright © 2018 Pearson Education, Inc.

3. $D = 7\cdot 1 - 3\cdot 2 = 1$

$$\begin{bmatrix} \frac{1}{1} & -\frac{2}{1} \\ -\frac{3}{1} & \frac{7}{1} \end{bmatrix} = \begin{bmatrix} 1 & -2 \\ -3 & 7 \end{bmatrix}$$

5. $D = 6\cdot 2 - 5\cdot 2 = 2$

$$\begin{bmatrix} \frac{2}{2} & -\frac{2}{2} \\ -\frac{5}{2} & \frac{6}{2} \end{bmatrix} = \begin{bmatrix} 1 & -1 \\ -\frac{5}{2} & 3 \end{bmatrix}$$

7. $D = 0.7\cdot 0.8 - 0.3\cdot 0.2 = 0.5$

$$\begin{bmatrix} \frac{.8}{.5} & -\frac{.2}{.5} \\ -\frac{.3}{.5} & \frac{.7}{.5} \end{bmatrix} = \begin{bmatrix} 1.6 & -.4 \\ -.6 & 1.4 \end{bmatrix}$$

9. For a 1×1 matrix $[a]$ $(a \neq 0)$, $[a]^{-1} = \begin{bmatrix} \frac{1}{a} \end{bmatrix}$.

$$\begin{bmatrix} \frac{1}{3} \end{bmatrix}$$

11. $$\begin{bmatrix} 1 & 2 \\ 2 & 6 \end{bmatrix}^{-1} \begin{bmatrix} 3 \\ 5 \end{bmatrix} = \begin{bmatrix} 3 & -1 \\ -1 & \frac{1}{2} \end{bmatrix} \begin{bmatrix} 3 \\ 5 \end{bmatrix} = \begin{bmatrix} 4 \\ -\frac{1}{2} \end{bmatrix}$$

$x = 4,\ y = -\frac{1}{2}$

13. $$\begin{bmatrix} \frac{1}{2} & 2 \\ 3 & 16 \end{bmatrix}^{-1} \begin{bmatrix} 4 \\ 0 \end{bmatrix} = \begin{bmatrix} 8 & -1 \\ -\frac{3}{2} & \frac{1}{4} \end{bmatrix} \begin{bmatrix} 4 \\ 0 \end{bmatrix} = \begin{bmatrix} 32 \\ -6 \end{bmatrix}$$

$x = 32, y = -6$

15. a. $$\begin{bmatrix} 0.8 & 0.3 \\ 0.2 & 0.7 \end{bmatrix} \begin{bmatrix} x \\ y \end{bmatrix} = \begin{bmatrix} m \\ s \end{bmatrix}$$

b. $$\begin{bmatrix} x \\ y \end{bmatrix} = \begin{bmatrix} 0.8 & 0.3 \\ 0.2 & 0.7 \end{bmatrix}^{-1} \begin{bmatrix} m \\ s \end{bmatrix} = \begin{bmatrix} 1.4 & -0.6 \\ -0.4 & 1.6 \end{bmatrix} \begin{bmatrix} m \\ s \end{bmatrix}$$

c. $$\begin{bmatrix} 1.4 & -0.6 \\ -0.4 & 1.6 \end{bmatrix} \begin{bmatrix} 100{,}000 \\ 50{,}000 \end{bmatrix} = \begin{bmatrix} 110{,}000 \\ 40{,}000 \end{bmatrix}$$

110,000 married; 40,000 single

d. $$\begin{bmatrix} 1.4 & -0.6 \\ -0.4 & 1.6 \end{bmatrix} \begin{bmatrix} 110{,}000 \\ 40{,}000 \end{bmatrix} = \begin{bmatrix} 130{,}000 \\ 20{,}000 \end{bmatrix}$$

130,000 married; 20,000 single

17. a. $$\begin{bmatrix} 0.7 & 0.1 \\ 0.3 & 0.9 \end{bmatrix} \begin{bmatrix} x \\ y \end{bmatrix} = \begin{bmatrix} u \\ v \end{bmatrix}$$

b. $$\begin{bmatrix} x \\ y \end{bmatrix} = \begin{bmatrix} 0.7 & 0.1 \\ 0.3 & 0.9 \end{bmatrix}^{-1} \begin{bmatrix} u \\ v \end{bmatrix} = \begin{bmatrix} \frac{3}{2} & -\frac{1}{6} \\ -\frac{1}{2} & \frac{7}{6} \end{bmatrix} \begin{bmatrix} u \\ v \end{bmatrix}$$

c. $$\begin{bmatrix} \frac{3}{2} & -\frac{1}{6} \\ -\frac{1}{2} & \frac{7}{6} \end{bmatrix} \begin{bmatrix} 6000 \\ 3000 \end{bmatrix} = \begin{bmatrix} 8500 \\ 500 \end{bmatrix}$$

$$\begin{bmatrix} 0.7 & 0.1 \\ 0.3 & 0.9 \end{bmatrix} \begin{bmatrix} 6000 \\ 3000 \end{bmatrix} = \begin{bmatrix} 4500 \\ 4500 \end{bmatrix}$$

8500; 4500

19. $$\begin{bmatrix} 5 & -2 & -2 \\ -1 & 1 & 0 \\ -1 & 0 & 1 \end{bmatrix} \begin{bmatrix} 1 \\ -1 \\ -1 \end{bmatrix} = \begin{bmatrix} 9 \\ -2 \\ -2 \end{bmatrix}$$

$x = 9, y = -2, z = -2$

21. $$\begin{bmatrix} 1 & 2 & 2 \\ 1 & 3 & 2 \\ 1 & 2 & 3 \end{bmatrix} \begin{bmatrix} 3 \\ 4 \\ 5 \end{bmatrix} = \begin{bmatrix} 21 \\ 25 \\ 26 \end{bmatrix}$$

$x = 21, y = 25, z = 26$

23. $$\begin{bmatrix} 1 & 0 & -2 & 0 \\ 0 & 1 & 0 & -5 \\ -4 & 0 & 9 & 0 \\ 0 & 2 & 1 & -9 \end{bmatrix} \begin{bmatrix} 1 \\ 0 \\ 0 \\ -1 \end{bmatrix} = \begin{bmatrix} 1 \\ 5 \\ -4 \\ 9 \end{bmatrix}$$

$x = 1, y = 5, z = -4, w = 9$

25. $$\begin{bmatrix} 9 & 0 & 2 & 0 \\ -20 & -9 & -5 & 5 \\ 4 & 0 & 1 & 0 \\ -4 & -2 & -1 & 1 \end{bmatrix} \begin{bmatrix} 0 \\ 1 \\ 2 \\ 0 \end{bmatrix} = \begin{bmatrix} 4 \\ -19 \\ 2 \\ -4 \end{bmatrix}$$

$x = 4, y = -19, z = 2, w = -4$

27. $D = a\cdot b - 0\cdot 0 = ab$

$$A^{-1} = \frac{1}{D}\begin{bmatrix} b & 0 \\ 0 & a \end{bmatrix} = \begin{bmatrix} \frac{b}{ab} & -\frac{0}{ab} \\ -\frac{0}{ab} & \frac{a}{ab} \end{bmatrix} = \begin{bmatrix} \frac{1}{a} & 0 \\ 0 & \frac{1}{b} \end{bmatrix}$$

29. a. $$\begin{cases} x + 2y = a \\ .9x = b \end{cases}$$

$$\begin{bmatrix} 1 & 2 \\ .9 & 0 \end{bmatrix} \begin{bmatrix} x \\ y \end{bmatrix} = \begin{bmatrix} a \\ b \end{bmatrix}$$

Copyright © 2018 Pearson Education, Inc.

b. $\begin{bmatrix} 1 & 2 \\ .9 & 0 \end{bmatrix}\begin{bmatrix} 450{,}000 \\ 360{,}000 \end{bmatrix} = \begin{bmatrix} 1{,}170{,}000 \\ 405{,}000 \end{bmatrix}$

After 1 year:
1,170,000 in group I,
405,000 in group II

$\begin{bmatrix} 1 & 2 \\ .9 & 0 \end{bmatrix}\begin{bmatrix} 1{,}170{,}000 \\ 405{,}000 \end{bmatrix} = \begin{bmatrix} 1{,}980{,}000 \\ 1{,}053{,}000 \end{bmatrix}$

After 2 years:
1,980,000 in group I,
1,053,000 in group II

c. $\begin{bmatrix} 1 & 2 \\ .9 & 0 \end{bmatrix}^{-1}\begin{bmatrix} 810{,}000 \\ 630{,}000 \end{bmatrix} = \begin{bmatrix} 700{,}000 \\ 55{,}000 \end{bmatrix}$

700,000 in group I, 55,000 in group II

31. If AB = 0 (zero matrix) and A has an inverse the B is a matrix of all zeros:
Proof : Assume AB = 0 and A has an inverse.

$A^{-1}(AB) = A^{-1}(0) \rightarrow$

$(A^{-1}A)B = A^{-1}(0) \rightarrow$

$(I_n)B = 0 \rightarrow$

$B = 0$

33. One example:
$AX = B$

Let $A = \begin{bmatrix} 1 & 1 \\ 1 & 1 \end{bmatrix}$ $X = \begin{bmatrix} x \\ y \end{bmatrix}$ $B = \begin{bmatrix} 2 \\ 3 \end{bmatrix}$

The system $\begin{cases} x + y = 2 \\ x + y = 3 \end{cases}$ has no solution.

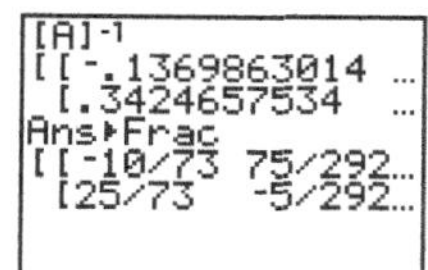

$\begin{bmatrix} -\frac{10}{73} & \frac{75}{292} \\ \frac{25}{73} & -\frac{5}{292} \end{bmatrix}$

37.

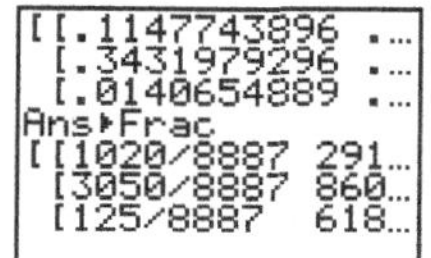

$\begin{bmatrix} \frac{1020}{8887} & \frac{2910}{8887} & -\frac{500}{8887} \\ \frac{3050}{8887} & \frac{860}{8887} & \frac{1990}{8887} \\ \frac{125}{8887} & \frac{618}{8887} & \frac{810}{8887} \end{bmatrix}$

39.

```
[A]-1[B]▸Frac
          [[-4/5]
           [28/5]
           [5   ]]
```

$x = -\frac{4}{5},\ y = \frac{28}{5},\ z = 5$

41.

```
[A]-1[B]▸Frac
            [[0]
             [2]
             [0]
             [2]]
```

$x = 0, y = 2, z = 0, w = 2$

43. Answer may vary. One possibility is with the message ERR:INVALID DIM

Exercises 2.5

1. $\left[\begin{array}{cc|cc} 7 & 3 & 1 & 0 \\ 5 & 2 & 0 & 1 \end{array}\right]$

$\left[\begin{array}{cc|cc} 1 & \frac{3}{7} & \frac{1}{7} & 0 \\ 0 & -\frac{1}{7} & -\frac{5}{7} & 1 \end{array}\right]$

$\left[\begin{array}{cc|cc} 1 & 0 & -2 & 3 \\ 0 & 1 & 5 & -7 \end{array}\right]$

$\begin{bmatrix} -2 & 3 \\ 5 & -7 \end{bmatrix}$

3. $\left[\begin{array}{cc|cc} 2 & 3 & 1 & 0 \\ -4 & -7 & 0 & 1 \end{array}\right]$

$\left[\begin{array}{cc|cc} 1 & \frac{3}{2} & \frac{1}{2} & 0 \\ 0 & -1 & 2 & 1 \end{array}\right]$

$\left[\begin{array}{cc|cc} 1 & 0 & \frac{7}{2} & \frac{3}{2} \\ 0 & 1 & -2 & -1 \end{array}\right]$

$\begin{bmatrix} \frac{7}{2} & \frac{3}{2} \\ -2 & -1 \end{bmatrix}$

Copyright © 2018 Pearson Education, Inc.

5. $\left[\begin{array}{cc|cc} \underline{2} & -4 & 1 & 0 \\ -1 & 2 & 0 & 1 \end{array}\right]$

$\left[\begin{array}{cc|cc} 1 & -2 & \frac{1}{2} & 0 \\ 0 & 0 & \frac{1}{2} & 1 \end{array}\right]$

No inverse

7. $\left[\begin{array}{ccc|ccc} \underline{1} & 2 & -2 & 1 & 0 & 0 \\ 1 & 1 & 1 & 0 & 1 & 0 \\ 0 & 0 & 1 & 0 & 0 & 1 \end{array}\right]$

$\left[\begin{array}{ccc|ccc} 1 & 2 & -2 & 1 & 0 & 0 \\ 0 & \underline{-1} & 3 & -1 & 1 & 0 \\ 0 & 0 & 1 & 0 & 0 & 1 \end{array}\right]$

$\left[\begin{array}{ccc|ccc} 1 & 0 & 4 & -1 & 2 & 0 \\ 0 & 1 & -3 & 1 & -1 & 0 \\ 0 & 0 & \underline{1} & 0 & 0 & 1 \end{array}\right]$

$\left[\begin{array}{ccc|ccc} 1 & 0 & 0 & -1 & 2 & -4 \\ 0 & 1 & 0 & 1 & -1 & 3 \\ 0 & 0 & 1 & 0 & 0 & 1 \end{array}\right]$

$\begin{bmatrix} -1 & 2 & -4 \\ 1 & -1 & 3 \\ 0 & 0 & 1 \end{bmatrix}$

9. $\left[\begin{array}{ccc|ccc} \underline{-2} & 5 & 2 & 1 & 0 & 0 \\ 1 & -3 & -1 & 0 & 1 & 0 \\ -1 & 2 & 1 & 0 & 0 & 1 \end{array}\right]$

$\left[\begin{array}{ccc|ccc} 1 & -\frac{5}{2} & -1 & -\frac{1}{2} & 0 & 0 \\ 0 & \underline{-\frac{1}{2}} & 0 & \frac{1}{2} & 1 & 0 \\ 0 & -\frac{1}{2} & 0 & -\frac{1}{2} & 0 & 1 \end{array}\right]$

$\left[\begin{array}{ccc|ccc} 1 & 0 & -1 & -3 & -5 & 0 \\ 0 & 1 & 0 & -1 & -2 & 0 \\ 0 & 0 & 0 & -1 & -1 & 1 \end{array}\right]$

No inverse

11. $\left[\begin{array}{cccc|cccc} \underline{1} & 6 & 0 & 0 & 1 & 0 & 0 & 0 \\ 1 & 5 & 0 & 0 & 0 & 1 & 0 & 0 \\ 0 & 0 & 4 & 2 & 0 & 0 & 1 & 0 \\ 0 & 0 & 50 & 2 & 0 & 0 & 0 & 1 \end{array}\right]$

$\left[\begin{array}{cccc|cccc} 1 & 6 & 0 & 0 & 1 & 0 & 0 & 0 \\ 0 & \underline{-1} & 0 & 0 & -1 & 1 & 0 & 0 \\ 0 & 0 & 4 & 2 & 0 & 0 & 1 & 0 \\ 0 & 0 & 50 & 2 & 0 & 0 & 0 & 1 \end{array}\right]$

$\left[\begin{array}{cccc|cccc} 1 & 0 & 0 & 0 & -5 & 6 & 0 & 0 \\ 0 & 1 & 0 & 0 & 1 & -1 & 0 & 0 \\ 0 & 0 & \underline{4} & 2 & 0 & 0 & 1 & 0 \\ 0 & 0 & 50 & 2 & 0 & 0 & 0 & 1 \end{array}\right]$

$\left[\begin{array}{cccc|cccc} 1 & 0 & 0 & 0 & -5 & 6 & 0 & 0 \\ 0 & 1 & 0 & 0 & 1 & -1 & 0 & 0 \\ 0 & 0 & 1 & \frac{1}{2} & 0 & 0 & \frac{1}{4} & 0 \\ 0 & 0 & 0 & \underline{-23} & 0 & 0 & -\frac{25}{2} & 1 \end{array}\right]$

$\left[\begin{array}{cccc|cccc} 1 & 0 & 0 & 0 & -5 & 6 & 0 & 0 \\ 0 & 1 & 0 & 0 & 1 & -1 & 0 & 0 \\ 0 & 0 & 1 & 0 & 0 & 0 & -\frac{1}{46} & \frac{1}{46} \\ 0 & 0 & 0 & 1 & 0 & 0 & \frac{25}{46} & -\frac{1}{23} \end{array}\right]$

$\begin{bmatrix} -5 & 6 & 0 & 0 \\ 1 & -1 & 0 & 0 \\ 0 & 0 & -\frac{1}{46} & \frac{1}{46} \\ 0 & 0 & \frac{25}{46} & -\frac{1}{23} \end{bmatrix}$

13. Find the inverse of $\begin{bmatrix} 1 & 1 & 2 \\ 3 & 2 & 2 \\ 1 & 1 & 3 \end{bmatrix}$.

$\left[\begin{array}{ccc|ccc} 1 & 1 & 2 & 1 & 0 & 0 \\ 3 & 2 & 2 & 0 & 1 & 0 \\ 1 & 1 & 3 & 0 & 0 & 1 \end{array}\right]$

$\left[\begin{array}{ccc|ccc} 1 & 1 & 2 & 1 & 0 & 0 \\ 0 & -1 & -4 & -3 & 1 & 0 \\ 0 & 0 & 1 & -1 & 0 & 1 \end{array}\right]$

$\left[\begin{array}{ccc|ccc} 1 & 0 & -2 & -2 & 1 & 0 \\ 0 & 1 & 4 & 3 & -1 & 0 \\ 0 & 0 & 1 & -1 & 0 & 1 \end{array}\right]$

Copyright © 2018 Pearson Education, Inc.

$$\left[\begin{array}{ccc|ccc} 1 & 0 & 0 & -4 & 1 & 2 \\ 0 & 1 & 0 & 7 & -1 & -4 \\ 0 & 0 & 1 & -1 & 0 & 1 \end{array}\right]$$

$$\begin{bmatrix} 1 & 1 & 2 \\ 3 & 2 & 2 \\ 1 & 1 & 3 \end{bmatrix}^{-1} \begin{bmatrix} 3 \\ 4 \\ 5 \end{bmatrix} = \begin{bmatrix} -4 & 1 & 2 \\ 7 & -1 & -4 \\ -1 & 0 & 1 \end{bmatrix} \begin{bmatrix} 3 \\ 4 \\ 5 \end{bmatrix} = \begin{bmatrix} 2 \\ -3 \\ 2 \end{bmatrix}$$

$x = 2, y = -3, z = 2$

15. Find the inverse of $\begin{bmatrix} 1 & 4 & 3 \\ 1 & 3 & 4 \\ 2 & 3 & 3 \end{bmatrix}$.

$$\left[\begin{array}{ccc|ccc} 1 & 4 & 3 & 1 & 0 & 0 \\ 1 & 3 & 4 & 0 & 1 & 0 \\ 2 & 3 & 3 & 0 & 0 & 1 \end{array}\right]$$

$$\left[\begin{array}{ccc|ccc} 1 & 4 & 3 & 1 & 0 & 0 \\ 0 & -1 & 1 & -1 & 1 & 0 \\ 0 & -5 & -3 & -2 & 0 & 1 \end{array}\right]$$

$$\left[\begin{array}{ccc|ccc} 1 & 0 & 7 & -3 & 4 & 0 \\ 0 & 1 & -1 & 1 & -1 & 0 \\ 0 & 0 & -8 & 3 & -5 & 1 \end{array}\right]$$

$$\left[\begin{array}{ccc|ccc} 1 & 0 & 0 & -\frac{3}{8} & -\frac{3}{8} & \frac{7}{8} \\ 0 & 1 & 0 & \frac{5}{8} & -\frac{3}{8} & -\frac{1}{8} \\ 0 & 0 & 1 & -\frac{3}{8} & \frac{5}{8} & -\frac{1}{8} \end{array}\right]$$

$$\begin{bmatrix} 1 & 4 & 3 \\ 1 & 3 & 4 \\ 2 & 3 & 3 \end{bmatrix}^{-1} \begin{bmatrix} 15 \\ 17 \\ 16 \end{bmatrix} = \begin{bmatrix} -\frac{3}{8} & -\frac{3}{8} & \frac{7}{8} \\ \frac{5}{8} & -\frac{3}{8} & -\frac{1}{8} \\ -\frac{3}{8} & \frac{5}{8} & -\frac{1}{8} \end{bmatrix} \begin{bmatrix} 15 \\ 17 \\ 16 \end{bmatrix} = \begin{bmatrix} 2 \\ 1 \\ 3 \end{bmatrix}$$

$x = 2, y = 1, z = 3$

17. Find the inverse of $\begin{bmatrix} 1 & 0 & -2 & -2 \\ 0 & 1 & 0 & -5 \\ -4 & 0 & 9 & 9 \\ 0 & 2 & 1 & -8 \end{bmatrix}$.

$$\left[\begin{array}{cccc|cccc} 1 & 0 & -2 & -2 & 1 & 0 & 0 & 0 \\ 0 & 1 & 0 & -5 & 0 & 1 & 0 & 0 \\ -4 & 0 & 9 & 9 & 0 & 0 & 1 & 0 \\ 0 & 2 & 1 & -8 & 0 & 0 & 0 & 1 \end{array}\right]$$

$$\left[\begin{array}{cccc|cccc} 1 & 0 & -2 & -2 & 1 & 0 & 0 & 0 \\ 0 & 1 & 0 & -5 & 0 & 1 & 0 & 0 \\ 0 & 0 & 1 & 1 & 4 & 0 & 1 & 0 \\ 0 & 2 & 1 & -8 & 0 & 0 & 0 & 1 \end{array}\right]$$

$$\left[\begin{array}{cccc|cccc} 1 & 0 & -2 & -2 & 1 & 0 & 0 & 0 \\ 0 & 1 & 0 & -5 & 0 & 1 & 0 & 0 \\ 0 & 0 & 1 & 1 & 4 & 0 & 1 & 0 \\ 0 & 0 & 1 & 2 & 0 & -2 & 0 & 1 \end{array}\right]$$

$$\left[\begin{array}{cccc|cccc} 1 & 0 & 0 & 0 & 9 & 0 & 2 & 0 \\ 0 & 1 & 0 & -5 & 0 & 1 & 0 & 0 \\ 0 & 0 & 1 & 1 & 4 & 0 & 1 & 0 \\ 0 & 0 & 0 & 1 & -4 & -2 & -1 & 1 \end{array}\right]$$

$$\left[\begin{array}{cccc|cccc} 1 & 0 & 0 & 0 & 9 & 0 & 2 & 0 \\ 0 & 1 & 0 & 0 & -20 & -9 & -5 & 5 \\ 0 & 0 & 1 & 0 & 8 & 2 & 2 & -1 \\ 0 & 0 & 0 & 1 & -4 & -2 & -1 & 1 \end{array}\right]$$

$$\begin{bmatrix} 1 & 0 & -2 & -2 \\ 0 & 1 & 0 & -5 \\ -4 & 0 & 9 & 9 \\ 0 & 2 & 1 & -8 \end{bmatrix}^{-1} \begin{bmatrix} 0 \\ 1 \\ 2 \\ 3 \end{bmatrix}$$

$$= \begin{bmatrix} 9 & 0 & 2 & 0 \\ -20 & -9 & -5 & 5 \\ 8 & 2 & 2 & -1 \\ -4 & -2 & -1 & 1 \end{bmatrix} \begin{bmatrix} 0 \\ 1 \\ 2 \\ 3 \end{bmatrix}$$

$$= \begin{bmatrix} 4 \\ -4 \\ 3 \\ -1 \end{bmatrix}$$

$x = 4, y = -4, z = 3, w = -1$

19. $A \cdot \begin{bmatrix} 2 & 5 \\ 1 & 3 \end{bmatrix} = \begin{bmatrix} -1 & 0 \\ 4 & 2 \end{bmatrix}$

$$A = A \cdot \begin{bmatrix} 2 & 5 \\ 1 & 3 \end{bmatrix} \begin{bmatrix} 2 & 5 \\ 1 & 3 \end{bmatrix}^{-1}$$

$$= \begin{bmatrix} -1 & 0 \\ 4 & 2 \end{bmatrix} \begin{bmatrix} 2 & 5 \\ 1 & 3 \end{bmatrix}^{-1}$$

$$= \begin{bmatrix} -1 & 0 \\ 4 & 2 \end{bmatrix} \begin{bmatrix} 3 & -5 \\ -1 & 2 \end{bmatrix}$$

$$= \begin{bmatrix} -3 & 5 \\ 10 & -16 \end{bmatrix}$$

Copyright © 2018 Pearson Education, Inc.

21. $\begin{cases} x+y+z=100 \\ x-2y=0 \\ x+y-z=26 \end{cases}$

$$\left[\begin{array}{ccc|ccc} 1 & 1 & 1 & 1 & 0 & 0 \\ 1 & -2 & 0 & 0 & 1 & 0 \\ 1 & 1 & -1 & 0 & 0 & 1 \end{array}\right]$$

$$\left[\begin{array}{ccc|ccc} 1 & 1 & 1 & 1 & 0 & 0 \\ 0 & -3 & -1 & -1 & 1 & 0 \\ 0 & 0 & -2 & -1 & 0 & 1 \end{array}\right]$$

$$\left[\begin{array}{ccc|ccc} 1 & 1 & 1 & 1 & 0 & 0 \\ 0 & 1 & \frac{1}{3} & \frac{1}{3} & -\frac{1}{3} & 0 \\ 0 & 0 & -2 & -1 & 0 & 1 \end{array}\right]$$

$$\left[\begin{array}{ccc|ccc} 1 & 0 & \frac{2}{3} & \frac{2}{3} & \frac{1}{3} & 0 \\ 0 & 1 & \frac{1}{3} & \frac{1}{3} & -\frac{1}{3} & 0 \\ 0 & 0 & -2 & -1 & 0 & 1 \end{array}\right]$$

$$\left[\begin{array}{ccc|ccc} 1 & 0 & \frac{2}{3} & \frac{2}{3} & \frac{1}{3} & 0 \\ 0 & 1 & \frac{1}{3} & \frac{1}{3} & -\frac{1}{3} & 0 \\ 0 & 0 & 1 & \frac{1}{2} & 0 & -\frac{1}{2} \end{array}\right]$$

$$\left[\begin{array}{ccc|ccc} 1 & 0 & 0 & \frac{1}{3} & \frac{1}{3} & \frac{1}{3} \\ 0 & 1 & 0 & \frac{1}{6} & -\frac{1}{3} & \frac{1}{6} \\ 0 & 0 & 1 & \frac{1}{2} & 0 & -\frac{1}{2} \end{array}\right]$$

$$\begin{bmatrix} \frac{1}{3} & \frac{1}{3} & \frac{1}{3} \\ \frac{1}{6} & -\frac{1}{3} & \frac{1}{6} \\ \frac{1}{2} & 0 & -\frac{1}{2} \end{bmatrix}\begin{bmatrix} 100 \\ 0 \\ 26 \end{bmatrix} = \begin{bmatrix} 42 \\ 21 \\ 37 \end{bmatrix}$$

$x = 42, y = 21, z = 37$

23. $\begin{cases} x+y+z=100 \\ x-5y+3z=0 \\ 6x-29y+z=0 \end{cases}$

$$\left[\begin{array}{ccc|ccc} 1 & 1 & 1 & 1 & 0 & 0 \\ 1 & -5 & 3 & 0 & 1 & 0 \\ 6 & -29 & 1 & 0 & 0 & 1 \end{array}\right]$$

$$\left[\begin{array}{ccc|ccc} 1 & 1 & 1 & 1 & 0 & 0 \\ 0 & -6 & 2 & -1 & 1 & 0 \\ 0 & -35 & -5 & -6 & 0 & 1 \end{array}\right]$$

$$\left[\begin{array}{ccc|ccc} 1 & 1 & 1 & 1 & 0 & 0 \\ 0 & 1 & -\frac{1}{3} & \frac{1}{6} & -\frac{1}{6} & 0 \\ 0 & -35 & -5 & -6 & 0 & 1 \end{array}\right]$$

$$\left[\begin{array}{ccc|ccc} 1 & 0 & \frac{4}{3} & \frac{5}{6} & \frac{1}{6} & 0 \\ 0 & 1 & -\frac{1}{3} & \frac{1}{6} & -\frac{1}{6} & 0 \\ 0 & 0 & -\frac{50}{3} & -\frac{1}{6} & -\frac{35}{6} & 1 \end{array}\right]$$

$$\left[\begin{array}{ccc|ccc} 1 & 0 & \frac{4}{3} & \frac{5}{6} & \frac{1}{6} & 0 \\ 0 & 1 & -\frac{1}{3} & \frac{1}{6} & -\frac{1}{6} & 0 \\ 0 & 0 & 1 & \frac{1}{100} & \frac{7}{20} & -\frac{3}{50} \end{array}\right]$$

$$\left[\begin{array}{ccc|ccc} 1 & 0 & 0 & \frac{41}{50} & -\frac{3}{10} & \frac{2}{25} \\ 0 & 1 & 0 & \frac{17}{100} & -\frac{1}{20} & -\frac{1}{50} \\ 0 & 0 & 1 & \frac{1}{100} & \frac{7}{20} & -\frac{3}{50} \end{array}\right]$$

$$\begin{bmatrix} \frac{41}{50} & -\frac{3}{10} & \frac{2}{25} \\ \frac{17}{100} & -\frac{1}{20} & -\frac{1}{50} \\ \frac{1}{100} & \frac{7}{20} & -\frac{3}{50} \end{bmatrix}\begin{bmatrix} 100 \\ 0 \\ 0 \end{bmatrix} = \begin{bmatrix} 82 \\ 17 \\ 1 \end{bmatrix}$$

$x = 82, y = 17, z = 1$

Exercises 2.6

1. $A_{21} = 0.2$, 20 cents of energy are required to produce \$1 worth of manufactured goods.

3. $A_{31} = 0.1, A_{32} = 0.2, A_{33} = 0.15$

$A_{32} > A_{33} > A_{31}$

The energy sector uses the greatest amount of services in order to produce \$1 worth of output.

5. $[10{,}000{,}000][0.2 \quad 0.25 \quad 0.15]$

$= [2{,}000{,}000 \quad 2{,}500{,}000 \quad 1{,}500{,}000]$

\$2 million for Manufacturing, \$2.5 million for Energy, \$1.5 million for Services for a total of \$6 million.

7. $A_{13} = 0.2, A_{23} = 0.15, A_{33} = 0.15$

$A_{13} > A_{23} = A_{33}$

The services sector is most dependent on manufacturing.

Copyright © 2018 Pearson Education, Inc.

9. The initial consumption matrix is

$$\begin{bmatrix} 0.3 & 0.1 & 0.2 \\ 0.2 & 0.25 & 0.15 \\ 0.1 & 0.2 & 0.15 \end{bmatrix}\begin{bmatrix} 10 \\ 20 \\ 30 \end{bmatrix} = \begin{bmatrix} 11.00 \\ 11.50 \\ 9.50 \end{bmatrix}$$

11. The production matrix is

$$\begin{bmatrix} 1.58 & 0.33 & 0.43 \\ 0.48 & 1.5 & 0.38 \\ 0.3 & 0.39 & 1.32 \end{bmatrix}\begin{bmatrix} 5 \\ 6 \\ 7 \end{bmatrix} = \begin{bmatrix} 12.89 \\ 14.06 \\ 13.08 \end{bmatrix}$$

13. $D_{\text{new}} = \begin{bmatrix} 4 \\ 1.5 \\ 9 \end{bmatrix}$

$$(I - A)^{-1} D_{new} = \begin{bmatrix} 1.01 & 0.20 & 0.50 \\ 0.02 & 1.05 & 0.23 \\ 0.01 & 0.09 & 1.08 \end{bmatrix}\begin{bmatrix} 4 \\ 1.5 \\ 9 \end{bmatrix} = \begin{bmatrix} 8.84 \\ 3.725 \\ 9.895 \end{bmatrix}$$

Coal: \$8.84 billion, steel: \$3.725 billion, electricity: \$9.895 billion

15. $D_{\text{new}} = \begin{bmatrix} 3 \\ 1 \\ 4 \end{bmatrix}$

$$(I - A)^{-1} D_{\text{new}} = \begin{bmatrix} 1.04 & 0.02 & 0.10 \\ 0.21 & 1.01 & 0.02 \\ 0.11 & 0.02 & 1.02 \end{bmatrix}\begin{bmatrix} 3 \\ 1 \\ 4 \end{bmatrix} = \begin{bmatrix} 3.54 \\ 1.72 \\ 4.43 \end{bmatrix}$$

Computers: \$354 million, semiconductors: \$172 million

17. $D_{\text{new}} = \begin{bmatrix} 4 \\ 2 \\ 5 \end{bmatrix}$

$$(I - A)D_{\text{new}} = \begin{bmatrix} 1 & -0.15 & -0.43 \\ -0.02 & 0.97 & -0.20 \\ -0.01 & -0.08 & 0.95 \end{bmatrix}\begin{bmatrix} 4 \\ 2 \\ 5 \end{bmatrix} = \begin{bmatrix} 1.55 \\ 0.86 \\ 4.55 \end{bmatrix}$$

\$1.55 billion worth of coal; \$.86 billion worth of steel; \$4.55 billion worth of electricity

19. **a.**

$$\begin{array}{c} \quad\quad T \quad\;\; E \\ A = \begin{bmatrix} 0.25 & 0.30 \\ 0.20 & 0.15 \end{bmatrix} \begin{matrix} T \\ E \end{matrix} \end{array}$$

b. $(I - A)^{-1} = \left(\begin{bmatrix} 1 & 0 \\ 0 & 1 \end{bmatrix} - \begin{bmatrix} 0.25 & 0.30 \\ 0.20 & 0.15 \end{bmatrix}\right)^{-1} \approx \begin{bmatrix} 1.47 & 0.52 \\ 0.35 & 1.30 \end{bmatrix}$

c. $(I - A)^{-1} D \approx \begin{bmatrix} 1.47 & 0.52 \\ 0.35 & 1.30 \end{bmatrix}\begin{bmatrix} 5 \\ 3 \end{bmatrix} = \begin{bmatrix} 8.91 \\ 5.65 \end{bmatrix}$

Transportation should produce \$8.91 billion worth of output and Energy should produce \$5.65 billion.

Copyright © 2018 Pearson Education, Inc.

d. $\begin{bmatrix} 0.25 & 0.30 \\ 0.20 & 0.15 \end{bmatrix}\begin{bmatrix} 8.91 \\ 5.65 \end{bmatrix} = \begin{bmatrix} 3.92 \\ 2.63 \end{bmatrix}$

Transportation should use \$3.92 billion worth of output and Energy should use \$2.63 billion.

21. $A = \begin{array}{c} \begin{array}{cc} P & I \end{array} \\ \begin{bmatrix} 0.02 & 0.01 \\ 0.10 & 0.05 \end{bmatrix} \end{array} \begin{array}{c} \\ P \\ I \end{array}$

$D = \begin{bmatrix} 930 \\ 465 \end{bmatrix}$

$(I - A)^{-1}D = \begin{bmatrix} 955 \\ 590 \end{bmatrix}$

Plastics: \$955,000,
industrial equipment: \$590,000

23. a. $A = \begin{array}{c} \\ \text{w} \\ \text{s} \\ \text{c} \end{array} \begin{array}{c} \begin{array}{ccc} \text{w} & \text{s} & \text{c} \end{array} \\ \begin{bmatrix} 0.3 & 0 & 0.1 \\ 0.2 & 0.3 & 0.2 \\ 0.1 & 0.2 & 0.05 \end{bmatrix} \end{array}, D = \begin{bmatrix} 1 \\ 4 \\ 2 \end{bmatrix}$

b. $(I - A)^{-1} = \left(\begin{bmatrix} 1 & 0 & 0 \\ 0 & 1 & 0 \\ 0 & 0 & 1 \end{bmatrix} - \begin{bmatrix} 0.3 & 0 & 0.1 \\ 0.2 & 0.3 & 0.2 \\ 0.1 & 0.2 & 0.05 \end{bmatrix} \right)^{-1} \approx \begin{bmatrix} 1.47 & 0.05 & 0.16 \\ 0.49 & 1.54 & 0.38 \\ 0.26 & 0.33 & 1.15 \end{bmatrix}$

c. $(I - A)^{-1}D = \begin{bmatrix} 1.47 & 0.05 & 0.16 \\ 0.49 & 1.54 & 0.38 \\ 0.26 & 0.33 & 1.15 \end{bmatrix}\begin{bmatrix} 1 \\ 4 \\ 2 \end{bmatrix} = \begin{bmatrix} 1.99 \\ 7.41 \\ 3.88 \end{bmatrix}$

Wood: \$1.99, steel: \$7.41, coal \$3.88

d. $\begin{bmatrix} 0.30 & 0 & 0.10 \\ 0.20 & 0.30 & 0.20 \\ 0.10 & 0.20 & 0.05 \end{bmatrix}\begin{bmatrix} 1.99 \\ 7.41 \\ 3.88 \end{bmatrix} = \begin{bmatrix} 0.99 \\ 3.40 \\ 1.88 \end{bmatrix}$

Wood should use \$0.99 worth of output, steel should use \$3.40, and coal should use \$1.88.

25. $(I - A)^{-1}\begin{bmatrix} 100 \\ 80 \\ 200 \end{bmatrix} \approx \begin{bmatrix} 398 \\ 313 \\ 452 \end{bmatrix}$

manufacturing: \$398 million,
transportation: \$313 million,
agriculture: \$452 million

 Copyright © 2018 Pearson Education, Inc.

27.
$$A = \begin{bmatrix} 0 & 0.50 & 0.30 \\ 0.30 & 0.10 & 0.20 \\ 0.40 & 0.30 & 0.30 \end{bmatrix} \begin{matrix} M \\ B \\ F \end{matrix} \quad (\text{columns } M, B, F)$$

$$D = \begin{bmatrix} 20 \\ 15 \\ 18 \end{bmatrix}$$

$$(I-A)^{-1}D \approx \begin{bmatrix} 85 \\ 68 \\ 103 \end{bmatrix}$$

Merchant: \$85,000, baker: \$68,000, farmer: \$103,000

29. $(I-A)^{-1}\begin{bmatrix} 3 \\ 1 \\ 3 \end{bmatrix} = (I-A)^{-1}\left(\begin{bmatrix} 2 \\ 1 \\ 3 \end{bmatrix} + \begin{bmatrix} 1 \\ 0 \\ 0 \end{bmatrix}\right)$ When the matrix $(I-A)^{-1}$ is multiplied by the column matrix $\begin{bmatrix} 1 \\ 0 \\ 0 \end{bmatrix}$, the entries give the first column of the matrix $(I-A)^{-1}$, which represents the additional amounts that must be produced by the three industries.

$$= (I-A)^{-1}D_{new} - (I-A)^{-1}D_{old}$$

$$= (I-A)^{-1}\left(\begin{bmatrix} 3 \\ 1 \\ 3 \end{bmatrix} - \begin{bmatrix} 2 \\ 1 \\ 3 \end{bmatrix}\right)$$

$$= (I-A)^{-1}\left(\begin{bmatrix} 1 \\ 0 \\ 0 \end{bmatrix}\right)$$

$$= \begin{bmatrix} 1.01 & .20 & .50 \\ .02 & 1.05 & .23 \\ .01 & .09 & 1.08 \end{bmatrix}\begin{bmatrix} 1 \\ 0 \\ 0 \end{bmatrix} = \begin{bmatrix} 1.01 \\ .02 \\ .01 \end{bmatrix}$$

31.

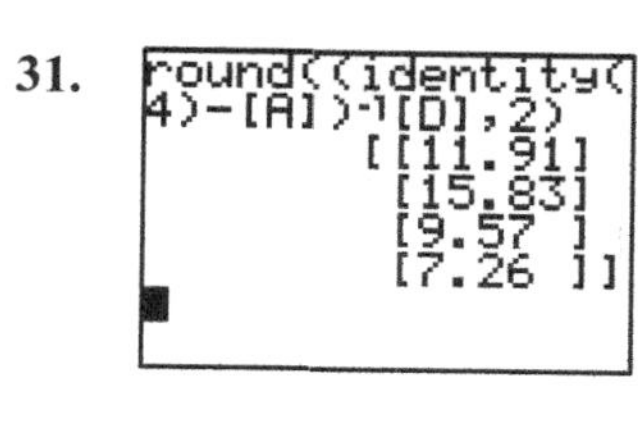

$$\begin{bmatrix} 11.91 \\ 15.83 \\ 9.57 \\ 7.26 \end{bmatrix}$$

Chapter 2 Review Exercises

1. $\left[\begin{array}{cc|c} 3 & -6 & 1 \\ 2 & 4 & 6 \end{array}\right] \xrightarrow{\frac{1}{3}R_1} \left[\begin{array}{cc|c} 1 & -2 & \frac{1}{3} \\ 2 & 4 & 6 \end{array}\right]$

$\xrightarrow{R_2+(-2)R_1} \left[\begin{array}{cc|c} 1 & -2 & \frac{1}{3} \\ 0 & 8 & \frac{16}{3} \end{array}\right]$

2. $\left[\begin{array}{cc|c} -5 & -3 & 1 \\ 4 & 2 & 0 \\ 0 & 6 & 7 \end{array}\right] \xrightarrow{\frac{1}{2}R_2} \left[\begin{array}{cc|c} -5 & -3 & 1 \\ 2 & 1 & 0 \\ 0 & 6 & 7 \end{array}\right]$

$\xrightarrow{R_3+(-6)R_2} \left[\begin{array}{cc|c} 1 & 0 & 1 \\ 2 & 1 & 0 \\ -12 & 0 & 7 \end{array}\right]$

3. $\left[\begin{array}{cc|c} \frac{1}{2} & -1 & -3 \\ 4 & -5 & -9 \end{array}\right]$

$\left[\begin{array}{cc|c} 1 & -2 & -6 \\ 0 & \underline{3} & 15 \end{array}\right]$

$\left[\begin{array}{cc|c} 1 & 0 & 4 \\ 0 & 1 & 5 \end{array}\right]$

$x = 4, y = 5$

4. $\left[\begin{array}{ccc|c} \underline{3} & 0 & 9 & 42 \\ 2 & 1 & 6 & 30 \\ -1 & 3 & -2 & -20 \end{array}\right]$

$\left[\begin{array}{ccc|c} 1 & 0 & 3 & 14 \\ 0 & \underline{1} & 0 & 2 \\ 0 & 3 & 1 & -6 \end{array}\right]$

$\left[\begin{array}{ccc|c} 1 & 0 & 3 & 14 \\ 0 & 1 & 0 & 2 \\ 0 & 0 & \underline{1} & -12 \end{array}\right]$

$\left[\begin{array}{ccc|c} 1 & 0 & 0 & 50 \\ 0 & 1 & 0 & 2 \\ 0 & 0 & 1 & -12 \end{array}\right]$

$x = 50, y = 2, z = -12$

5. $\left[\begin{array}{ccc|c} 3 & -6 & 6 & -5 \\ -2 & 3 & -5 & \frac{7}{3} \\ 1 & 1 & 10 & 3 \end{array}\right]$

$\left[\begin{array}{ccc|c} 1 & -2 & 2 & -\frac{5}{3} \\ 0 & -1 & -1 & -1 \\ 0 & 3 & 8 & \frac{14}{3} \end{array}\right]$

$\left[\begin{array}{ccc|c} 1 & 0 & 4 & \frac{1}{3} \\ 0 & 1 & 1 & 1 \\ 0 & 0 & 5 & \frac{5}{3} \end{array}\right]$

$\left[\begin{array}{ccc|c} 1 & 0 & 0 & -1 \\ 0 & 1 & 0 & \frac{2}{3} \\ 0 & 0 & 1 & \frac{1}{3} \end{array}\right]$

$x = -1,\ y = \frac{2}{3},\ z = \frac{1}{3}$

6. $\left[\begin{array}{ccc|c} \underline{3} & 6 & -9 & 1 \\ 2 & 4 & -6 & 1 \\ 3 & 4 & 5 & 0 \end{array}\right]$

$\left[\begin{array}{ccc|c} 1 & 2 & -3 & \frac{1}{3} \\ 0 & 0 & 0 & \frac{1}{3} \\ 0 & -2 & 14 & -1 \end{array}\right]$

$\left[\begin{array}{ccc|c} 1 & 2 & -3 & \frac{1}{3} \\ 0 & \underline{-2} & 14 & -1 \\ 0 & 0 & 0 & \frac{1}{3} \end{array}\right]$

$\left[\begin{array}{ccc|c} 1 & 0 & 11 & -\frac{2}{3} \\ 0 & 1 & -7 & \frac{1}{2} \\ 0 & 0 & 0 & \frac{1}{3} \end{array}\right]$

No solution

7. $\left[\begin{array}{cccc|c} \underline{1} & 2 & -5 & 3 & 16 \\ -5 & -7 & 13 & -9 & -50 \\ -1 & 1 & -7 & 2 & 9 \\ 3 & 4 & -7 & 6 & 33 \end{array}\right]$

$\left[\begin{array}{cccc|c} 1 & 2 & -5 & 3 & 16 \\ 0 & \underline{3} & -12 & 6 & 30 \\ 0 & 3 & -12 & 5 & 25 \\ 0 & -2 & 8 & -3 & -15 \end{array}\right]$

$\left[\begin{array}{cccc|c} 1 & 0 & 3 & -1 & -4 \\ 0 & 1 & -4 & 2 & 10 \\ 0 & 0 & 0 & \underline{-1} & -5 \\ 0 & 0 & 0 & 1 & 5 \end{array}\right]$

$\left[\begin{array}{cccc|c} 1 & 0 & 3 & 0 & 1 \\ 0 & 1 & -4 & 0 & 0 \\ 0 & 0 & 0 & 1 & 5 \\ 0 & 0 & 0 & 0 & 0 \end{array}\right]$

$$\begin{cases} x + 3z = 1 \\ y - 4z = 0 \\ w = 5 \\ 0 = 0 \end{cases}$$

z = any value, $x = 1 - 3z$, $y = 4z$, $w = 5$

Copyright © 2018 Pearson Education, Inc.

8. $\left[\begin{array}{rr|r} 5 & -10 & 5 \\ 3 & -8 & -3 \\ -3 & 7 & 0 \end{array}\right]$

$\left[\begin{array}{rr|r} 1 & -2 & 1 \\ 0 & -2 & -6 \\ 0 & 1 & 3 \end{array}\right]$

$\left[\begin{array}{rr|r} 1 & 0 & 7 \\ 0 & 1 & 3 \\ 0 & 0 & 0 \end{array}\right]$

$x = 7, y = 3$

9. $\begin{bmatrix} 2+3 \\ -1+4 \\ 0+7 \end{bmatrix} = \begin{bmatrix} 5 \\ 3 \\ 7 \end{bmatrix}$

10. $\begin{bmatrix} 1\cdot 3+3\cdot 1+(-2)0 & 1\cdot 5+3\cdot 0+(-2)(-6) \\ 4\cdot 3+0\cdot 1+(-1)0 & 4\cdot 5+0\cdot 0+(-1)(-6) \end{bmatrix}$

$= \begin{bmatrix} 6 & 17 \\ 12 & 26 \end{bmatrix}$

11. $\begin{bmatrix} \frac{3}{4}\cdot 8 & \frac{3}{4}\cdot(-6) \\ \frac{3}{4}\cdot\frac{2}{3} & \frac{3}{4}\cdot 0 \end{bmatrix} = \begin{bmatrix} 6 & \frac{-9}{2} \\ \frac{1}{2} & 0 \end{bmatrix}$

12. $\begin{bmatrix} 1.4-.8 & -3-7 \\ 8.2-1.6 & 0-(-2) \\ 4-0 & 5.5-(-5.5) \end{bmatrix} = \begin{bmatrix} .6 & -10 \\ 6.6 & 2 \\ 4 & 11 \end{bmatrix}$

13. $AB = \begin{bmatrix} 23 & -10+5k \\ -9 & 15+k \end{bmatrix}$

$BA = \begin{bmatrix} 23 & 15 \\ 6-3k & 15+k \end{bmatrix}$

Therefore,

$$-10+5k = 15$$
$$5k = 25$$
$$k = 5$$
$$6-3k = -9$$
$$-3k = -15$$
$$k = 5$$

Since any value of k will work in the $a_{2,2}$ position $k = 5$ is the only solution.

14. $AB = \begin{bmatrix} 3k+5 & -10+k \\ 9 & 17 \end{bmatrix}$

$BA = \begin{bmatrix} 17 & -10+k \\ 9 & 3k+5 \end{bmatrix}$

Therefore,

$$3k+5 = 17$$
$$3k = 12$$
$$k = 4$$
$$3k+5 = 17$$
$$3k = 12$$
$$k = 4$$

Since any value of k will work in the $a_{1,2}$ position, $k = 4$ is the only solution

15. $\begin{bmatrix} 3 & 2 \\ 5 & 4 \end{bmatrix}^{-1} = \begin{bmatrix} \frac{4}{2} & -\frac{2}{2} \\ -\frac{5}{2} & \frac{3}{2} \end{bmatrix} = \begin{bmatrix} 2 & -1 \\ -\frac{5}{2} & \frac{3}{2} \end{bmatrix}$

$\begin{bmatrix} 2 & -1 \\ -\frac{5}{2} & \frac{3}{2} \end{bmatrix}\begin{bmatrix} 0 \\ 2 \end{bmatrix} = \begin{bmatrix} -2 \\ 3 \end{bmatrix}$

$x = -2, y = 3$

16. a. $\begin{bmatrix} 4 & -2 & 3 \\ 8 & -3 & 5 \\ 7 & -2 & 4 \end{bmatrix}\begin{bmatrix} 1 \\ 0 \\ 3 \end{bmatrix} = \begin{bmatrix} 13 \\ 23 \\ 19 \end{bmatrix}$

$x = 13, y = 23, z = 19$

b. $\begin{bmatrix} -2 & 2 & -1 \\ 3 & -5 & 4 \\ 5 & -6 & 4 \end{bmatrix}\begin{bmatrix} 0 \\ -1 \\ 2 \end{bmatrix} = \begin{bmatrix} -4 \\ 13 \\ 14 \end{bmatrix}$

$x = -4, y = 13, z = 14$

17. $\left[\begin{array}{rr|rr} 2 & 6 & 1 & 0 \\ 1 & 2 & 0 & 1 \end{array}\right]$

$\left[\begin{array}{rr|rr} 1 & 3 & \frac{1}{2} & 0 \\ 0 & -1 & -\frac{1}{2} & 1 \end{array}\right]$

$\left[\begin{array}{rr|rr} 1 & 0 & -1 & 3 \\ 0 & 1 & \frac{1}{2} & -1 \end{array}\right]$

$\begin{bmatrix} -1 & 3 \\ \frac{1}{2} & -1 \end{bmatrix}$

Copyright © 2018 Pearson Education, Inc.

18. $\left[\begin{array}{ccc|ccc} \underline{1} & 1 & 1 & 1 & 0 & 0 \\ 3 & 4 & 3 & 0 & 1 & 0 \\ 1 & 1 & 2 & 0 & 0 & 1 \end{array}\right]$

$\left[\begin{array}{ccc|ccc} 1 & 1 & 1 & 1 & 0 & 0 \\ 0 & \underline{1} & 0 & -3 & 1 & 0 \\ 0 & 0 & 1 & -1 & 0 & 1 \end{array}\right]$

$\left[\begin{array}{ccc|ccc} 1 & 0 & 1 & 4 & -1 & 0 \\ 0 & 1 & 0 & -3 & 1 & 0 \\ 0 & 0 & \underline{1} & -1 & 0 & 1 \end{array}\right]$

$\left[\begin{array}{ccc|ccc} 1 & 0 & 0 & 5 & -1 & -1 \\ 0 & 1 & 0 & -3 & 1 & 0 \\ 0 & 0 & 1 & -1 & 0 & 1 \end{array}\right]$

$\begin{bmatrix} 5 & -1 & -1 \\ -3 & 1 & 0 \\ -1 & 0 & 1 \end{bmatrix}$

19. $\begin{cases} c+w+s=1000 \\ 357c+137w+181s=269{,}000 \\ c-w-s=0 \end{cases}$

$\begin{bmatrix} 1 & 1 & 1 \\ 357 & 137 & 181 \\ 1 & -1 & -1 \end{bmatrix}^{-1} \begin{bmatrix} 1000 \\ 269{,}000 \\ 0 \end{bmatrix} = \begin{bmatrix} 500 \\ 0 \\ 500 \end{bmatrix}$

Corn: 500 acres, wheat: 0 acres, soybeans: 500 acres

20. a. $AC = \begin{bmatrix} 15 & 20 & 8 \\ 10 & 17 & 12 \end{bmatrix} \begin{bmatrix} 165 \\ 65 \\ 210 \end{bmatrix} = \begin{bmatrix} 5455 \\ 5275 \end{bmatrix}$

The total cost to make the equipment in the first store was \$5455 and in the second store \$5275.

b. $AS = \begin{bmatrix} 15 & 20 & 8 \\ 10 & 17 & 12 \end{bmatrix} \begin{bmatrix} 200 \\ 80 \\ 250 \end{bmatrix} = \begin{bmatrix} 6600 \\ 6360 \end{bmatrix}$

The total revenue of the equipment in the first store was \$6600 and in the second store \$6360.

c. $S-C = \begin{bmatrix} 200 \\ 80 \\ 250 \end{bmatrix} - \begin{bmatrix} 165 \\ 65 \\ 210 \end{bmatrix} = \begin{bmatrix} 35 \\ 15 \\ 40 \end{bmatrix}$

The entries represent the profit per unit of each item.

d. $A(S-C) = \begin{bmatrix} 15 & 20 & 8 \\ 10 & 17 & 12 \end{bmatrix} \begin{bmatrix} 35 \\ 15 \\ 40 \end{bmatrix} = \begin{bmatrix} 1145 \\ 1085 \end{bmatrix}$

The total profit for the first store was \$1145 and for the second store \$1085.

21. a. $BA = \begin{bmatrix} 5000 & 8000 & 10{,}000 \end{bmatrix} \begin{bmatrix} 0.50 & 0.43 & 0.07 \\ 0.45 & 0.26 & 0.29 \\ 0.40 & 0.40 & 0.20 \end{bmatrix} = \begin{bmatrix} 10{,}100 & 8230 & 4670 \end{bmatrix}$

Total amount invested in bonds, stocks , and the conservative fixed income fund, respectively.

b. $BC = \begin{bmatrix} 5000 & 8000 & 10{,}000 \end{bmatrix} \begin{bmatrix} 0.0032 & 0.1119 \\ 0.0233 & 0.0976 \\ 0.0320 & 0.0467 \end{bmatrix} = \begin{bmatrix} 522.40 & 1807.30 \end{bmatrix}$

total return on the investments for one year and five years, respectively.

c. $2B = \begin{bmatrix} 2(5000) & 2(8000) & 2(10{,}000) \end{bmatrix} = \begin{bmatrix} 10{,}000 & 16{,}000 & 20{,}000 \end{bmatrix}$ The result of doubling the amounts invested

d. \$8230 is the total amount invested in stocks.

e. \$522.40 is the total return after one year.

Copyright © 2018 Pearson Education, Inc.

22. a. $AB = \begin{bmatrix} 11 & 7 & 12 \\ 9 & 5 & 16 \\ 13 & 8 & 9 \\ 13 & 7 & 10 \end{bmatrix}\begin{bmatrix} 11 \\ 9 \\ 12 \end{bmatrix} = \begin{bmatrix} 328 \\ 336 \\ 323 \\ 326 \end{bmatrix}\begin{matrix} Sara \\ Quinn \\ Tamia \\ Zack \end{matrix}$

total amount earned by each person for the week.

b. Quinn earned the most and Tamia earned the least.

c. $AB = \begin{bmatrix} 11 & 7 & 12 \\ 9 & 5 & 16 \\ 13 & 8 & 9 \\ 13 & 7 & 10 \end{bmatrix}\begin{bmatrix} 12 \\ 9 \\ 11 \end{bmatrix} = \begin{bmatrix} 327 \\ 329 \\ 327 \\ 329 \end{bmatrix}\begin{matrix} Sara \\ Quinn \\ Tamia \\ Zack \end{matrix}$

Quinn and Zack both earn $329.

d. Sara worked 11 hours in concessions, 7 hours at the front desk, and 12 hours cleaning for a total of 30 hours.

23. Let x = number of apples
y = number of bananas
z = number of oranges

$$x+y+z=18$$
$$0.85x+0.20y+0.76z=9$$
$$-x+y-z=0$$

$$\left[\begin{array}{ccc|c} 1 & 1 & 1 & 18 \\ 0.85 & 0.20 & 0.76 & 9 \\ -1 & 1 & -1 & 0 \end{array}\right]$$

$$\xrightarrow{R_3+R_1}\left[\begin{array}{ccc|c} 1 & 1 & 1 & 18 \\ 0.85 & 0.20 & 0.76 & 9 \\ 0 & 2 & 0 & 18 \end{array}\right]$$

$$\xrightarrow{R_2+(-0.85)R_1}\left[\begin{array}{ccc|c} 1 & 1 & 1 & 18 \\ 0 & -0.65 & -0.09 & -6.3 \\ 0 & 2 & 0 & 18 \end{array}\right]$$

$$\xrightarrow{R_2 \leftrightarrow R_3}\left[\begin{array}{ccc|c} 1 & 1 & 1 & 18 \\ 0 & 2 & 0 & 18 \\ 0 & -0.65 & -0.09 & -6.3 \end{array}\right]$$

$$\xrightarrow{\frac{1}{2}R_2}\left[\begin{array}{ccc|c} 1 & 1 & 1 & 18 \\ 0 & 1 & 0 & 9 \\ 0 & -0.65 & -0.09 & -6.3 \end{array}\right]$$

$$\xrightarrow{R_1+(-1)R_2}\left[\begin{array}{ccc|c} 1 & 0 & 1 & 9 \\ 0 & 1 & 0 & 9 \\ 0 & -0.65 & -0.09 & -6.3 \end{array}\right]$$

$$\xrightarrow{R_3+(0.65)R_2}\left[\begin{array}{ccc|c} 1 & 0 & 1 & 9 \\ 0 & 1 & 0 & 9 \\ 0 & 0 & -0.09 & -0.45 \end{array}\right]$$

$$\xrightarrow{-\frac{1}{0.09}R_3}\left[\begin{array}{ccc|c} 1 & 0 & 1 & 9 \\ 0 & 1 & 0 & 9 \\ 0 & 0 & 1 & 5 \end{array}\right]$$

$$\xrightarrow{R_1+(-1)R_3}\left[\begin{array}{ccc|c} 1 & 0 & 0 & 4 \\ 0 & 1 & 0 & 9 \\ 0 & 0 & 1 & 5 \end{array}\right]$$

4 apples, 9 bananas, 5 oranges

24. $x = A$'s current stockpile,
$y = B$'s current stockpile,
$a = A$'s next-year's stockpile,
$b = B$'s next-year's stockpile
$0.8x + 0.2y = a$
$0.1x + 0.9y = b$

a. Next year: $\begin{bmatrix} 0.8 & 0.2 \\ 0.1 & 0.9 \end{bmatrix}\begin{bmatrix} 10,000 \\ 7000 \end{bmatrix} = \begin{bmatrix} 9400 \\ 7300 \end{bmatrix}$

Two years: $\begin{bmatrix} 0.8 & 0.2 \\ 0.1 & 0.9 \end{bmatrix}^2\begin{bmatrix} 10,000 \\ 7000 \end{bmatrix} = \begin{bmatrix} 8980 \\ 7510 \end{bmatrix}$

A: 9400, 8980; B: 7300, 7510

b. Previous year:

$\begin{bmatrix} 0.8 & 0.2 \\ 0.1 & 0.9 \end{bmatrix}^{-1}\begin{bmatrix} 10,000 \\ 7000 \end{bmatrix} \approx \begin{bmatrix} 10,857 \\ 6571 \end{bmatrix}$

Two years ago:

Copyright © 2018 Pearson Education, Inc.

$$\left(\begin{bmatrix} 0.8 & 0.2 \\ 0.1 & 0.9 \end{bmatrix}^{-1}\right)^2 \begin{bmatrix} 10,000 \\ 7000 \end{bmatrix} \approx \begin{bmatrix} 12,082 \\ 5959 \end{bmatrix}$$

A: 10,857, 12,082; B: 6571, 5959

c. $a - b = (0.8x + 0.2y) - (0.1x + 0.9y)$
$= 0.7x - 0.7y = 0.7(x - y)$
Thus, the "missile gap" of the following year is 70% of the previous, so the "missile gap" decreases 30% each year.
$a + b = 0.9x + 1.1y < x + y$
when $0.1y < 0.1x$ or $y < x$.
$a + b = 0.9x + 1.1y > x + y$ when $y > x$.

25. $$\left(\begin{bmatrix} 1 & 0 \\ 0 & 1 \end{bmatrix} - \begin{bmatrix} 0.4 & 0.2 \\ 0.1 & 0.3 \end{bmatrix}\right)^{-1} \begin{bmatrix} 8 \\ 12 \end{bmatrix} = \begin{bmatrix} 20 \\ 20 \end{bmatrix}$$

Industry I: 20; industry II: 20

26. Let x = number of nickels
y = number of dimes
z = number of quarters

$$\begin{cases} x + y + z = 30 \\ 0.05x + 0.10y + 0.25z = 3.30 \\ y = 5z \end{cases}$$

$$\left[\begin{array}{ccc|c} \underline{1} & 1 & 1 & 30 \\ 0.05 & 0.10 & 0.25 & 3.30 \\ 0 & 1 & -5 & 0 \end{array}\right]$$

$$\left[\begin{array}{ccc|c} 1 & 1 & 1 & 30 \\ 0 & \underline{0.05} & 0.20 & 1.8 \\ 0 & 1 & -5 & 0 \end{array}\right]$$

$$\left[\begin{array}{ccc|c} 1 & 1 & 1 & 30 \\ 0 & \underline{1} & 4 & 36 \\ 0 & 1 & -5 & 0 \end{array}\right]$$

$$\left[\begin{array}{ccc|c} 1 & 0 & -3 & -6 \\ 0 & 1 & 4 & 36 \\ 0 & 0 & \underline{-9} & -36 \end{array}\right]$$

$$\left[\begin{array}{ccc|c} 1 & 0 & -3 & -6 \\ 0 & 1 & 4 & 36 \\ 0 & 0 & \underline{1} & 4 \end{array}\right]$$

$$\left[\begin{array}{ccc|c} 1 & 0 & 0 & 6 \\ 0 & 1 & 0 & 20 \\ 0 & 0 & 1 & 4 \end{array}\right]$$

Joe has 4 quarters.

27. **a.** True; a system of equations has no solution, exactly one solution, or infinitely many solutions.

b. False; a system of equations could have two or more equations that are multiples of each other.

c. True; At least one variable must be dependent on another when there are less equations than variables.

28. a. True; The matrix and itself will have the same dimensions and therefore can be added.

b. False; In order to be multiplied by itself, the number of rows must be the same as the number of columns. In other words, only a square matrix can be multiplied by itself.

29. One possible system with infinitely many solutions is: $\begin{cases} 2x + 3y = 4 \\ 4x + 6y = 8 \end{cases}$.

30. One possible system with no solution is: $\begin{cases} 2x + 3y = 4 \\ 2x + 3y = 5 \end{cases}$.

31. Answers may vary, but two possible matrices are: $A = \begin{bmatrix} 1 & 1 \\ 1 & 1 \end{bmatrix}$ and $B = \begin{bmatrix} -1 & 1 \\ 1 & -1 \end{bmatrix}$.

32. If the matrix has no inverse, there will be a row of zeros in the resulting Gauss-Jordan matrix.

33. The column values should add to less than one in an input-output matrix because they reflect the amount of input required from each industry and all the industries are dependent on each other.

Copyright © 2018 Pearson Education, Inc.

Chapter 3

Exercises 3.1

1. False

3. True

5. $2x - 5 \geq 3$
$2x \geq 8$
$x \geq 4$

7. $-5x + 13 \leq -2$
$-5x \leq -15$
$x \geq 3$

9. $2x + y \leq 5$
$y \leq -2x + 5$

11. $5x - \frac{1}{3}y \leq 6$

$-\frac{1}{3}y \leq -5x + 6$

$y \geq 15x - 18$

13. $4x \geq -3$

$x \geq -\frac{3}{4}$

15. $3(2) + 5(1) \leq 12$
$6 + 5 \leq 12$
$11 \leq 12$
Yes

17. $0 \geq -2(3) + 7$
$0 \geq -6 + 7$
$0 \geq 1$
No

19. $5 \leq 3(3) - 4$
$5 \leq 9 - 4$
$5 \leq 5$
Yes

21. $7 \geq 5$
Yes

23.

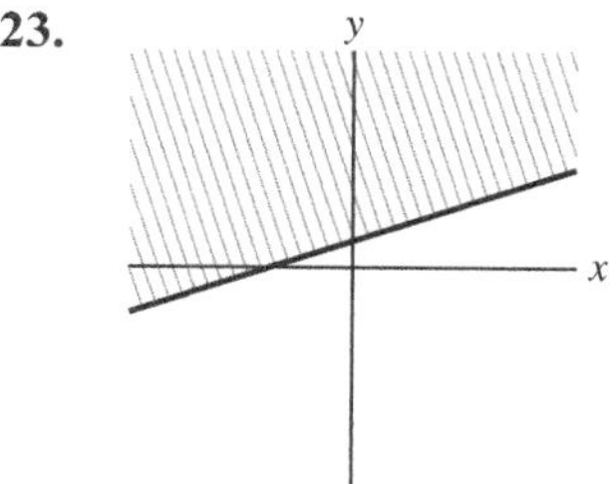

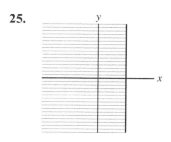

25.

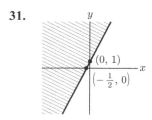

27. Use both intercept points to find the slope of the given line, the points are (3, 0) and (0, 3). The slope is: $m = \frac{3-0}{0-3} = \frac{3}{-3} = -1$. Using the slope, the y intercept point, the solid line, and the shading of the graph, the equation of the inequality is $y \leq -x + 3$.

29. Use both intercept points to find the slope of the given line, the points are $(4, 0)$ and $(0, -2)$. The slope is: $m = \frac{0+2}{4-0} = \frac{2}{4} = \frac{1}{2}$. Using the slope, the y intercept point, the solid line, and the shading of the graph, the equation of the inequality is $y \geq \frac{1}{2}x - 2$.

31.

33.

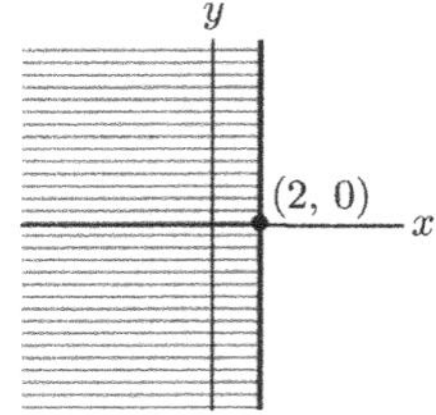

35. $x+4y\geq 12$

$$y\geq -\frac{1}{4}x+3$$

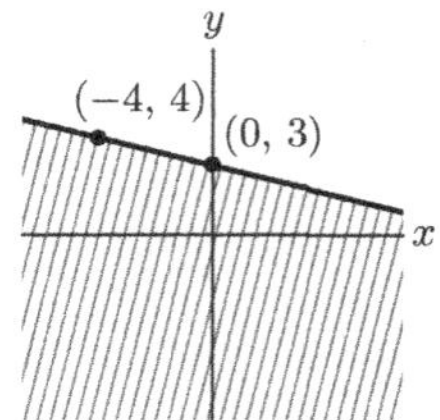

37. $4x-5y+25\geq 0$

$$y\leq \frac{4}{5}x+5$$

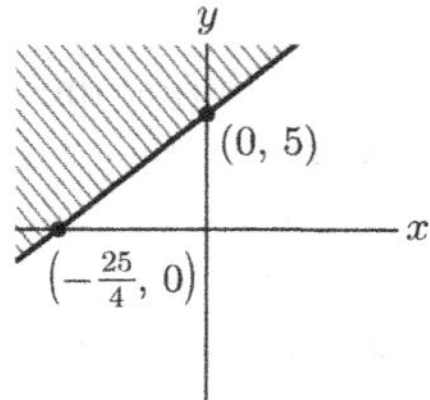

39. $\frac{1}{2}x-\frac{1}{3}y\leq 1$

$$y\geq \frac{3}{2}x-3$$

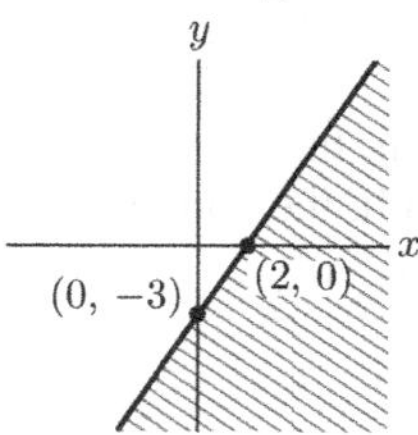

41. $.5x+.4y\leq 2$

$$y\leq -1.25x+5$$

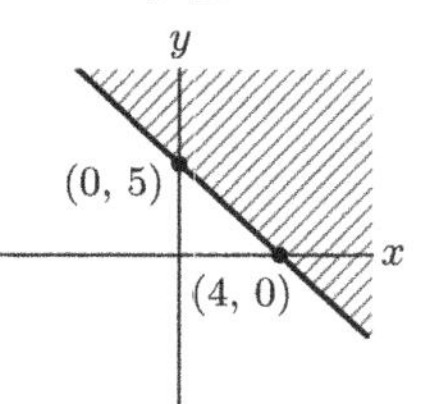

43. $\begin{cases} y\leq 2x-4 \\ y\geq 0 \end{cases}$

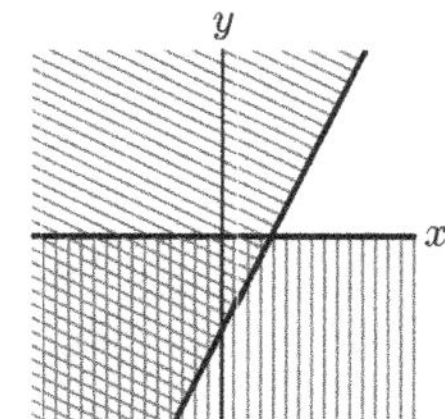

45. $\begin{cases} x+2y\geq 2 \\ 3x-y\geq 3 \end{cases}$

$$\begin{cases} y\geq -\frac{1}{2}x+1 \\ y\leq 3x-3 \end{cases}$$

y
x

 Copyright © 2018 Pearson Education, Inc.

47. $\begin{cases} x+5y \le 10 \\ x+y \le 3 \\ x \ge 0,\ y \ge 0 \end{cases}$

$\begin{cases} y \le -\frac{1}{5}x+2 \\ y \le -x+3 \\ x \ge 0,\ y \ge 0 \end{cases}$

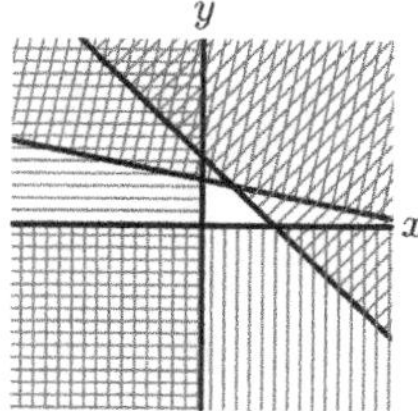

49. $\begin{cases} 6(8)+3(7) \le 96 \\ 8+7 \le 18 \\ 2(8)+6(7) \le 72 \\ 8 \ge 0,\ 7 \ge 0 \end{cases}$

$\begin{cases} 69 \le 96 \\ 15 \le 18 \\ 58 \le 72 \\ 8 \ge 0,\ 7 \ge 0 \end{cases}$

Yes

51. $\begin{cases} 6(9)+3(10) \le 96 \\ 9+10 \le 18 \\ 2(9)+6(10) \le 72 \\ 9 \ge 0,\ 10 \ge 0 \end{cases}$

$\begin{cases} 84 \le 96 \\ 19 \le 18 \\ 78 \le 72 \\ 9 \ge 0,\ 10 \ge 0 \end{cases}$

No

53. For $x = 3$, $y = 2(3) + 5 = 11$.
So (3, 9) is below.

55. $7-4x+5y=0$

$$y=\frac{4}{5}x-\frac{7}{5}$$

For $x = 0$, $y=\frac{4}{5}(0)-\frac{7}{5}=-\frac{7}{5}$.

So (0, 0) is above.

57. $8x-4y=4$

$y=2x-1$

$8x-4y=0$

$y=2x$

$\begin{cases} y \ge 2x-1 \\ y \le 2x \end{cases}$

59. d

61. $4x-2y=7$

$$y=2x-\frac{7}{2}$$

a.

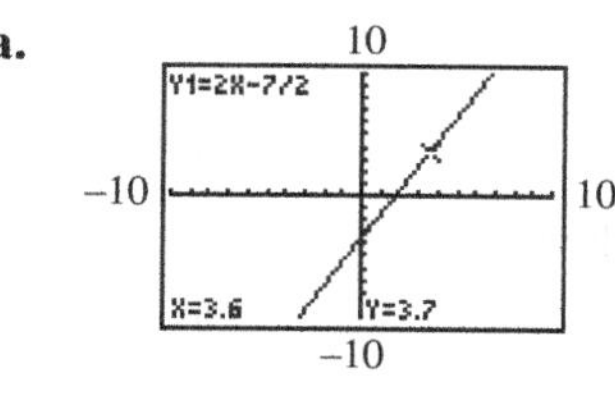

(3.6, 3.7)

b. Below, because (3.6, 3.7) is on the line.

63.

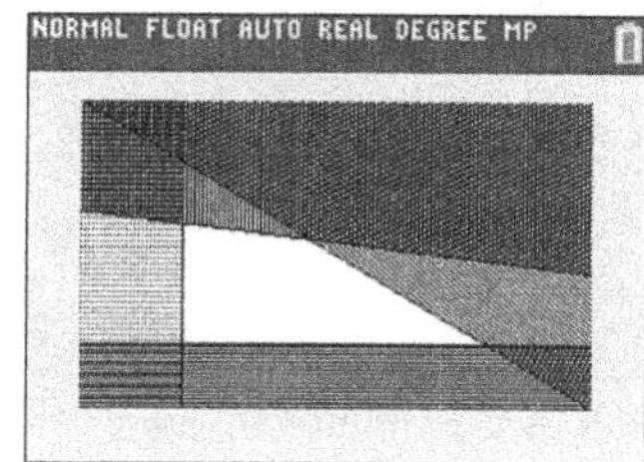

Exercises 3.2

1. (8, 7)

$\begin{cases} 6(8)+3(7) \le 96 \\ 8+7 \le 18 \\ 2(8)+6(7) \le 72 \\ 8 \ge 0,\ 7 \ge 0 \end{cases}$

$\begin{cases} 69 \le 96 & \text{true} \\ 15 \le 18 & \text{true} \\ 58 \le 72 & \text{true} \\ 8 \ge 0,\ 7 \ge 0 & \text{true} \end{cases}$

Yes

Copyright © 2018 Pearson Education, Inc.

3. (9, 10)

$$\begin{cases} 6(9)+3(10)\le 96 \\ 9+10\le 18 \\ 2(9)+6(10)\le 72 \\ 9\ge 0,\ 3\ge 0 \end{cases}$$

$$\begin{cases} 84\le 96 & \text{true} \\ 19\le 18 & \text{false} \\ 78\le 72 & \text{false} \\ 9\ge 0,\ 3\ge 0 & \text{true} \end{cases}$$

No

5. $\begin{cases} 6x\le 96 \\ x\le 18 \\ 2x\le 72 \\ x\ge 0 \end{cases} \Rightarrow \begin{cases} x\le 16 \\ x\le 18 \\ x\le 36 \\ x\ge 0 \end{cases}$

Therefore, 16 chairs could be produced

7. a.

	A	*B*	*Available*
Candy bars	2	1	500
Suckers	2	2	600
Profit (in cents)	40	30	

b. Candy Bars: $2x+y\le 500$
Suckers: $2x+2y\le 600$

c. $x\ge 0,\ y\ge 0$

d. $40x+30y$

e.

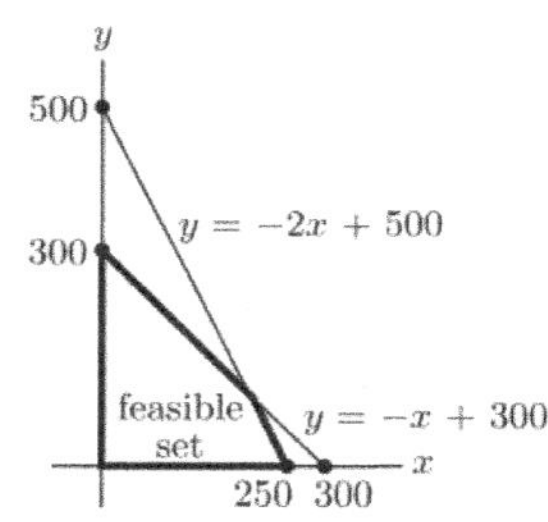

9. a.

	A	*B*	*Truck capacity*
Volume	4 cubic feet	3 cubic feet	300 cubic feet
Weight	100 pounds	200 pounds	10,000 pounds
Earnings	\$13	\$9	

b. Volume: $4x+3y\le 300$
Weight: $100x+200y\le 10{,}000$

c. $y\le 2x,\ x\ge 0,\ y\ge 0$

d. $13x+9y$

Copyright © 2018 Pearson Education, Inc.

e. In standard form, the inequalities from (b) and (c) are:

$$\begin{cases} y \le -\frac{4}{3}x + 100 \\ y \le -\frac{1}{2}x + 50 \\ y \le 2x \\ x \ge 0,\ y \ge 0 \end{cases}$$

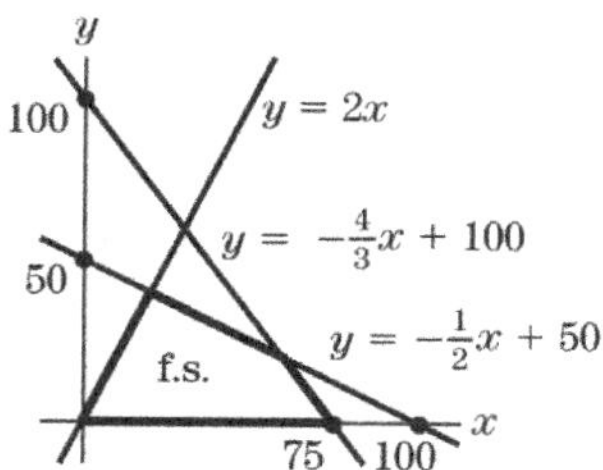

11. a.

	Essay questions	*Short-answer Questions*	*Available*
Time to answer	10 minutes	2 minutes	90 minutes
Quantity	10	50	
Required	3	10	
Worth	20 points	5 points	

b. $10x + 2y \le 90$

c. $3 \le x \le 10,\ 10 \le y \le 50$

d. $20x + 5y$

e. In standard form, the inequality from (b) is $y \le -5x + 45$. Graph the system:

$$\begin{cases} y \le -5x + 45 \\ 3 \le x \le 10 \\ 10 \le y \le 50 \end{cases}$$

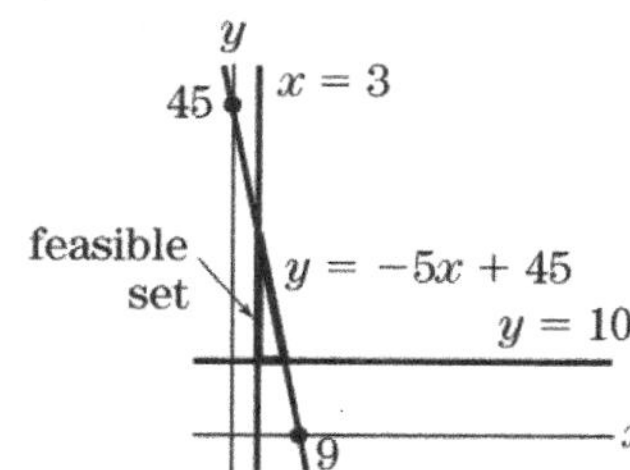

Note that the conditions $x \le 10$ and $y \le 50$ are superfluous because they are automatically assured if the other inequalities hold.

Copyright © 2018 Pearson Education, Inc.

13. a.

	Alfalfa	*Corn*	*Requirements*
Protein	0.13 pound	0.065 pound	4550 pounds
TDN	0.48 pound	0.96 pound	26,880 pounds
Vitamin A	2.16 IUs	0 IUs	43,200 IUs
Cost/lb	\$0.08	\$0.13	

b. Protein: $0.13x + 0.065y \geq 4550$
TDN: $0.48x + 0.96y \geq 26{,}880$
Vitamin A: $2.16x \geq 43{,}200$
Other: $y \geq 0$
(The condition $x \geq 0$ is unnecessary because it is automatically assured if the inequality for Vitamin A holds.)

c. In standard form, the inequalities are:

$$\begin{cases} y \geq -2x + 70{,}000 \\ y \geq -0.5x + 28{,}000 \\ x \geq 20{,}000 \\ y \geq 0 \end{cases}$$

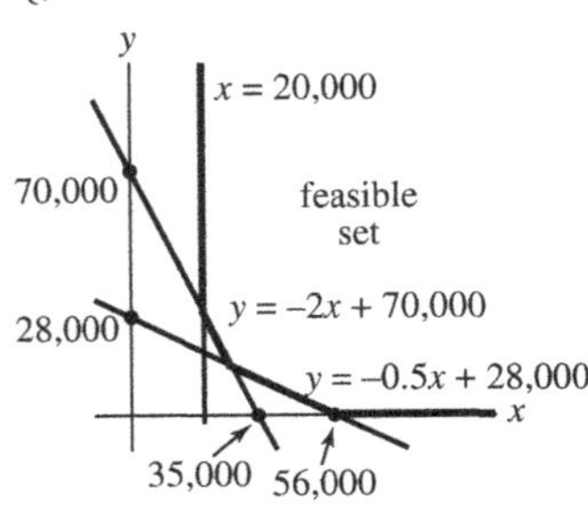

d. $.08x + .13y$

Exercises 3.3

1.

Vertex	$4x + 3y$
(0, 0)	$4(0) + 3(0) = 0$
(0, 20)	$4(0) + 3(20) = 60$
(20, 0)	$4(20) + 3(0) = 80$

The objective function is maximized at (20, 0).

3. Find the vertices.
Lower left corner: (0, 0)

y-intercept of $y = -\frac{1}{2}x + 4$: (0, 4)

Intersection of $y = -\frac{1}{2}x + 4$ and $y = -x + 6$:

$$-\frac{1}{2}x + 4 = -x + 6$$

$$\frac{1}{2}x = 2$$

$$x = 4$$

$$y = -\frac{1}{2}x + 4 = -\frac{1}{2}(4) + 4 = 2$$

(4, 2)

x-intercept of $y = -x + 6$: (6, 0)

Copyright © 2018 Pearson Education, Inc.

Vertex	$4x+3y$
(0, 0)	$4(0)+3(0)=0$
(0, 4)	$4(0)+3(4)=12$
(4, 2)	$4(4)+3(2)=22$
(6, 0)	$4(6)+3(0)=24$

The objective function is maximized at (6, 0).

5.

Vertex	$x+2y$
(0, 0)	$0+2(0)=0$
(0, 5)	$0+2(5)=10$
(3, 3)	$3+2(3)=9$
(4, 0)	$4+2(0)=4$

The objective function is maximized at (0, 5).

7.

Vertex	$2x+y$
(0, 0)	$2(0)+0=0$
(0, 5)	$2(0)+5=5$
(3, 3)	$2(3)+3=9$
(4, 0)	$2(4)+0=8$

The objective function is maximized at (3, 3).

9.

Vertex	$8x+y$
(0, 7)	$8(0)+7=7$
(1, 2)	$8(1)+2=10$
(2, 1)	$8(2)+1=17$
(6, 0)	$8(6)+0=48$

The objective function is minimized at (0, 7).

11.

Vertex	$2x+3y$
(0, 7)	$2(0)+3(7)=21$
(1, 2)	$2(1)+3(2)=8$
(2, 1)	$2(2)+3(1)=7$
(6, 0)	$2(6)+3(0)=12$

The objective function is minimized at (2, 1).

13. $\begin{cases} 15x \geq 90 \\ 810x \geq 1620 \\ \frac{1}{9}x \geq 1 \\ x \geq 0 \end{cases} \Rightarrow \begin{cases} x \geq 6 \\ x \geq 2 \\ x \geq 9 \\ x \geq 0 \end{cases}$

Therefore, 9 cups would be required

15. Find the vertices.

y-intercept of $y=-x+300$ (0, 300)

Intersection of

$y=-x+300$ and $y=-2x+500$

$-x+300=-2x+500$

$x=200$

$y=-200+300$

$y=100$

(200,100)

x-intercept of $y=-2x+500$: (250, 0)

Vertex	*Profit* $=40x+30y$
(0, 300)	$40(0)+30(300)=9000$
(200, 100)	$40(200)+30(100)=$ 11000
(250, 0)	$40(250)+30(0)$ $=10000$

The profit is maximized at (200, 100).
Joe should make 200 Assortment A's and 100 Assortment B's for a profit of \$110.

17. Find the vertices.
Lower left corner: (0, 0)

Intersection of $y=2x$ and $y=-\frac{1}{2}x+50$:

$2x=-\frac{1}{2}x+50$

$\frac{5}{2}x=50$

$x=20$

$y=2x=2(20)=40$

(20, 40)

Copyright © 2018 Pearson Education, Inc.

Intersection of $y=-\frac{1}{2}x+50$ and

$y=-\frac{4}{3}x+100$:

$-\frac{1}{2}x+50=-\frac{4}{3}x+100$

$\frac{5}{6}x=50$

$x=60$

$y=-\frac{1}{2}x+50=-\frac{1}{2}(60)+50=20$

(60, 20)

x-intercept of $y=-\frac{4}{3}x+100$: (75, 0)

Vertex	*Earnings* = 13*x* + 9*y*
(0, 0)	13(0) + 9(0) = 0
(20, 40)	13(20) + 9(40) = 620
(60, 20)	13(60) + 9(20) = 960
(75, 0)	13(75) + 9(0) = 975

The earnings are maximized at (75, 0).
Ship 75 crates of cargo A and no crates of cargo B.

19.

Vertex	*Score* = 20*x* + 5*y*
(3, 10)	20(3) + 5(10) = 110
(3, 30)	20(3) + 5(30) = 210
(7, 10)	20(7) + 5(10) = 190

The score is maximized at (3, 30).
Answer 3 essay questions and 30 short-answer questions.

21. Find the vertices.
x-intercept of $y = -0.5x + 28{,}000$: (56,000, 0)

Intersection of $y = -0.5x + 28{,}000$ and $y = -2x + 70{,}000$:

$-0.5x+28{,}000=-2x+70{,}000$

$1.5x=42{,}000$

$x=28{,}000$

$y = -2x + 70{,}000 = -2(28{,}000) + 70{,}000 = 14{,}000$

(28,000, 14,000)

Intersection of $y = -2x + 70{,}000$ and $x = 20{,}000$:
$y = -2x + 70{,}000 = -2(20{,}000) + 70{,}000 = 30{,}000$
(20,000, 30,000)

Vertex	*cost* = 0.08*x* +0 .13*y*
(56,000, 0)	0.08(56,000) + 0.13(0) = 4480
(28,000, 14,000)	0.08(28,000) + 0.13(14,000) = 4060
(20,000, 30,000)	0.08(20,000) + 0.13(30,000) = 5500

The cost is minimized at (28,000, 14,000). Buy 28,000 pounds of alfalfa and 14,000 pounds of corn at a cost of $4060.

23.

Vertex	*Profit* = 150*x* + 70*y*
(0, 0)	150(0) + 70(0) = 0
(0, 12)	150(0) + 70(12) = 840
(9, 9)	150(9) + 70(9) = 1980
(14, 4)	150(14) + 70(4) = 2380
(16, 0)	150(16) + 70(0) = 2400

The profit is maximized at (16, 0).
Make 16 chairs and no sofas.

25. In standard form the inequalities are:

$$\begin{cases} y\ge -2x+10 \\ y\ge -\frac{1}{2}x+7 \\ x\ge 0,\ y\ge 0 \end{cases}$$

y
feasible set
(0, 10)
$y=-2x+10$
$y=-\frac{1}{2}x+7$
(2, 6)
x
(14, 0)

Find the vertices.
y-intercept of $y = -2x + 10$: (0, 10)

Intersection of $y = -2x + 10$ and $y=-\frac{1}{2}x+7$:

$-\frac{1}{2}x+7=-2x+10$

$\frac{3}{2}x=3$

$x=2$

 Copyright © 2018 Pearson Education, Inc.

$y = -2x + 10 = -2(2) + 10 = 6$
(2, 6)

x-intercept of $y = -\frac{1}{2}x + 7$: (14, 0)

Vertex	$3x + 4y$
(0, 10)	$3(0) + 4(10) = 40$
(2, 6)	$3(2) + 4(6) = 30$
(14, 0)	$3(14) + 4(0) = 42$

The minimum value is 30 and occurs at (2, 6).

27. In standard form the inequalities are:

$$\begin{cases} y \le -\frac{1}{2}x + 10 \\ y \ge -\frac{3}{2}x + 12 \\ x \le 6 \\ x \ge 0,\ y \ge 0 \end{cases}$$

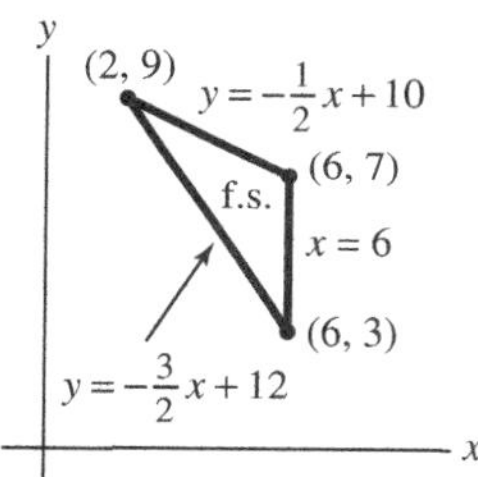

Find the vertices.

Intersection of $y = -\frac{1}{2}x + 10$ and

$y = -\frac{3}{2}x + 12$:

$-\frac{1}{2}x + 10 = -\frac{3}{2}x + 12$

$x = 2$

$y = -\frac{1}{2}x + 10 = -\frac{1}{2}(2) + 10 = 9$

(2, 9)

Intersection of $y = -\frac{1}{2}x + 10$ and $x = 6$:

$x = 6$

$y = -\frac{1}{2}x + 10 = -\frac{1}{2}(6) + 10 = 7$

(6, 7)

Intersection of $y = -\frac{3}{2}x + 12$ and $x = 6$:

$x = 6$

$y = -\frac{3}{2}x + 12 = -\frac{3}{2}(6) + 12 = 3$

(6, 3)

Vertex	$2x + 5y$
(2, 9)	$2(2) + 5(9) = 49$
(6, 7)	$2(6) + 5(7) = 47$
(6, 3)	$2(6) + 5(3) = 27$

The maximum value is 49 and occurs at (2, 9).

29. In standard form the inequalities are:

$$\begin{cases} y \le -\frac{1}{3}x + 40 \\ y \le -\frac{7}{2}x + 78 \\ x \le 20 \\ x \ge 0,\ y \ge 0 \end{cases}$$

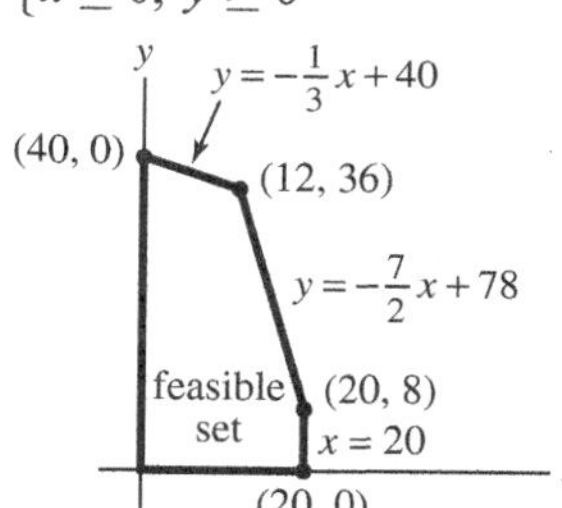

Find the vertices.
Lower left corner: (0, 0)

y-intercept of $y = -\frac{1}{3}x + 40$: (0, 40)

Intersection of $y = -\frac{1}{3}x + 40$ and

$y = -\frac{7}{2}x + 78$:

$-\frac{1}{3}x + 40 = -\frac{7}{2}x + 78$

$\frac{19}{6}x = 38$

$x = 12$

$y = -\frac{1}{3}x + 40 = -\frac{1}{3}(12) + 40 = 36$

(12, 36)

Copyright © 2018 Pearson Education, Inc.

Intersection of $y=-\frac{7}{2}x+78$ and $x=20$:

$x=20$

$y=-\frac{7}{2}x+78=-\frac{7}{2}(20)+78=8$

(20, 8)

Intersection of $x=20$ and $y=0$: (20, 0)

Vertex	$100x+150y$
(0, 0)	100(0) + 150(0) = 0
(0, 40)	100(0) + 150(40) = 6000
(12, 36)	100(12) + 150(36) = 6600
(20, 8)	100(20) + 150(8) = 3200
(20, 0)	100(20) + 150(0) = 2000

The maximum value is 6600 and occurs at (12, 36).

31. The inequalities are given in standard form.

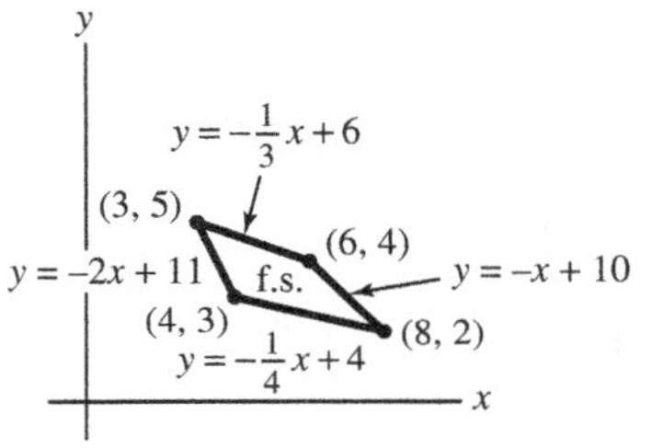

Find the vertices.

Intersection of $y=-2x+11$ and $y=-\frac{1}{3}x+6$:

$-\frac{1}{3}x+6=-2x+11$

$\frac{5}{3}x=5$

$x=3$

$y=-\frac{1}{3}x+6=-\frac{1}{3}(3)+6=5$

(3, 5)

Intersection of $y=-\frac{1}{3}x+6$ and $y=-x+10$:

$-\frac{1}{3}x+6=-x+10$

$\frac{2}{3}x=4$

$x=6$

$y=-x+10=-6+10=4$

(6, 4)

Intersection of $y=-x+10$ and $y=-\frac{1}{4}x+4$:

$-\frac{1}{4}x+4=-x+10$

$\frac{3}{4}x=6$

$x=8$

$y=-x+10=-8+10=2$

(8, 2)

Intersection of $y=-\frac{1}{4}x+4$ and $y=-2x+11$:

$-\frac{1}{4}x+4=-2x+11$

$\frac{7}{4}x=7$

$x=4$

$y=-2x+11=-2(4)+11=3$

(4, 3)

Vertex	$7x+4y$
(3, 5)	7(3) + 4(5) = 41
(6, 4)	7(6) + 4(4) = 58
(8, 2)	7(8) + 4(2) = 64
(4, 3)	7(4) + 4(3) = 40

The minimum value is 40 and occurs at (4, 3).

Copyright © 2018 Pearson Education, Inc.

33.

	Hockey games	*Soccer games*	*Available*
Assembly	2 labor-hours	3 labor-hours	42 labor-hours
Testing	2 labor-hours	1 labor-hour	26 labor-hours

Let x be the number of hockey games produced each day, and let y be the number of soccer games produced each day.

Assembly: $2x + 3y \le 42$

Testing: $2x + y \le 26$

In standard form the equations are:

$$\begin{cases} y \le -\frac{2}{3}x + 14 \\ y \le -2x + 26 \\ x \ge 0,\ y \ge 0 \end{cases}$$

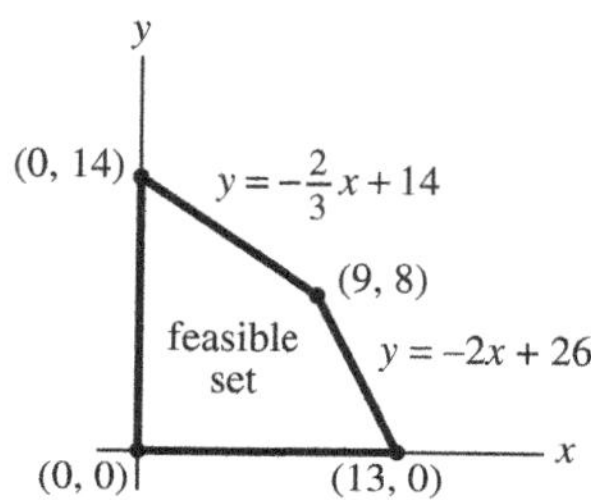

Find the vertices.

Lower left corner: (0, 0)

y-intercept of $y = -\frac{2}{3}x + 14$: (0, 14)

Intersection of $y = -\frac{2}{3}x + 14$ and $y = -2x + 26$:

$$-\frac{2}{3}x + 14 = -2x + 26$$

$$\frac{4}{3}x = 12$$

$$x = 9$$

$y = -2x + 26 = -2(9) + 26 = 8$

(9, 8)

x-intercept of $y = -2x + 26$: (13, 0)

The total daily output is simply the total number of games produced, or $x + y$.

Vertex	*Output* $= x + y$
(0, 0)	$0 + 0 = 0$
(0, 14)	$0 + 14 = 14$
(9, 8)	$9 + 8 = 17$
(13, 0)	$13 + 0 = 13$

The maximum output occurs at (9, 8). Produce 9 hockey games and 8 soccer games each day.

Copyright © 2018 Pearson Education, Inc.

35. Since the farmer can spend \$2400 for labor at \$8 per hour, the available labor is $2400 \div 8 = 300$ hours.

	Oats	*Corn*	*Available*
Capital	\$18	\$36	\$2100
Labor	2 hours	6 hours	300 hours
Land	1 acre	1 acre	100 acres
Revenue	\$55	\$125	

Let x be the number of acres of oats, and let y be the number of acres of corn.
Capital: $18x + 36y \le 2100$
Labor: $2x + 6y \le 300$
Land: $x + y \le 100$
In standard form the inequalities are:

$$\begin{cases} y \le -\frac{1}{2}x + \frac{175}{3} \\ y \le -\frac{1}{3}x + 50 \\ y \le -x + 100 \\ x \ge 0,\ y \ge 0 \end{cases}$$

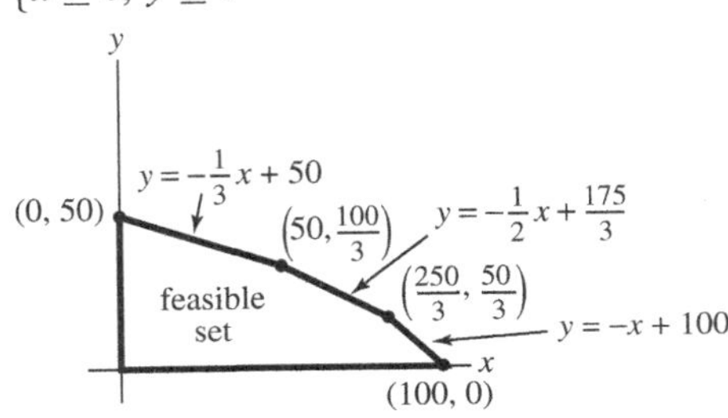

Find the vertices.
Lower left corner: (0, 0)

y-intercept of $y = -\frac{1}{3}x + 50$: (0, 50)

Intersection of $y = -\frac{1}{3}x + 50$ and $y = -\frac{1}{2}x + \frac{175}{3}$:

$$-\frac{1}{3}x + 50 = -\frac{1}{2}x + \frac{175}{3}$$

$$\frac{1}{6}x = \frac{25}{3}$$

$$x = 50$$

$$y = -\frac{1}{3}x + 50 = -\frac{1}{3}(50) + 50 = \frac{100}{3}$$

$$\left(50, \frac{100}{3}\right)$$

Copyright © 2018 Pearson Education, Inc.

Intersection of $y = -\frac{1}{2}x + \frac{175}{3}$ and $y = -x + 100$:

$$-\frac{1}{2}x + \frac{175}{3} = -x + 100$$
$$\frac{1}{2}x = \frac{125}{3}$$
$$x = \frac{250}{3}$$
$$y = -x + 100 = -\frac{250}{3} + 100 = \frac{50}{3}$$
$$\left(\frac{250}{3}, \frac{50}{3}\right)$$

x-intercept of $y = -x + 100$: (100, 0)
Find the objective function for the profit.
Revenue: $55x + 125y$
Leftover capital: $2100 - 18x - 36y$
Leftover labor cash reserve: $2400 - 8(2x + 6y) = 2400 - 16x - 48y$
The profit is the sum of the above: $21x + 41y + 4500$

Vertex	*Profit* $= 21x + 41y + 4500$
(0, 0)	$21(0) + 41(0) + 4500 = 4500$
(0, 50)	$21(0) + 41(50) + 4500 = 6550$
$\left(50, \frac{100}{3}\right)$	$21(50) + 41\left(\frac{100}{3}\right) + 4500 \approx 6916.67$
$\left(\frac{250}{3}, \frac{50}{3}\right)$	$21\left(\frac{250}{3}\right) + 41\left(\frac{50}{3}\right) + 4500 \approx 6933.33$
(100, 0)	$21(100) + 41(0) + 4500 = 6600$

The maximum profit is achieved at $\left(\frac{250}{3}, \frac{50}{3}\right)$, or $\left(83\frac{1}{3}, 16\frac{2}{3}\right)$. The farmer should plant $83\frac{1}{3}$ acres of oats and $16\frac{2}{3}$ acres of corn to make a profit of $6933.33.

37.

	Regular	Deluxe	*Available*
Capital	$32	$38	$2100
Labor	4 hours	6 hours	280 hours
Price	$46	$55	

Let x be the number of regular bags made each day, and let y be the number of deluxe

$capital: 32x + 38y \le 2100$

$Labor: 4x + 6y \le 280$

$x \ge 0, y \ge 0$

Copyright © 2018 Pearson Education, Inc.

In standard form the inequalities are:

$$\begin{cases} y \le -\dfrac{16}{19}x + \dfrac{1050}{19} \\ y \le -\dfrac{2}{3}x + \dfrac{140}{3} \\ x \ge 0,\ y \ge 0 \end{cases}$$

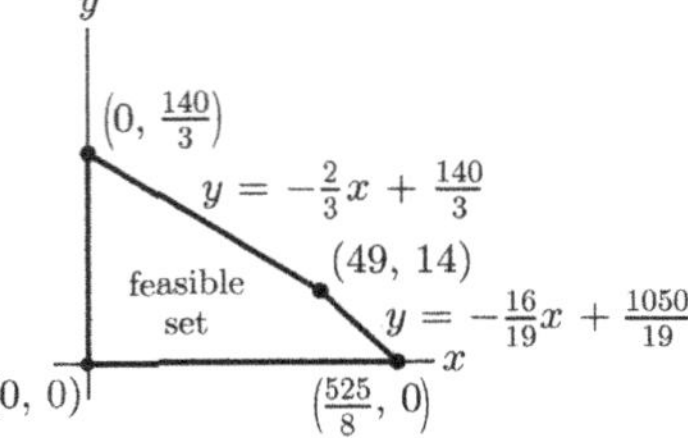

Find the vertices:

y-intercept of $y = -\dfrac{2}{3}x + \dfrac{140}{3}$: $\left(0, \dfrac{140}{3}\right)$

Intersection of $y = -\dfrac{16}{19}x + \dfrac{1050}{19}$ and $y = -\dfrac{2}{3}x + \dfrac{140}{3}$:

$$-\frac{16}{19}x + \frac{1050}{19} = -\frac{2}{3}x + \frac{140}{3}$$

$$\frac{10}{57}x = \frac{490}{57}$$

$$x = 49$$

$$y = -\frac{16}{19}x + \frac{1050}{19} = -\frac{16}{19}(49) + \frac{1050}{19} = 14$$

(49, 14)

x-intercept of $y = -\dfrac{16}{19}x + \dfrac{1050}{19}$: $\left(\dfrac{525}{8}, 0\right)$

Vertex	*Revenue* = 46*x* + 55*y*
$\left(0, \dfrac{140}{3}\right)$	$46(0) + 55(46.67) \approx 2566.67$
(49, 14)	$46(49) + 55(14) = 3024$
$\left(\dfrac{525}{8}, 0\right)$	$46(65.625) + 55(0) = 3018.75$

The maximum profit is achieved at (49, 14). Make 49 regular bags and 14 deluxe.

 Copyright © 2018 Pearson Education, Inc.

39.

	Food A	*Food B*	*Requirement*
Protein	4 units	3 units	42 units
Carbohydrates	2 units	6 units	30 units
Fat	2 units	1 unit	18 units
Weight	3 pounds	2 pounds	

Let x be the number of tubes of food A, and let y be the number of tubes of food B.
Protein: $4x + 3y \geq 42$
Carbohydrates: $2x + 6y \geq 30$
Fat: $2x + y \geq 18$
In standard form the inequalities are:

$$\begin{cases} y \geq -\frac{4}{3}x + 14 \\ y \geq -\frac{1}{3}x + 5 \\ y \geq -2x + 18 \\ x \geq 0,\ y \geq 0 \end{cases}$$

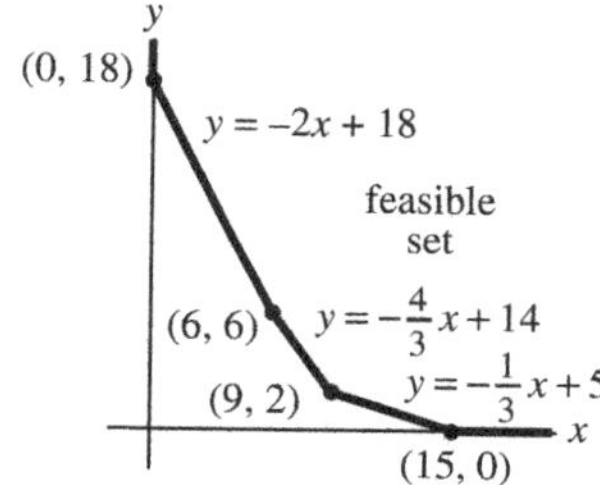

Find the vertices.
y-intercept of $y = -2x + 18$: (0, 18)

Intersection of $y = -2x + 18$ and $y = -\frac{4}{3}x + 14$:

$$-\frac{4}{3}x + 14 = -2x + 18$$
$$\frac{2}{3}x = 4$$
$$x = 6$$

$y = -2x + 18 = -2(6) + 18 = 6$
(6, 6)

Intersection of $y = -\frac{4}{3}x + 14$ and $y = -\frac{1}{3}x + 5$:

$$-\frac{1}{3}x + 5 = -\frac{4}{3}x + 14$$
$$x = 9$$
$$y = -\frac{1}{3}x + 5 = -\frac{1}{3}(9) + 5 = 2$$

(9, 2)

x-intercept of $y = -\frac{1}{3}x + 5$: (15, 0)

Copyright © 2018 Pearson Education, Inc.

Vertex	*Weight* $= 3x + 2y$
(0, 18)	$3(0) + 2(18) = 36$
(6, 6)	$3(6) + 2(6) = 30$
(9, 2)	$3(9) + 2(2) = 31$
(15, 0)	$3(15) + 2(0) = 45$

The minimum weight is achieved at (6, 6).
Send 6 tubes of food A and 6 tubes of food B.

41.

	Fruit Delight	*Heavenly Punch*	*Available*
Pineapple juice	10 ounces	10 ounces	9000 ounces
Orange juice	3 ounces	2 ounces	2400 ounces
Apricot juice	1 ounce	2 ounces	1400 ounces
Profit	$0.40	$0.60	

Let x be the number of cans of Fruit Delight, and let y be the number of cans of Heavenly Punch produced each week.
Pineapple: $10x + 10y \le 9000$
Orange: $3x + 2y \le 2400$
Apricot: $x + 2y \le 1400$
In standard form the inequalities are:

$$\begin{cases} y \le -x + 900 \\ y \le -\frac{3}{2}x + 1200 \\ y \le -\frac{1}{2}x + 700 \\ x \ge 0,\ y \ge 0 \end{cases}$$

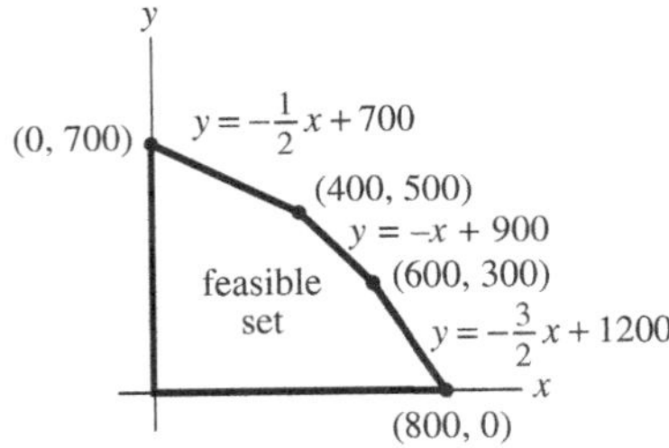

Find the vertices.
Lower left corner: (0, 0)

y-intercept of $y = -\frac{1}{2}x + 700$: (0, 700)

Intersection of $y = -\frac{1}{2}x + 700$ and $y = -x + 900$:

$$-\frac{1}{2}x + 700 = -x + 900$$

$$\frac{1}{2}x = 200$$

$$x = 400$$

Copyright © 2018 Pearson Education, Inc.

$y = -x + 900 = -400 + 900 = 500$
(400, 500)

Intersection of $y = -x + 900$ and $y = -\frac{3}{2}x + 1200$:

$$-x + 900 = -\frac{3}{2}x + 1200$$
$$\frac{1}{2}x = 300$$
$$x = 600$$

$y = -x + 900 = -600 + 900 = 300$
(600, 300)

x-intercept of $y = -\frac{3}{2}x + 1200$: (800, 0)

Vertex	*Profit* = 0.4*x* + 0.6*y*
(0, 0)	0.4(0) + 0.6(0) = 0
(0, 700)	0.4(0) + 0.6(700) = 420
(400, 500)	0.4(400) + 0.6(500) = 460
(600, 300)	0.4(600) + 0.6(300) = 420
(800, 0)	0.4(800) + 0.6(0) = 320

The maximum profit is achieved at (400, 500).
Make 400 cans of Fruit Delight and 500 cans of Heavenly Punch.

43. a. Since the conditions are less restrictive than they were in Exercise 35, the optimal solution found in Exercise 35 will still be in the feasible set.

b. The total cost to plant an acre of oats is \$18 capital plus \$16 for labor, or \$34.
The total cost to plant an acre of corn is \$36 capital plus \$48 for labor, or \$84.

	Oats	*Corn*	*Available*
Cap. And Lab.	\$34	\$84	\$4500
Land	1 acre	1 acre	100 acres
Revenue	\$55	\$125	

Let x be the number of acres of oats, and let y be the number of acres of corn.
Capital and Labor: $34x + 84y \le 4500$
Land: $x + y \le 100$
In standard form the inequalities are:

$$\begin{cases} y \le -\frac{17}{42}x + \frac{375}{7} \\ y \le -x + 100 \\ x \ge, y \ge 0 \end{cases}$$

Copyright © 2018 Pearson Education, Inc.

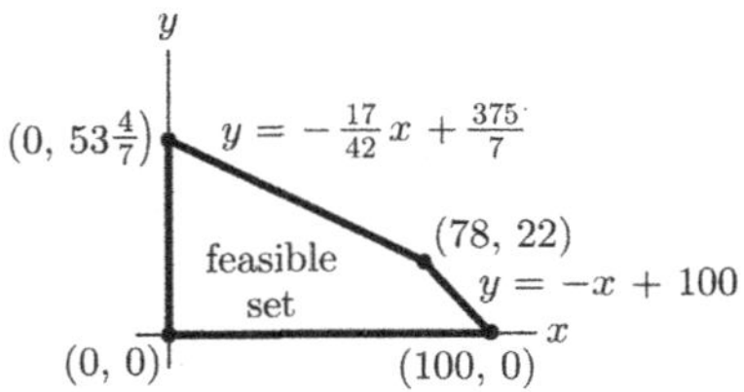

Find the vertices.

Lower left corner: (0, 0)

y-intercept of $y = -\dfrac{17}{42}x + \dfrac{375}{7}$: $\left(0, 53\dfrac{4}{7}\right)$

Intersection of $y = -\dfrac{17}{42}x + \dfrac{375}{7}$ and $y = -x + 100$:

$$-\frac{17}{42}x + \frac{375}{7} = -x + 100$$
$$\frac{25}{42}x = \frac{325}{7}$$
$$x = 78$$
$$y = -x + 100 = -(78) + 100 = 22$$

$(78, 22)$

x-intercept of $y = -x + 100$: (100, 0)

Note that the objective function for the profit is the same as in Exercise 35.

Find the objective function for the profit.

Revenue: $55x + 125y$

Leftover capital: $2100 - 18x - 36y$

Leftover labor cash reserve: $2400 - 8(2x + 6y) = 2400 - 16x - 48y$

The profit is the sum of the above: $21x + 41y + 4500$

Vertex	*Profit* $= 21x + 41y + 4500$
(0, 0)	$21(0) + 41(0) + 4500 = 4500$
(78, 22)	$21(78) + 41(22) + 4500 = 7040$
$\left(0, \frac{375}{7}\right)$	$21(0) + 41\left(\frac{375}{7}\right) + 4500 \approx 6696.43$
(100, 0)	$21(100) + 41(0) + 4500 = 6600$

The maximum profit is achieved at $(78, 22)$. The farmer should plant 78 acres of oats and 22 acres of corn to make a profit of \$7040. Yes, it provides more profit.

45.

	Cupid	*Patriotic*	*Available*
Red	60	30	2400
White	40	35	1750
Blue	0	35	1470
Profit	8	6	

Let x be the number of Cupid assortments produced, and let y be the number of Patriotic assortments produced.

Copyright © 2018 Pearson Education, Inc.

Red: $60x + 30y \le 2400$
White: $40x + 35y \le 1750$

Blue: $35y \le 1470$
In standard form the inequalities are:

$$\begin{cases} y \le -2x + 80 \\ y \le -\frac{8}{7}x + 50 \\ y \le 42 \\ x \ge 0,\ y \ge 0 \end{cases}$$

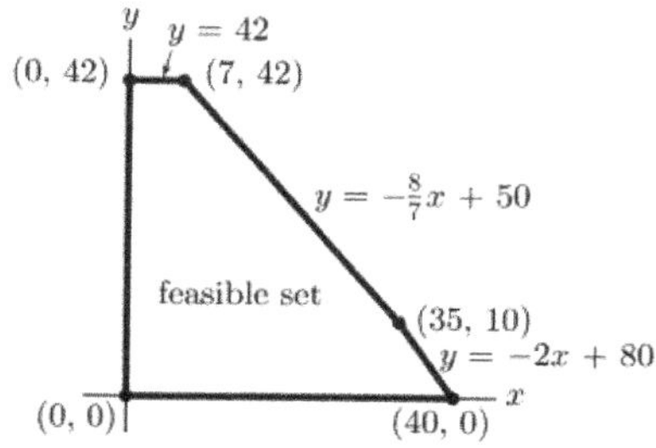

Find the vertices.
Intersection of $y = 0$ and $x = 0$ (0, 0)

Intersection of $y = 42$ and $x = 0$ (0, 42)

Intersection of $y = 42$ and $y = -\frac{8}{7}x + 50$

$$42 = -\frac{8}{7}x + 50$$
$$\frac{8}{7}x = 8$$
$$x = 7$$

(7, 42)

Intersection of $y = -2x + 80$ and $y = -\frac{8}{7}x + 50$

$$-2x + 80 = -\frac{8}{7}x + 50$$
$$30 = \frac{6}{7}x$$
$$x = 35$$

$$y = -2(35) + 80$$
$$y = 10$$

(35, 10)

Copyright © 2018 Pearson Education, Inc.

Intersection of $y=0$ and $y=-2x+80$

$0=-2x+80$

$2x=80$

$x=40$

(40, 0)

Vertex	*Profit* = 8*x* + 6*y*
(0, 0)	8(0) + 6(0) = 0
(0, 42)	8(0) + 6(42) = 252
(7, 42)	8(7) + 6(42) = 308
(35, 10)	8(35) + 6(10) = 340
(40, 0)	8(40) + 6(0) = 320

The maximum profit occurs at (35, 10).
Produce 35 Cupid assortments and 10 Patriotic assortments.

47.

	Pamper Me	*Best Friends*	*Available*
Shower Gel	1	2	400
Bubble Bath	2	2	550
Candles	2	0	400
Profit	15	12	

Let x be the number of Pamper Me baskets produced, and let y be the number of Best Friends baskets produced.
Shower Gel: $x+2y \le 400$

Bubble Bath: $2x+2y \le 550$

Candles: $2x \le 400$

In standard form the inequalities are:

$$\begin{cases} y \le -0.5x+200 \\ y \le -x+275 \\ x \le 200 \\ x \ge 0,\ y \ge 0 \end{cases}$$

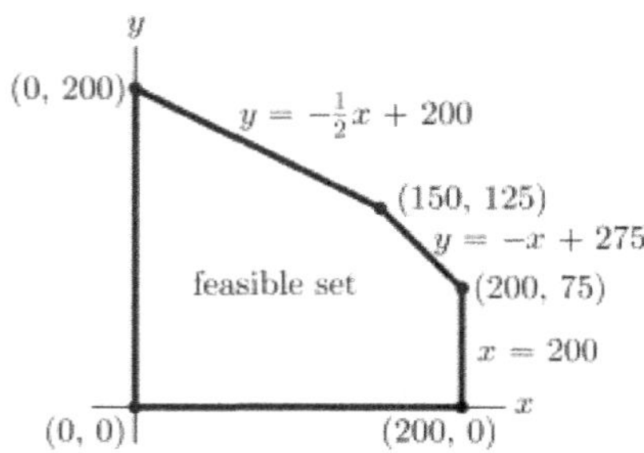

Find the vertices.
Intersection of $y=0$ and $x=0$ (0, 0)

 Copyright © 2018 Pearson Education, Inc.

Intersection of $y = -0.5x + 200$ and $x = 0$

$y = -0.5(0) + 200$

$y = 200$

(0, 200)

Intersection of $y = -0.5x + 200$ and $y = -x + 275$

$-0.5x + 200 = -x + 275$

$0.5x = 75$

$x = 150$

$y = -150 + 275$

$y = 125$

(150, 125)

Intersection of $y = -x + 275$ and $x = 200$

$y = -200 + 275$

$y = 75$

(200, 75)

Intersection of $y = 0$ and $x = 200$

(200, 0)

Vertex	*Profit* = 15*x* + 12*y*
(0, 0)	15(0) + 12(0) = 0
(0, 200)	15(0) + 12(200) = 2400
(150, 125)	15(150) + 12(125) = 3750
(200, 75)	15(200) + 12(75) = 3900
(200, 0)	15(200) + 12(0) = 3000

The maximum profit occurs at (200, 75).
Produce 200 Pamper Me baskets and 75 Best Friends baskets.

49. Since $x \geq 0$ and $y \geq 0$, $4x + 3y \geq x + y$. Therefore, the inequality $x + y \geq 6$ implies that $4x + 3y \geq 6$, which contradicts $4x + 3y \leq 4$.
The feasible set contains no points.

51.

	A	B
1	2	10
2	6	14
3		2
4		6
5		30
6		
7	Cell	Content
8	B1	=2*x+y
9	B2	=x+2*y
10	B3	=x
11	B4	=y
12	B5	=3*x+4*y

Copyright © 2018 Pearson Education, Inc.

Exercises 3.4

1. a. $21x + 14y = c$

$$y = -\frac{3}{2}x + \frac{c}{14}$$

b. Up

c. B

3. The objective function $3x + 2y$ has constant value on any line of slope $-\frac{3}{2}$. Since this is between –1 and –4, the objective function is maximized at C.

5. The objective function $10x + 2y$ has constant value on any line of slope –5. Since this is less (steeper) than –4, the objective function is maximized at D.

7. The objective function $2x + 10y$ has constant value on any line of slope $-\frac{1}{5}$. Since this is between 0 and $-\frac{1}{4}$, the objective function is minimized at D.

9. The objective function $2x + 3y$ has constant value on any line of slope $-\frac{2}{3}$. Since this is between $-\frac{1}{4}$ and –1, the objective function is minimized at C.

11. The objective function $x + ky$ has constant value on any line of slope $-\frac{1}{k}$. The slope of the line containing (0, 5) and (3, 4) is $\frac{4-5}{3-0} = -\frac{1}{3}$, and the slope of the line containing (3, 4) and (4, 0) is $\frac{0-4}{4-3} = -4$. We require $-4 \le -\frac{1}{k} \le -\frac{1}{3}$. This is equivalent to $4 \ge \frac{1}{k} \ge \frac{1}{3}$, or $\frac{1}{4} \le k \le 3$.

13.

	Brand A	*Brand B*	*Requirement*
Protein	3 units	1 unit	6 units
Carbohydrate	1 unit	1 unit	4 units
Fat	2 units	6 units	12 units
Cost	$0.80	$0.50	

Let x be the number of units of brand A, and let y be the number of units of brand B.
Protein: $3x + y \ge 6$
Carbohydrate: $x + y \ge 4$
Fat: $2x + 6y \ge 12$
In standard form the inequalities are:

Copyright © 2018 Pearson Education, Inc.

$$\begin{cases} y \geq -3x+6 \\ y \geq -x+4 \\ y \geq -\frac{1}{3}x+2 \\ x \geq 0,\ y \geq 0 \end{cases}$$

Vertex	*Cost* = .8*x* + .5*y*
(0, 6)	3
(1, 3)	2.3
(3, 1)	2.9
(6, 0)	4.8

The minimum cost of $2.30 is obtained at (1, 3).
Feed 1 can of brand A and 3 cans of brand B.

15. Let the variables represent amounts in thousands of dollars.

	Low-risk	*Medium-risk*	*High-risk*
Yield	0.06	0.07	0.08
Variables	x	y	$9-x-y$

The required inequalities are:

$$\begin{cases} x \leq y+1 \\ x+y \geq 5 \\ y+(9-x-y) \leq 7 \\ x \geq 0,\ y \geq 0 \\ 9-x-y \geq 0 \end{cases} \text{ or } \begin{cases} y \geq x-1 \\ y \geq -x+5 \\ x \geq 2 \\ x \geq 0,\ y \geq 0 \\ y \leq -x+9 \end{cases}$$

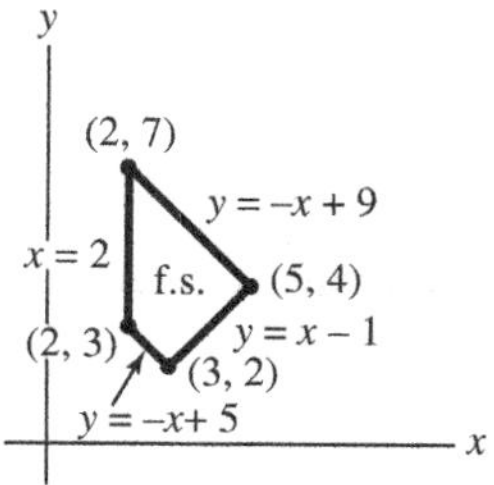

(The inequalities $x \geq 0$ and $y \geq 0$ do not appear in the graph, as they are assured by the other inequalities.)
The objective function for the expected yield is $0.06x + 0.07y + 0.08(9 - x - y)$, or $0.72 - 0.02x - 0.01y$.

Copyright © 2018 Pearson Education, Inc.

Vertex	*Yield* = .72 − .02*x* − .01*y*
(2, 3)	.65
(2, 7)	.61
(5, 4)	.58
(3, 2)	.64

The maximum yield of \$650 is achieved at (2, 3). Then $9 - x - y = 9 - 2 - 3 = 4$.
Invest \$2000 in low-risk stocks, \$3000 in medium-risk stocks, and \$4000 in high-risk stocks.

17.

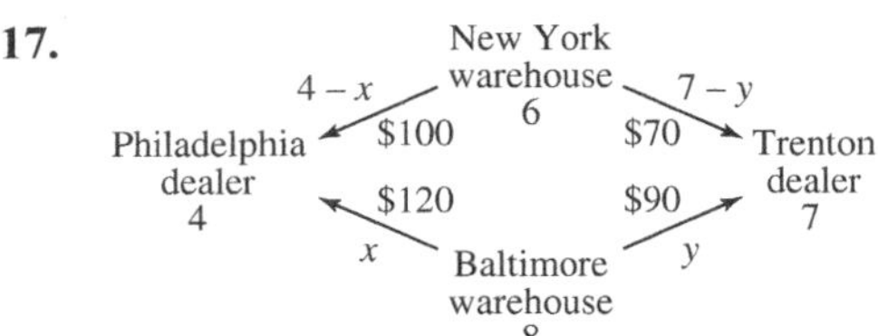

The required inequalities are:

$$\begin{cases} x \geq 0,\ y \geq 0 \\ 4 - x \geq 0 \\ 7 - y \geq 0 \\ x + y \leq 8 \\ (4-x) + (7-y) \leq 6 \end{cases} \quad \text{or} \quad \begin{cases} x \geq 0,\ y \geq 0 \\ x \leq 4 \\ y \leq 7 \\ y \leq -x + 8 \\ y \geq -x + 5 \end{cases}$$

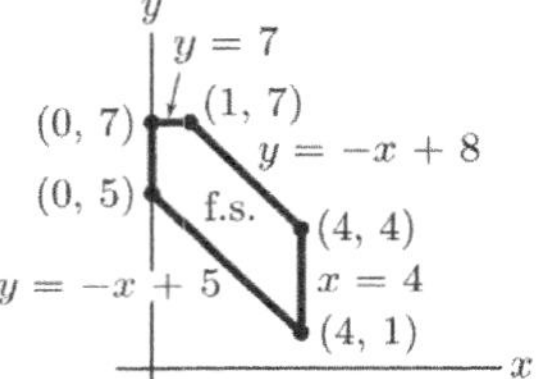

(The inequality $y \geq 0$ does not appear in the graph, as it is assured by the other inequalities.)

The objective function for the cost is $120x + 90y + 100(4 - x) + 70(7 - y)$, or $890 + 20x + 20y$.

Vertex	*Cost* = 890 + 20*x* + 20*y*
(0, 5)	990
(0, 7)	1030
(1, 7)	1050
(4, 4)	1050
(4, 1)	990

The minimum cost of \$990 is achieved at (0, 5) or (4, 1)—or anywhere along the segment connecting these points.

 Copyright © 2018 Pearson Education, Inc.

There are several ways to minimize costs, as summarized below.

Baltimore to Philadelphia	*Baltimore to Trenton*	*New York to Philadelphia*	*New York to Trenton*
0	5	4	2
1	4	3	3
2	3	2	4
3	2	1	5
4	1	0	6

19. Let the variables represent the number of thousands of gallons.

	Gasoline	*Jet fuel*	*Diesel fuel*
Profit	\$0.15	\$0.12	\$0.10
Variables	x	y	$100 - x - y$

The required inequalities are:

$$\begin{cases} x \geq 5,\ y \geq 5 \\ 100 - x - y \geq 5 \\ x + y \geq 20 \\ x + (100 - x - y) \geq 50 \end{cases} \quad \text{or} \quad \begin{cases} x \geq 5,\ y \geq 5 \\ y \geq -x + 95 \\ y \geq -x + 20 \\ y \leq 50 \end{cases}$$

(Note that the inequalities above have ignored the somewhat subtle issue of whether there will be enough gasoline for *both* the airline and the trucking firm at the same time. This is only a potential issue if both suppliers require gasoline—that is, if $y \leq 20$ and $100 - x - y \leq 50$. In this case, we require that the gasoline requirements of each firm be less than the amount of gasoline actually produced—that is,
$(20 - y) + (50 - (100 - x - y)) \leq x$. This inequality is equivalent to $-30 \leq 0$, so it is always satisfied.)

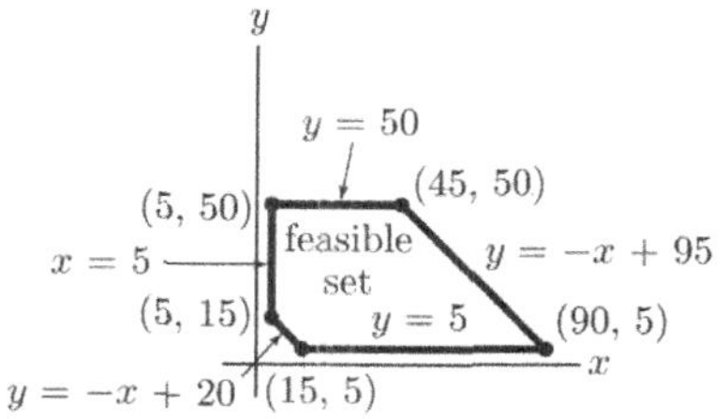

y
y = 50
(45, 50)
(5, 50)
feasible set
x = 5
y = −x + 95
(5, 15)
y = 5
(90, 5)
x
y = −x + 20
(15, 5)

The objective function for the profit is $0.15x + 0.12y + 0.1(100 - x - y)$, or $10 + 0.05x + 0.02y$.

Vertex	*Profit* = $10 + 0.05x + 0.02y$
(5, 50)	11.25
(45, 50)	13.25
(90, 5)	14.6
(15, 5)	10.85
(5, 15)	10.55

The maximum profit of \$14,600 is achieved at (90, 5). Then $100 - x - y = 100 - 90 - 5 = 5$.
Produce 90,000 gallons of gasoline, 5000 gallons of jet fuel, and 5000 gallons of diesel fuel.

Copyright © 2018 Pearson Education, Inc.

21.

	High-capacity	*Low-capacity*	*Available*
Cost ($thousands)	50	30	1080
Drivers	1	1	30
Capacity	320 cases	200 cases	

Let x be the number of high-capacity trucks, and let y be the number of low-capacity trucks. The required inequalities are:

$$\begin{cases} 50x+30y\le 1080 \\ x+y\le 30 \\ x\le 15 \\ x\ge 0,\ y\ge 0 \end{cases} \quad \text{or} \quad \begin{cases} y\le -\frac{5}{3}x+36 \\ y\le -x+30 \\ x\le 15 \\ x\ge 0,\ y\ge 0 \end{cases}$$

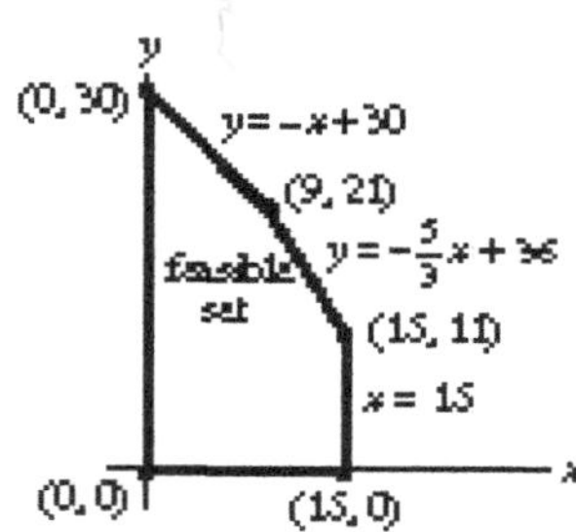

Vertex	*Capacity* = 320x + 200y
(0, 0)	0
(0, 30)	6000
(9, 21)	7080
(15, 11)	7000
(15, 0)	4800

The maximum capacity of 7080 cases is achieved at (9, 21).
Buy 9 high-capacity trucks and 21 low-capacity trucks.

23. Let x = pounds of coffee shipped from San Jose to Salt Lake City, and
y = pounds of coffee shipped from San Jose to Reno.

The required inequalities are:

$$\begin{cases} x+y\le 500 \\ (400-x)+(350-y)\le 700 \\ x\ge 0, y\ge 0 \\ 400-x\ge 0, 350-y\ge 0 \end{cases} \quad \text{or} \quad \begin{cases} y\le -x+500 \\ y\ge -x+50 \\ x\ge 0, y\ge 0 \\ x\le 400, y\le 350 \end{cases}$$

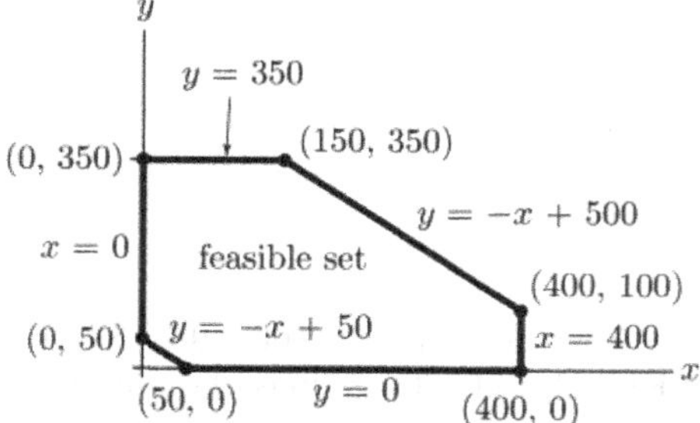

 Copyright © 2018 Pearson Education, Inc.

Objective function: [cost] = $2x+2.5y+2.5(400-x)+3.5(350-y)=2225-.5x-y$.

Vertex	*Cost* = $2225-.5x-y$
(0, 50)	2175
(0, 350)	1875
(150, 350)	1800
(400, 100)	1925
(400, 0)	2025
(50, 0)	2200

The minimum cost of $1800 is at (150, 350).
Ship 150 pounds of coffee from San Jose to Salt Lake City, 350 pounds from San Jose to Reno, and 250 pounds from Seattle to Salt Lake City.

25.

	Kit I	*Kit II*	*Kit III*	*Available*
Filters	1	2	1	54
Gravel (pounds)	2	2	3	100
Fish food	1	0	2	53
Profit	$7	$10	$13	

Let x be the number of Kit I's, and let y be the number of Kit II's. Then $x-y$ is the number of Kit III's. The required inequalities are:

$$\begin{cases} x+2y+(x-y)\leq 54 \\ 2x+2y+3(x-y)\leq 100 \\ x+2(x-y)\leq 53 \\ x-y\geq 0 \\ x\geq 0, y\geq 0 \end{cases} \quad \text{or} \quad \begin{cases} y\leq -2x+54 \\ y\geq 5x-100 \\ y\geq \frac{3}{2}x-\frac{53}{2} \\ y\leq x \\ x\geq 0, y\geq 0 \end{cases}$$

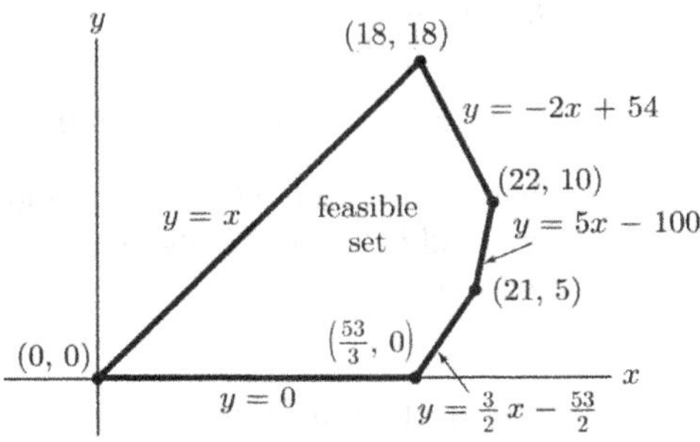

The objective function is $[\text{profit}]=7x+10y+13(x-y)$
$$=20x-3y$$

Vertex	*Profit* = $20x - 3y$
(0, 0)	0
(18, 18)	306
(22, 10)	410
(21, 5)	405
$\left(\frac{53}{3}, 0\right)$	$353\frac{1}{3}$

The maximum profit of \$410 is achieved at (22, 10). Create 22 Kit I, 10 Kit II; and 12 Kit III.

27.a. $Ax+70y=p$ or $y=-\frac{A}{70}x+\frac{p}{70}$

The slope of the line is $-\frac{A}{70}$.

b. $-2 \le -\frac{A}{70} \le -1$

$-140 \le -A \le -70$

$140 \ge A \ge 70$

c. $80x + Ay = p$

$y = -\frac{80}{A}x + \frac{p}{A}$

$-2 \le -\frac{80}{A} \le -1$

$-\frac{1}{2} \ge -\frac{A}{80} \ge -1$

$40 \le A \le 80$

29. The objective function $ax + by$ has constant value on any line of slope $-\frac{a}{b}$. The slope of the line containing (8, 3) and (9, 0) is $\frac{0-3}{9-8} = -3$. Therefore, if $-\frac{a}{b} < 0$, we require $-\frac{a}{b} \le -3$, or $\frac{a}{b} \ge 3$, where $a > 0$ and $b > 0$. One possibility is $a = 5$ and $b = 1$, giving the objective function $5x + y$. (We could also have chosen an objective function that is constant on lines of nonnegative or undefined slope of the form $ax + by$ with $a \ge 0$ and $b \le 0$.)

31. The objective function $ax + by$ has constant value on any line of slope $-\frac{a}{b}$. Since the slope of the line containing (3, 8) and (8, 3) is $\frac{3-8}{8-3} = -1$, and the slope of the line containing (8, 3) and (9, 0) is $\frac{0-3}{9-8} = -3$, we require

$-1 \ge -\frac{a}{b} \ge -3$, or $1 \le \frac{a}{b} \le 3$.

One possibility is $a = 2$ and $b = 1$, giving the objective function $2x + y$.

33. The objective function $ax + by$ has constant value on any line of slope $-\frac{a}{b}$. The slope of the line containing (6, 1) and (9, 0) is $\frac{0-1}{9-6} = -\frac{1}{3}$.

We require $0 \ge -\frac{a}{b} \ge -\frac{1}{3}$, or $0 \le \frac{a}{b} \le \frac{1}{3}$, where $a \ge 0$ and $b > 0$. One possibility is $a = 1$ and $b = 5$, giving the objective function $x + 5y$.

35. The objective function $ax + by$ has constant value on any line of slope $-\frac{a}{b}$. The slope of the line containing (1, 6) and (6, 1) is $\frac{1-6}{6-1} = -1$, and the slope of the line containing (6, 1) and (9, 0) is $\frac{0-1}{9-6} = -\frac{1}{3}$.

Copyright © 2018 Pearson Education, Inc.

We require $-\frac{1}{3} \geq -\frac{a}{b} \geq -1$, or

$\frac{1}{3} \leq \frac{a}{b} \leq 1$, where $a > 0$ and $b > 0$. One possibility is $a = 2$ and $b = 3$, giving the objective function $2x + 3y$.

37.

Cell	*Name*	*Final Value*	*Reduced Cost*	*Objective Coefficient*	*Allowable Increase*	*Allowable Decrease*
\$A\$1	*x*	.857142857	0	21	21	11.66666667
\$A\$2	*y*	3.428571429	0	14	17.5	7

The range of optimality for the cost of rice is [21 – 11.67, 21 + 21] or [9.33, 42].
The range of optimality for the cost of soybeans is [14 – 7, 14 + 17.5] or [7, 31.5].

Chapter 3 Review Exercises

1. $3(1) + 4(2) \geq 11$

$3 + 8 \geq 11$

$11 \geq 11$

Yes

2. $x - 3y \geq 12$

$y \leq \frac{1}{3}x - 4$

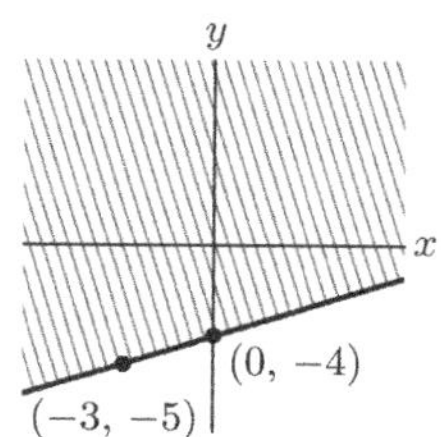

3. $\begin{cases} 2x - y = 1 \\ x + 2y = 13 \end{cases}$

$\begin{cases} y = 2x - 1 \\ y = -\frac{1}{2}x + \frac{13}{2} \end{cases}$

$2x - 1 = -\frac{1}{2}x + \frac{13}{2}$

$\frac{5}{2}x = \frac{15}{2}$

$x = 3$

$y = 2(3) - 1 = 5$

(3, 5)

Copyright © 2018 Pearson Education, Inc.

4.

	Type A	*Type B*	*Required or available*
Passengers	50	300	1400
Flight attendants	3	4	42
Cost	\$14,000	\$90,000	

Let x be the number of type A planes, and let y be the number of type B planes.
The required inequalities are:

$$\begin{cases} 50x+300y \geq 1400 \\ 3x+4y \leq 42 \\ x \geq y \\ x \geq 0,\ y \geq 0 \end{cases} \quad \text{or} \quad \begin{cases} y \geq -\frac{1}{6}x+\frac{14}{3} \\ y \leq -\frac{3}{4}x+\frac{21}{2} \\ y \leq x \\ x \geq 0,\ y \geq 0 \end{cases}$$

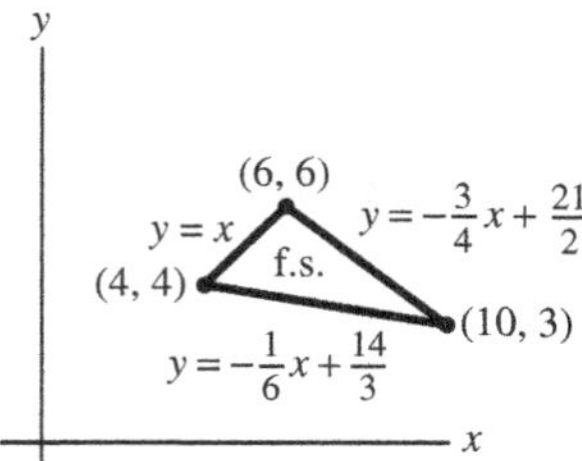

(The inequalities $x \geq 0$ and $y \geq 0$ are not shown in the graph because they are assured by the other inequalities.)

Vertex	*Cost* = 14,000x + 90,000y
(4, 4)	416,000
(6, 6)	624,000
(10, 3)	410,000

The minimum cost of \$410,000 is achieved at (10, 3).
Use 10 type A planes and 3 type B planes.

5.

	Wheat germ	*Enriched oat flour*	*Required*
Niacin	2 milligrams	3 milligrams	7 milligrams
Iron	3 milligrams	3 milligrams	9 milligrams
Thiamin	0.5 milligram	0.25 milligram	1 milligram
Cost	6 cents	8 cents	

Let x be the number of ounces of wheat germ, and let y be the number of ounces of enriched oat flour.

Copyright © 2018 Pearson Education, Inc.

The required inequalities are:

$$\begin{cases} 2x+3y \geq 7 \\ 3x+3y \geq 9 \\ .5x+.25y \geq 1 \\ x \geq 0,\ y \geq 0 \end{cases} \text{ or } \begin{cases} y \geq -\frac{2}{3}x+\frac{7}{3} \\ y \geq -x+3 \\ y \geq -2x+4 \\ x \geq 0,\ y \geq 0 \end{cases}$$

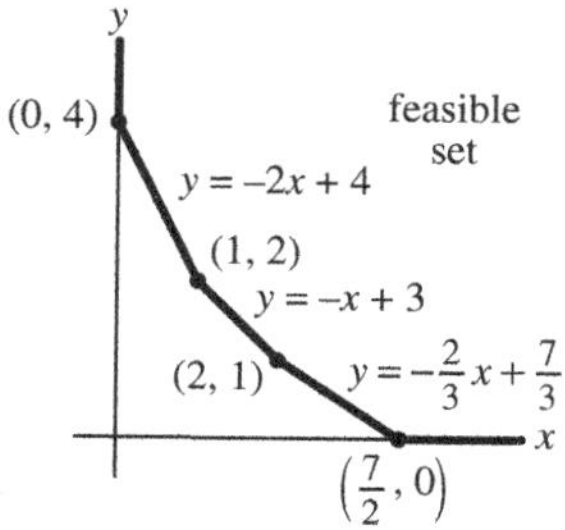

Vertex	*Cost* = 6x + 8y
(0, 4)	32
(1, 2)	22
(2, 1)	20
$\left(\frac{7}{2}, 0\right)$	21

The minimum cost of 20 cents is achieved at (2, 1).
Use 2 ounces of wheat germ and 1 ounce of enriched oat flour.

6.

	Hardtops	*Sports cars*	*Available*
Assemble	8 labor-hours	18 labor-hours	360 labor-hours
Paint	2 labor-hours	2 labor-hours	50 labor-hours
Upholster	2 labor-hours	1 labor-hour	40 labor-hours
Profit	\$90	\$100	

Let x be the number of hardtops and let y be the number of Sports cars
The required inequalities are:

$$\begin{cases} 8x+18y \leq 360 \\ 2x+2y \leq 50 \\ 2x+y \leq 40 \\ x \geq 0,\ y \geq 0 \end{cases} \text{ or } \begin{cases} y \leq -\frac{4}{9}x+20 \\ y \leq -x+25 \\ y \leq -2x+40 \\ x \geq 0,\ y \geq 0 \end{cases}$$

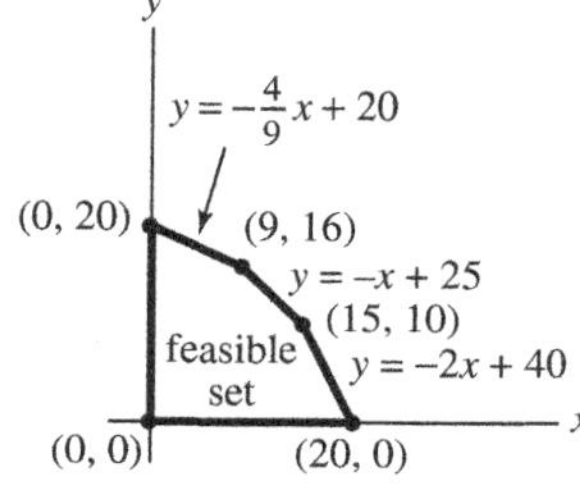

Copyright © 2018 Pearson Education, Inc.

Vertex	*Profit* = 90*x* + 100*y*
(0, 0)	0
(0, 20)	2000
(9, 16)	2410
(15, 10)	2350
(20, 0)	1800

The maximum profit of $2410 is achieved at (9, 16).
Produce 9 hardtops and 16 sports cars.

7.

	Type A	*Type B*	*Available*
Peanuts	6 oz	12 oz	5400 oz
Raisins	1 oz	3 oz	1200 oz
Cashews	4 oz	2 oz	2400 oz
Revenue	$4.25	$6.55	

Let x be the amount of mixture A, and let y be the amount of mixture B.

Peanuts : $6x+12y \le 5400$

Raisins : $x+3y \le 1200$

Cashews : $4x+2y \le 2400$

In standard form the inequalities are:

$$\begin{cases} y \le -\frac{1}{2}x+450 \\ y \le -\frac{1}{3}x+400 \\ y \le -2x+1200 \\ x \ge 0,\ y \ge 0 \end{cases}$$

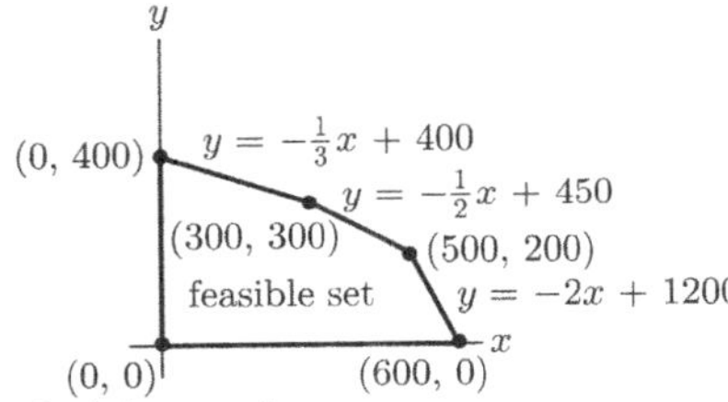

Find the vertices.
Lower left corner: (0, 0)

y-intercept of $y=-\frac{1}{3}x+400$: (0, 400)

Intersection of $y=-\frac{1}{3}x+400$ and $y=-\frac{1}{2}x+450$:

$$-\frac{1}{3}x+400=-\frac{1}{2}x+450$$

$$\frac{1}{6}x=50$$

$$x=300$$

Copyright © 2018 Pearson Education, Inc.

$y = -\frac{1}{3}x + 400 = -\frac{1}{3}(300) + 400 = 300$
(300, 300)

Intersection of $y = -\frac{1}{2}x + 450$ and $y = -2x + 1200$:

$-\frac{1}{2}x + 450 = -2x + 1200$

$\frac{3}{2}x = 750$

$x = 500$

$y = -\frac{1}{2}x + 450 = -\frac{1}{2}(500) + 450 = 200$
(500, 200)
x-intercept of $y = -2x + 1200$: (600, 0)

Vertex	*Revenue* = 4.25x + 6.55y
(0, 0)	4.25(0) + 6.55(0) = 0
(0, 400)	4.25(0) + 6.55(400) = 2620
(300, 300)	4.25(300) + 6.55(300) = 3240
(500, 200)	4.25(500) + 6.55(200) = 3435
(600, 0)	4.25(600) + 6.55(0) = 2550

The maximum revenue is achieved at (500, 200).
Make 500 boxes of mixture A and 200 boxes of mixture B.

8.

	Elementary	*Intermediate*	*Advanced*
Profit	$8000	$7000	$1000
Variables	x	y	$72 - x - y$

The required inequalities are:

$$\begin{cases} 72 - x - y \geq 4 \\ x \geq 3y \\ y \geq 2(72 - x - y) \\ x \geq 0,\ y \geq 0 \end{cases} \quad \text{or} \quad \begin{cases} y \leq -x + 68 \\ y \leq \frac{1}{3}x \\ y \leq -\frac{2}{3}x + 48 \\ x \geq 0,\ y \geq 0 \end{cases}$$

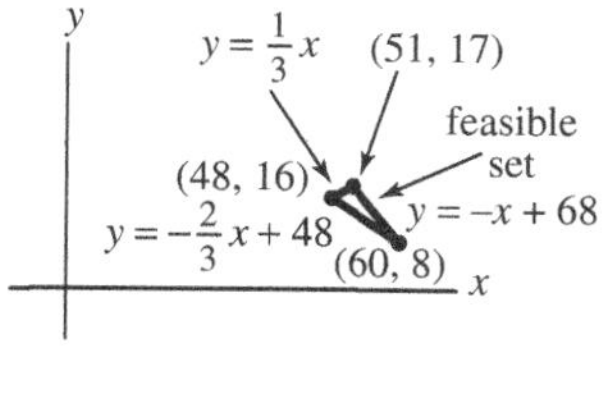

y
$y = \frac{1}{3}x$
(51, 17)
feasible set
(48, 16)
$y = -x + 68$
$y = -\frac{2}{3}x + 48$
(60, 8)
x

The objective function for the annual profit is $8000x + 7000y + 1000(72 - x - y)$, or $72{,}000 + 7000x + 6000y$.

Copyright © 2018 Pearson Education, Inc.

Vertex	*Profit* = 72,000 + 7000*x* + 6000*y*
(48, 16)	504,000
(51, 17)	531,000
(60, 8)	540,000

The maximum annual profit of $540,000 is achieved at (60, 8).
Then $72 - x - y = 72 - 60 - 8 = 4$.
Publish 60 elementary books, 8 intermediate books, and 4 advanced books.

9.

	Rochester	*Queens*	*Available*
Transport time	15 hours	20 hours	2100 hours
Cost	$15	$30	$3000
Profit	$40	$30	

Let x be the number of computers sent from Rochester, and let y be the number of computers sent from Queens. The required inequalities are:

$$\begin{cases} 15x+20y \le 2100 \\ 15x+30y \le 3000 \\ x \le 80 \\ y \le 120 \\ x \ge 0,\ y \ge 0 \end{cases} \quad \text{or} \quad \begin{cases} y \le -\frac{3}{4}x+105 \\ y \le -\frac{1}{2}x+100 \\ x \le 80 \\ y \le 120 \\ x \ge 0,\ y \ge 0 \end{cases}$$

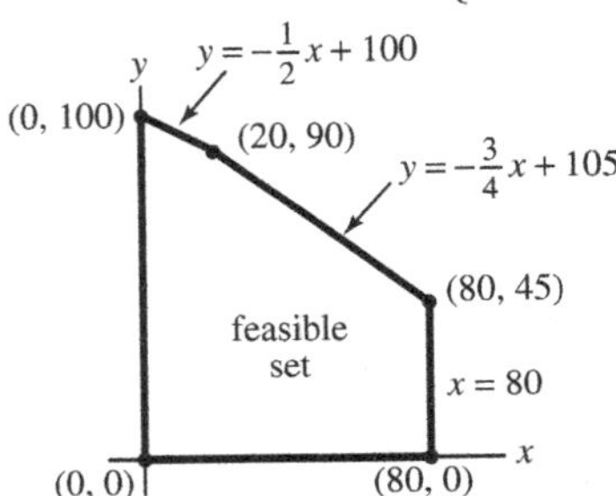

(The inequality $y \le 120$ does not appear in the graph, as it is assured by the other inequalities.)

Vertex	*Profit* = 40*x* + 30*y*
(0, 0)	0
(0, 100)	3000
(20, 90)	3500
(80, 45)	4550
(80, 0)	3200

The maximum profit of $4550 is achieved at (80, 45).
Package 80 computers at Rochester and 45 computers at Queens.

 Copyright © 2018 Pearson Education, Inc.

10.

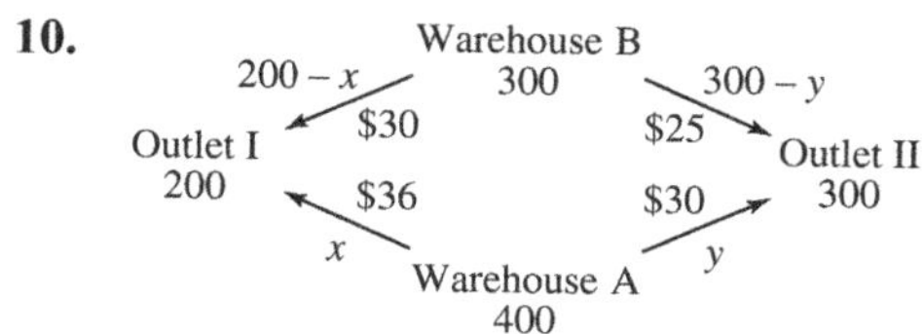

The required inequalities are:

$$\begin{cases} x \geq 0,\ y \geq 0 \\ 200 - x \geq 0 \\ 300 - y \geq 0 \\ x + y \leq 400 \\ (200 - x) + (300 - y) \leq 300 \end{cases} \quad \text{or} \quad \begin{cases} x \geq 0,\ y \geq 0 \\ x \leq 200 \\ y \leq 300 \\ y \leq -x + 400 \\ y \geq -x + 200 \end{cases}$$

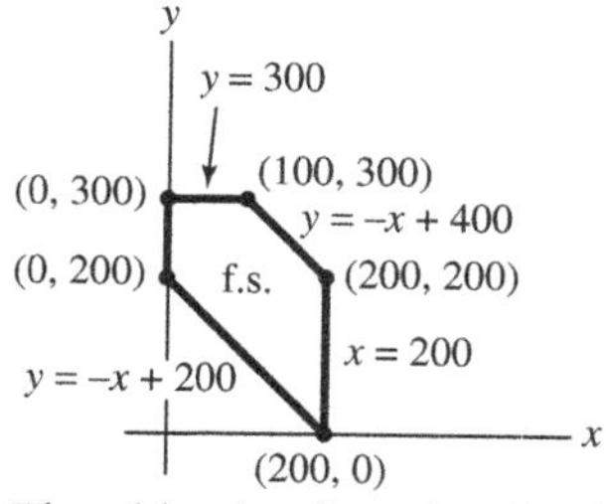

The objective function for the cost is $36x + 30y + 30(200 - x) + 25(300 - y)$, or $13{,}500 + 6x + 5y$.

Vertex	*Cost* = 13,500 + 6*x* + 5*y*
(0, 200)	14,500
(0, 300)	15,000
(100, 300)	15,600
(200, 200)	15,700
(200, 0)	14,700

The minimum cost of $14,500 is achieved at (0, 200). Then $200 - x = 200 - 0 = 200$, and $300 - y = 300 - 200 = 100$.

Transport no refrigerators from warehouse A to outlet I, 200 refrigerators from warehouse A to outlet II, 200 refrigerators from warehouse B to outlet I, and 100 refrigerators from warehouse B to outlet II.

Copyright © 2018 Pearson Education, Inc.

11.

	CD	*Mutual fund*	*Stocks*
Yield	0.05	0.07	0.09
Variables	x	y	$10{,}000-(x+y)$

The required inequalities are:

$$\begin{cases} y \le x + 10{,}000 - (x+y) \\ y + 10{,}000 - (x+y) \le 8000 \\ x \ge 0,\ y \ge 0 \\ x \le 10{,}000,\ y \le 10{,}000 \end{cases} \quad \text{or} \quad \begin{cases} y \le 5000 \\ x \ge 2000 \\ y \ge 0 \\ x \le 10{,}000 \end{cases}$$

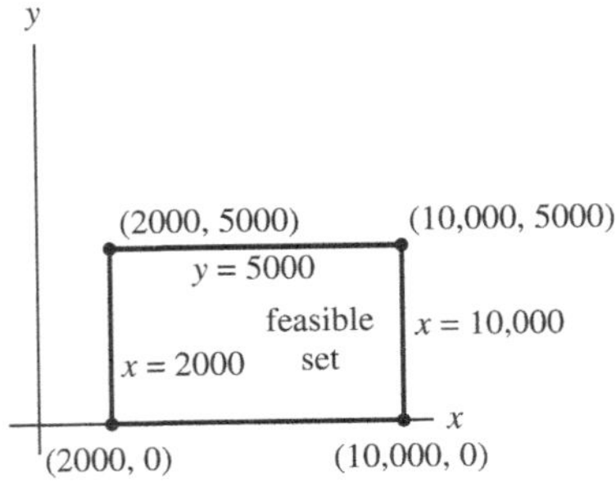

The objective function is [return] = $0.05x + 0.07y + 0.09[10{,}000 - (x+y)]$ or $900 - 0.04x - 0.02y$.

Vertex	*Return* = $900 - 0.04x - 0.02y$
(2000, 0)	820
(2000, 5000)	720
(10,000, 5000)	400
(10,000, 0)	500

The maximum return of $820 is achieved at (2000, 0). Invest $2000 in the CD, $0 in mutual funds; and $8000 in stocks.

12. a. Yes; The added constraint may add to the original cost therefore it may increase the optimal cost.

b. No; The added constraint will not decrease the original cost.

13. a. Yes; The removed constraint may increase the profit therefore it may increase the optimal profit.

b. No; The removed constraint will not decrease profit.

14. Answers will vary.

15. One of the boundary points will always be a maximum or a minimum of the problem.

16. Every point on a line (or line segment) always has the same objective function value, provided the line (or line segment) has the same slope as a line with constant objective function.

 Copyright © 2018 Pearson Education, Inc.

Chapter 4

Exercises 4.1

1. $\begin{cases} 20x + 30y + u = 3500 \\ 50x + 10y + v = 5000 \\ -8x - 13y + M = 0 \end{cases}$

Maximize M given $x \geq 0$, $y \geq 0$, $u \geq 0$, $v \geq 0$.

3. $\begin{cases} x + y + z + u = 100 \\ 3x + z + v = 200 \\ 5x + 10y + w = 100 \\ -x - 2y + 3z + M = 0 \end{cases}$

Maximize M given $x \geq 0$, $y \geq 0$, $z \geq 0$, $u \geq 0$, $v \geq 0$, $w \geq 0$.

5. $\begin{cases} 4x + 6y - 7z + u = 16 \\ 3x + 2y + v = 11 \\ 9y + 3z + w = 21 \\ -3x - 5y - 12z + M = 0 \end{cases}$

Maximize M given $x \geq 0$, $y \geq 0$, $z \geq 0$, $u \geq 0$, $v \geq 0$, $w \geq 0$.

7. a.

$$\begin{array}{ccccc|c} x & y & u & v & M & \\ 20 & 30 & 1 & 0 & 0 & 3500 \\ 50 & 10 & 0 & 1 & 0 & 5000 \\ -8 & -13 & 0 & 0 & 1 & 0 \end{array}$$

b. $x = 0, y = 0, u = 3500, v = 5000, M = 0$

9. a.

$$\begin{array}{ccccccc|c} x & y & z & u & v & w & M & \\ 1 & 1 & 1 & 1 & 0 & 0 & 0 & 100 \\ 3 & 0 & 1 & 0 & 1 & 0 & 0 & 200 \\ 5 & 10 & 0 & 0 & 0 & 1 & 0 & 100 \\ -1 & -2 & 3 & 0 & 0 & 0 & 1 & 0 \end{array}$$

b. $x = 0, y = 0, z = 0, u = 100, v = 200, w = 100, M = 0$

11. a.

$$\begin{array}{ccccccc|c} x & y & z & u & v & w & M & \\ 4 & 6 & -7 & 1 & 0 & 0 & 0 & 16 \\ 3 & 2 & 0 & 0 & 1 & 0 & 0 & 11 \\ 0 & 9 & 3 & 0 & 0 & 1 & 0 & 21 \\ -3 & -5 & -12 & 0 & 0 & 0 & 1 & 0 \end{array}$$

b. $x = 0, y = 0, z = 0, u = 16, v = 11, w = 21, M = 0$

13. $x = 15, y = 0, u = 10, v = 0, M = 20$

15. $x = 14, y = 0, z = 17, u = 0, v = 0, M = 56$

17. $x = 42, y = 54, z = 61, u = 0, v = 0, w = 0,$ $M = 604$

19. $x = 10, y = 0, z = 15, u = 23, v = 0, w = 0,$ $M = -11$

Copyright © 2018 Pearson Education, Inc.

21. a. Divide the first row by 2, then use matrix operations to change the remaining values in column 1 to zeros.

$$\begin{array}{c|ccccc|c} & x & y & u & v & M & \\ \hline x & 1 & \frac{3}{2} & \frac{1}{2} & 0 & 0 & 6 \\ v & 0 & -\frac{1}{2} & -\frac{1}{2} & 1 & 0 & 4 \\ M & 0 & -5 & 5 & 0 & 1 & 60 \end{array}$$

$x = 6, y = 0, u = 0, v = 4, M = 60$

b. Divide the first row by 3, then use matrix operations to change the remaining values in column 2 to zeros.

$$\begin{array}{c|ccccc|c} & x & y & u & v & M & \\ \hline y & \frac{2}{3} & 1 & \frac{1}{3} & 0 & 0 & 4 \\ v & \frac{1}{3} & 0 & -\frac{1}{3} & 1 & 0 & 6 \\ M & \frac{10}{3} & 0 & \frac{20}{3} & 0 & 1 & 80 \end{array}$$

$x = 0, y = 4, u = 0, v = 6, M = 80$

c. Use matrix operations to change the remaining values in column 1 to zeros.

$$\begin{array}{c|ccccc|c} & x & y & u & v & M & \\ \hline u & 0 & 1 & 1 & -2 & 0 & -8 \\ x & 1 & 1 & 0 & 1 & 0 & 10 \\ M & 0 & -10 & 0 & 10 & 1 & 100 \end{array}$$

$x = 10, y = 0, u = -8, v = 0, M = 100$

d. Use matrix operations to change the remaining values in column 2 to zeros.

$$\begin{array}{c|ccccc|c} & x & y & u & v & M & \\ \hline u & -1 & 0 & 1 & -3 & 0 & -18 \\ y & 1 & 1 & 0 & 1 & 0 & 10 \\ M & 10 & 0 & 0 & 20 & 1 & 200 \end{array}$$

$x = 0, y = 10, u = -18, v = 0, M = 200$

e. M changes the most after the pivot operation in part (d).

23. a. Group I variables are x and y.

Group II variables are u, v, and M.

b. (i) Divide the first row by 2, then use matrix operations to change the remaining values in column 1 to zeros.

$$\begin{array}{ccccc|c} x & y & u & v & M & \\ \hline 1 & \frac{5}{2} & \frac{1}{2} & 0 & 0 & 50 \\ 0 & -\frac{13}{2} & -\frac{3}{2} & 1 & 0 & 150 \\ 0 & 18 & 5 & 0 & 1 & 500 \end{array}$$

Feasible:

Group I variables: u and y Group II variables: x, v, M

(ii) Divide the first row by 5, then use matrix operations to change the remaining values in column 2 to zeros.

$$\begin{array}{ccccc|c} x & y & u & v & M & \\ \hline \frac{2}{5} & 1 & \frac{1}{5} & 0 & 0 & 20 \\ \frac{13}{5} & 0 & -\frac{1}{5} & 1 & 0 & 280 \\ -\frac{36}{5} & 0 & \frac{7}{5} & 0 & 1 & 140 \end{array}$$

Feasible:

Group I variables: x and u Group II variables: y, v, M

(iii) Divide the second row by 3, then use matrix operations to change the remaining values in column 1 to zeros.

$$\begin{array}{ccccc|c} x & y & u & v & M & \\ \hline 0 & \frac{13}{3} & 1 & -\frac{2}{3} & 0 & -100 \\ 1 & \frac{1}{3} & 0 & \frac{1}{3} & 0 & 100 \\ 0 & -\frac{11}{3} & 0 & \frac{10}{3} & 1 & 1000 \end{array}$$

Not Feasible: Group I variables: y and v Group II variables: x, u, M

(iv) Use matrix operations to change the remaining value in column 2 to zeros.

$$\begin{array}{ccccc|c} x & y & u & v & M & \\ \hline -13 & 0 & 1 & -5 & 0 & -1400 \\ 3 & 1 & 0 & 1 & 0 & 300 \\ 11 & 0 & 0 & 7 & 1 & 2100 \end{array}$$

Not Feasible: Group I variables: x and v Group II variables: y, u, M

c. M becomes greatest after the pivot operation in part (i).

Exercises 4.2

1. 2; second row, first column

3. $\frac{1}{4}$; first row, second column

5. 1; third row, second column

Copyright © 2018 Pearson Education, Inc.

7. a. –12 is the most negative entry in the last row, and $\frac{6}{3} < \frac{10}{2}$. Pivot about the 3.

b.

$$\begin{array}{c|ccccc|c} & x & y & u & v & M & \\ \hline u & \frac{16}{3} & 0 & 1 & -\frac{2}{3} & 0 & 6 \\ y & \frac{1}{3} & 1 & 0 & \frac{1}{3} & 0 & 2 \\ \hline M & 0 & 0 & 0 & 4 & 1 & 24 \end{array}$$

c. $x = 0, y = 2, u = 6, v = 0, M = 24$

9. a. –2 is the only negative entry in the last row, and $\frac{5}{10} < \frac{12}{12}$. Pivot about the 10.

b.

$$\begin{array}{c|ccccc|c} & x & y & u & v & M & \\ \hline u & -13 & 0 & 1 & -\frac{6}{5} & 0 & 6 \\ y & \frac{3}{2} & 1 & 0 & \frac{1}{10} & 0 & \frac{1}{2} \\ \hline M & 7 & 0 & 0 & \frac{1}{5} & 1 & 1 \end{array}$$

c. $x = 0,\ y = \frac{1}{2},\ u = 6, v = 0, M = 1$

In Exercises 11–31, the pivot elements are underlined.

11.

$$\begin{array}{c|ccccc|c} & x & y & u & v & M & \\ \hline u & 1 & 1 & 1 & 0 & 0 & 7 \\ v & 1 & \underline{2} & 0 & 1 & 0 & 10 \\ \hline M & -1 & -3 & 0 & 0 & 1 & 0 \end{array}$$

$$\begin{array}{c|ccccc|c} & x & y & u & v & M & \\ \hline u & \frac{1}{2} & 0 & 1 & -\frac{1}{2} & 0 & 2 \\ y & \frac{1}{2} & 1 & 0 & \frac{1}{2} & 0 & 5 \\ \hline M & \frac{1}{2} & 0 & 0 & \frac{3}{2} & 1 & 15 \end{array}$$

$x = 0, y = 5; M = 15$

13.

$$\begin{array}{c|ccccc|c} & x & y & u & v & M & \\ \hline u & \underline{5} & 1 & 1 & 0 & 0 & 80 \\ v & 3 & 2 & 0 & 1 & 0 & 76 \\ \hline M & -4 & -2 & 0 & 0 & 1 & 0 \end{array}$$

$$\begin{array}{c|ccccc|c} & x & y & u & v & M & \\ \hline x & 1 & \frac{1}{5} & \frac{1}{5} & 0 & 0 & 16 \\ v & 0 & \underline{\frac{7}{5}} & -\frac{3}{5} & 1 & 0 & 28 \\ \hline M & 0 & -\frac{6}{5} & \frac{4}{5} & 0 & 1 & 64 \end{array}$$

$$\begin{array}{c|ccccc|c} & x & y & u & v & M & \\ \hline x & 1 & 0 & \frac{2}{7} & -\frac{1}{7} & 0 & 12 \\ y & 0 & 1 & -\frac{3}{7} & \frac{5}{7} & 0 & 20 \\ \hline M & 0 & 0 & \frac{2}{7} & \frac{6}{7} & 1 & 88 \end{array}$$

$x = 12, y = 20; M = 88$

15.

$$\begin{array}{c|cccccc|c} & x & y & z & u & v & M & \\ \hline u & 1 & 0 & \underline{2} & 1 & 0 & 0 & 10 \\ v & 0 & 3 & 1 & 0 & 1 & 0 & 24 \\ \hline M & -1 & -3 & -5 & 0 & 0 & 1 & 0 \end{array}$$

$$\begin{array}{c|cccccc|c} & x & y & z & u & v & M & \\ \hline z & \frac{1}{2} & 0 & 1 & \frac{1}{2} & 0 & 0 & 5 \\ v & -\frac{1}{2} & \underline{3} & 0 & -\frac{1}{2} & 1 & 0 & 19 \\ \hline M & \frac{3}{2} & -3 & 0 & \frac{5}{2} & 0 & 1 & 25 \end{array}$$

$$\begin{array}{c|cccccc|c} & x & y & z & u & v & M & \\ \hline z & \frac{1}{2} & 0 & 1 & \frac{1}{2} & 0 & 0 & 5 \\ y & -\frac{1}{6} & 1 & 0 & -\frac{1}{6} & \frac{1}{3} & 0 & \frac{19}{3} \\ \hline M & 1 & 0 & 0 & 2 & 1 & 1 & 44 \end{array}$$

$x = 0,\ y = \frac{19}{3},\ z = 5; M = 44$

17.

$$\begin{array}{c|ccccccc|c} & x & y & u & v & w & M & \\ \hline u & 5 & \underline{1} & 1 & 0 & 0 & 0 & 30 \\ v & 3 & 2 & 0 & 1 & 0 & 0 & 60 \\ w & 1 & 1 & 0 & 0 & 1 & 0 & 50 \\ \hline M & -2 & -3 & 0 & 0 & 0 & 1 & 0 \end{array}$$

Copyright © 2018 Pearson Education, Inc.

$$\begin{array}{c|cccccc|c} & x & y & u & v & w & M & \\ \hline y & 5 & 1 & 1 & 0 & 0 & 0 & 30 \\ v & -7 & 0 & -2 & 1 & 0 & 0 & 0 \\ w & -4 & 0 & -1 & 0 & 1 & 0 & 20 \\ \hline M & 13 & 0 & 3 & 0 & 0 & 1 & 90 \end{array}$$

$x = 0, y = 30; M = 90$
Pivoting about the 2 instead gives the same solution although the tableau is different.

19.

$$\begin{array}{c|ccccc|c} & x & y & u & v & M & \\ \hline u & 2 & \underline{3} & 1 & 0 & 0 & 400 \\ v & 1 & 1 & 0 & 1 & 0 & 150 \\ \hline M & -6 & -7 & 0 & 0 & 1 & 300 \end{array}$$

$$\begin{array}{c|ccccc|c} & x & y & u & v & M & \\ \hline y & \frac{2}{3} & 1 & \frac{1}{3} & 0 & 0 & \frac{400}{3} \\ v & \underline{\frac{1}{3}} & 0 & -\frac{1}{3} & 1 & 0 & \frac{50}{3} \\ \hline M & -\frac{4}{3} & 0 & \frac{7}{3} & 0 & 1 & \frac{3700}{3} \end{array}$$

$$\begin{array}{c|ccccc|c} & x & y & u & v & M & \\ \hline y & 0 & 1 & 1 & -2 & 0 & 100 \\ x & 1 & 0 & -1 & 3 & 0 & 50 \\ \hline M & 0 & 0 & 1 & 4 & 1 & 1300 \end{array}$$

$x = 50, y = 100; M = 1300$

21. Let b be the number of large basketballs and f be the number of footballs manufactured. Maximize $2.50b + 2.00f$ subject to the constraints:

$$\begin{cases} 4b + 3f \le 768 \\ 20b + 30f \le 7200 \\ b \ge 0, f \ge 0 \end{cases}$$

(Note that 48 pounds was converted to ounces and 120 hours to minutes)

$$\begin{array}{c|ccccc|c} & b & f & u & v & M & \\ \hline u & \underline{4} & 3 & 1 & 0 & 0 & 768 \\ v & 20 & 30 & 0 & 1 & 0 & 7200 \\ \hline M & -2.5 & -2 & 0 & 0 & 1 & 0 \end{array}$$

$$\begin{array}{c|ccccc|c} & b & f & u & v & M & \\ \hline b & 1 & \frac{3}{4} & \frac{1}{4} & 0 & 0 & 192 \\ v & 0 & \underline{15} & -5 & 1 & 0 & 3360 \\ \hline M & 0 & -\frac{1}{8} & \frac{5}{8} & 0 & 1 & 480 \end{array}$$

$$\begin{array}{c|ccccc|c} & b & f & u & v & M & \\ \hline b & 1 & 0 & \frac{1}{2} & -\frac{1}{20} & 0 & 24 \\ f & 0 & 1 & -\frac{1}{3} & \frac{1}{15} & 0 & 224 \\ \hline M & 0 & 0 & \frac{7}{12} & \frac{1}{120} & 1 & 508 \end{array}$$

Therefore, 24 basketballs and 224 footballs should be manufactured.

 Copyright © 2018 Pearson Education, Inc.

23. Let c be the number of chairs, s be the number of sofas, and t be the number of tables manufactured each day. Maximize $80c + 70s + 120t$ subject to the constraints

$$\begin{cases} 6c+3s+8t \le 768 \\ c+s+2t \le 144 \\ 2c+5s \le 216 \\ c \ge 0,\ s \ge 0,\ t \ge 0 \end{cases}$$

	c	s	t	u	v	w	M	
u	6	3	8	1	0	0	0	768
v	1	1	$\underline{2}$	0	1	0	0	144
w	2	5	0	0	0	1	0	216
M	−80	−70	−120	0	0	0	1	0

	c	s	t	u	v	w	M	
u	$\underline{2}$	−1	0	1	−4	0	0	192
t	$\frac{1}{2}$	$\frac{1}{2}$	1	0	$\frac{1}{2}$	0	0	72
w	2	5	0	0	0	1	0	216
M	−20	−10	0	0	60	0	1	8640

	c	s	t	u	v	w	M	
c	1	$-\frac{1}{2}$	0	$\frac{1}{2}$	−2	0	0	96
t	0	$\frac{3}{4}$	1	$-\frac{1}{4}$	$\frac{3}{2}$	0	0	24
w	0	$\underline{6}$	0	−1	4	1	0	24
M	0	−20	0	10	20	0	1	10,560

	c	s	t	u	v	w	M	
c	1	0	0	$\frac{5}{12}$	$-\frac{5}{3}$	$\frac{1}{12}$	0	98
t	0	0	1	$-\frac{1}{8}$	1	$-\frac{1}{8}$	0	21
s	0	1	0	$-\frac{1}{6}$	$\frac{2}{3}$	$\frac{1}{6}$	0	4
M	0	0	0	$\frac{20}{3}$	$\frac{100}{3}$	$\frac{10}{3}$	1	10,640

98 chairs, 21 tables, 4 sofas.

25. Let b be the number of hours spent bicycling, j the number spent jogging, and s the number spent swimming each month. Maximize $200b + 475j + 275s$ subject to the constraints:

$$\begin{cases} b+j+s \le 30 \\ s \le 4 \\ -b+j-s \le 0 \\ b \ge 0,\ j \ge 0,\ s \ge 0 \end{cases}$$

Copyright © 2018 Pearson Education, Inc.

$$\begin{array}{c} \\ u \\ v \\ w \\ M \end{array}\begin{array}{c} \begin{array}{ccccccc|c} b & j & s & u & v & w & M & \end{array} \\ \left[\begin{array}{ccccccc|c} 1 & 1 & 1 & 1 & 0 & 0 & 0 & 30 \\ 0 & 0 & 1 & 0 & 1 & 0 & 0 & 4 \\ -1 & \underline{1} & -1 & 0 & 0 & 1 & 0 & 0 \\ \hline -200 & -475 & -275 & 0 & 0 & 0 & 1 & 0 \end{array}\right] \end{array}$$

$$\begin{array}{c} \\ u \\ v \\ j \\ M \end{array}\begin{array}{c} \begin{array}{ccccccc|c} b & j & s & u & v & w & M & \end{array} \\ \left[\begin{array}{ccccccc|c} 2 & 0 & 2 & 1 & 0 & -1 & 0 & 30 \\ 0 & 0 & \underline{1} & 0 & 1 & 0 & 0 & 4 \\ -1 & 1 & -1 & 0 & 0 & 1 & 0 & 0 \\ \hline -675 & 0 & -750 & 0 & 0 & 475 & 1 & 0 \end{array}\right] \end{array}$$

$$\begin{array}{c} \\ u \\ s \\ j \\ M \end{array}\begin{array}{c} \begin{array}{ccccccc|c} b & j & s & u & v & w & M & \end{array} \\ \left[\begin{array}{ccccccc|c} \underline{2} & 0 & 0 & 1 & -2 & -1 & 0 & 22 \\ 0 & 0 & 1 & 0 & 1 & 0 & 0 & 4 \\ -1 & 1 & 0 & 0 & 1 & 1 & 0 & 4 \\ \hline -675 & 0 & 0 & 0 & 750 & 475 & 1 & 3000 \end{array}\right] \end{array}$$

$$\begin{array}{c} \\ b \\ s \\ j \\ M \end{array}\begin{array}{c} \begin{array}{ccccccc|c} b & j & s & u & v & w & M & \end{array} \\ \left[\begin{array}{ccccccc|c} 1 & 0 & 0 & \frac{1}{2} & -1 & -\frac{1}{2} & 0 & 11 \\ 0 & 0 & 1 & 0 & 1 & 0 & 0 & 4 \\ 0 & 1 & 0 & \frac{1}{2} & 0 & \frac{1}{2} & 0 & 15 \\ \hline 0 & 0 & 0 & \frac{675}{2} & 75 & \frac{275}{2} & 1 & 10,425 \end{array}\right] \end{array}$$

11 hours bicycling, 4 hours swimming, 15 hours jogging.

27. Let a be the number of type A restaurants, b the number of type B restaurants, and c the number of type C restaurants. Maximize $40a + 30b + 25c$ subject to the constraints

$$\begin{cases} 600a + 400b + 300c \le 48,000 \\ 15a + 9b + 5c \le 1000 \\ a + b + c \le 70 \\ a \ge 0, b \ge 0, c \ge 0 \end{cases} \quad \text{(dollars in thousands)}$$

$$\begin{array}{c} \\ u \\ v \\ w \\ M \end{array}\begin{array}{c} \begin{array}{ccccccc|c} a & b & c & u & v & w & M & \end{array} \\ \left[\begin{array}{ccccccc|c} 600 & 400 & 300 & 1 & 0 & 0 & 0 & 48,000 \\ \underline{15} & 9 & 5 & 0 & 1 & 0 & 0 & 1000 \\ 1 & 1 & 1 & 0 & 0 & 1 & 0 & 70 \\ \hline -40 & -30 & -25 & 0 & 0 & 0 & 1 & 0 \end{array}\right] \end{array}$$

 Copyright © 2018 Pearson Education, Inc.

$$\begin{array}{c} \\ u \\ a \\ w \\ M \end{array}\begin{array}{c} \begin{array}{ccccccc|c} a & b & c & u & v & w & M & \end{array} \\ \left[\begin{array}{ccccccc|c} 0 & 40 & 100 & 1 & -40 & 0 & 0 & 8000 \\ 1 & \frac{3}{5} & \frac{1}{3} & 0 & \frac{1}{15} & 0 & 0 & \frac{200}{3} \\ 0 & \frac{2}{5} & \frac{2}{3} & 0 & -\frac{1}{15} & 1 & 0 & \frac{10}{3} \\ \hline 0 & -6 & -\frac{35}{3} & 0 & \frac{8}{3} & 0 & 1 & \frac{8000}{3} \end{array}\right] \end{array}$$

$$\begin{array}{c} \\ u \\ a \\ c \\ M \end{array}\begin{array}{c} \begin{array}{ccccccc|c} a & b & c & u & v & w & M & \end{array} \\ \left[\begin{array}{ccccccc|c} 0 & -20 & 0 & 1 & -30 & -150 & 0 & 7500 \\ 1 & \frac{2}{5} & 0 & 0 & \frac{1}{10} & -\frac{1}{2} & 0 & 65 \\ 0 & \frac{3}{5} & 1 & 0 & -\frac{1}{10} & \frac{3}{2} & 0 & 5 \\ \hline 0 & 1 & 0 & 0 & \frac{3}{2} & \frac{35}{2} & 1 & 2725 \end{array}\right] \end{array}$$

65 type A restaurants and 5 type C restaurants.

29. Let a be the number of bags of mix A, b the number of bags of mix B, and c the number of bags of mix C. Maximize $3a + 5b + 6c$ subject to the constraints

$$\begin{cases} 12a + 10b + 8c \le 1200 \\ 5a + 6b + 8c \le 800 \\ 3a + 4b + 4c \le 600 \\ a \ge 0, b \ge 0, c \ge 0 \end{cases}$$

$$\begin{array}{c} \\ u \\ v \\ w \\ M \end{array}\begin{array}{c} \begin{array}{ccccccc|c} a & b & c & u & v & w & M & \end{array} \\ \left[\begin{array}{ccccccc|c} 12 & 10 & 8 & 1 & 0 & 0 & 0 & 1200 \\ 5 & 6 & \underline{8} & 0 & 1 & 0 & 0 & 800 \\ 3 & 4 & 4 & 0 & 0 & 1 & 0 & 600 \\ \hline -3 & -5 & -6 & 0 & 0 & 0 & 1 & 0 \end{array}\right] \end{array}$$

$$\begin{array}{c} \\ u \\ c \\ w \\ M \end{array}\begin{array}{c} \begin{array}{ccccccc|c} a & b & c & u & v & w & M & \end{array} \\ \left[\begin{array}{ccccccc|c} 7 & \underline{4} & 0 & 1 & -1 & 0 & 0 & 400 \\ \frac{5}{8} & \frac{3}{4} & 1 & 0 & \frac{1}{8} & 0 & 0 & 100 \\ \frac{1}{2} & 1 & 0 & 0 & -\frac{1}{2} & 1 & 0 & 200 \\ \hline \frac{3}{4} & -\frac{1}{2} & 0 & 0 & \frac{3}{4} & 0 & 1 & 600 \end{array}\right] \end{array}$$

$$\begin{array}{c} \\ b \\ c \\ w \\ M \end{array}\begin{array}{c} \begin{array}{ccccccc|c} a & b & c & u & v & w & M & \end{array} \\ \left[\begin{array}{ccccccc|c} \frac{7}{4} & 1 & 0 & \frac{1}{4} & -\frac{1}{4} & 0 & 0 & 100 \\ -\frac{11}{16} & 0 & 1 & -\frac{3}{16} & \frac{5}{16} & 0 & 0 & 25 \\ -\frac{5}{4} & 0 & 0 & -\frac{1}{4} & -\frac{1}{4} & 1 & 0 & 100 \\ \hline \frac{13}{8} & 0 & 0 & \frac{1}{8} & \frac{5}{8} & 0 & 1 & 650 \end{array}\right] \end{array}$$

No bags of Mix A, 100 bags of Mix B, and 25 bags of Mix C.

Copyright © 2018 Pearson Education, Inc.

31. Let x be the number of type A widgets, y be the number of type B widgets, and z be the number of type C widgets produced. Maximize $10x+20y+30z$ subject to the constraints:

$$\begin{cases} 12x+8y+4z \le 800 \\ 2x+6y+4z \le 600 \\ 2x+2y+8z \le 300 \\ x \ge 0, y \ge 0, z \ge 0 \end{cases}$$

	x	y	z	u	v	w	M	
u	12	8	4	1	0	0	0	800
v	2	6	4	0	1	0	0	600
w	2	2	8	0	0	1	0	300
M	−10	−20	−30	0	0	0	1	0

	x	y	z	u	v	w	M	
u	11	7	0	1	0	$-\frac{1}{2}$	0	650
v	1	5	0	0	0	$-\frac{1}{2}$	0	450
z	$\frac{1}{4}$	$\frac{1}{4}$	1	0	0	$\frac{1}{8}$	0	$\frac{75}{2}$
M	$-\frac{5}{2}$	$-\frac{25}{2}$	0	0	0	$\frac{5}{2}$	1	2250

	x	y	z	u	v	w	M	
u	$\frac{48}{5}$	0	0	1	0	$\frac{1}{5}$	0	20
y	$\frac{1}{5}$	1	0	0	0	$-\frac{1}{10}$	0	90
z	$\frac{1}{5}$	0	1	0	0	$\frac{3}{20}$	0	15
M	0	0	0	0	0	$\frac{5}{2}$	1	2250

Since the last row has all positive values, the manufacturer should produce no type A widgets, 90 type B widgets, and 15 type C widgets, for a maximum profit of $2,250.

33.

	x	y	z	u	v	w	M	
u	1	1	1	1	0	0	0	600
v	1	3	0	0	1	0	0	600
w	2	0	1	0	0	1	0	900
M	−60	−90	−300	0	0	0	1	0

	x	y	z	u	v	w	M	
z	1	1	1	1	0	0	0	600
v	1	3	0	0	1	0	0	600
w	1	−1	0	−1	0	1	0	300
M	240	210	0	300	0	0	1	180,000

$x = 0, y = 0, z = 600; M = 180,000$

 Copyright © 2018 Pearson Education, Inc.

35.

$$\begin{array}{c} \\ u \\ v \\ M \end{array}\begin{array}{c} \begin{array}{cccccc} x & y & u & v & M & \end{array} \\ \left[\begin{array}{ccccc|c} 5 & 1 & 1 & 0 & 0 & 8 \\ 1 & \underline{2} & 0 & 1 & 0 & 10 \\ \hline -2 & -4 & 0 & 0 & 1 & 0 \end{array}\right] \end{array}$$

$$\begin{array}{c} \\ u \\ y \\ M \end{array}\begin{array}{c} \begin{array}{cccccc} x & y & u & v & M & \end{array} \\ \left[\begin{array}{ccccc|c} \frac{9}{2} & 0 & 1 & -\frac{1}{2} & 0 & 3 \\ \frac{1}{2} & 1 & 0 & \frac{1}{2} & 0 & 5 \\ \hline 0 & 0 & 0 & 2 & 1 & 20 \end{array}\right] \end{array}$$

$x = 0, y = 5; M = 20$

37.

$$\begin{array}{c} \\ u \\ v \\ w \\ M \end{array}\begin{array}{c} \begin{array}{cccccccc} x & y & z & u & v & w & M & \end{array} \\ \left[\begin{array}{ccccccc|c} 1 & 1 & 2 & 1 & 0 & 0 & 0 & 6 \\ 1 & 5 & 2 & 0 & 1 & 0 & 0 & 20 \\ \underline{2} & 1 & 1 & 0 & 0 & 1 & 0 & 4 \\ \hline -3 & -2 & -2 & 0 & 0 & 0 & 1 & 0 \end{array}\right] \end{array}$$

$$\begin{array}{c} \\ u \\ v \\ x \\ M \end{array}\begin{array}{c} \begin{array}{cccccccc} x & y & z & u & v & w & M & \end{array} \\ \left[\begin{array}{ccccccc|c} 0 & \frac{1}{2} & \frac{3}{2} & 1 & 0 & -\frac{1}{2} & 0 & 4 \\ 0 & \frac{9}{2} & \frac{3}{2} & 0 & 1 & -\frac{1}{2} & 0 & 18 \\ 1 & \frac{1}{2} & \frac{1}{2} & 0 & 0 & \frac{1}{2} & 0 & 2 \\ \hline 0 & -\frac{1}{2} & -\frac{1}{2} & 0 & 0 & \frac{3}{2} & 1 & 6 \end{array}\right] \end{array}$$

$$\begin{array}{c} \\ u \\ v \\ y \\ M \end{array}\begin{array}{c} \begin{array}{cccccccc} x & y & z & u & v & w & M & \end{array} \\ \left[\begin{array}{ccccccc|c} -1 & 0 & 1 & 1 & 0 & -1 & 0 & 2 \\ -9 & 0 & -3 & 0 & 1 & -5 & 0 & 0 \\ 2 & 1 & 1 & 0 & 0 & 1 & 0 & 4 \\ \hline 1 & 0 & 0 & 0 & 0 & 2 & 1 & 8 \end{array}\right] \end{array}$$

$x = 0, y = 4, z = 0; M = 8$

Pivoting about the $\frac{9}{2}$ in step 2 leads to the same solution. Pivoting about the $\frac{3}{2}$, first row, third column in step 2 leads to a different solution: $x = 0, y = 2, z = 2$. M is still 8.

Exercises 4.3

1. Maximize $3x + 4y$ subject to the constraints:

$$\begin{cases} -5x - 3y \le -6 \\ 2x - 3y \le 7 \\ x \ge 0, y \ge 0 \end{cases}$$

3. Maximize $-x - y - z$ subject to the constraints:

$$\begin{cases} -2x - 3y - z \le -7 \\ -5x - 6y - 7z \le -8 \\ x \ge 0, y \ge 0, z \ge 0 \end{cases}$$

5. -3; second row, second column

7. $-\frac{1}{2}$; third row, second column

9.

$$\begin{array}{c} \\ u \\ v \\ M \end{array}\begin{array}{c} \begin{array}{cccccc} x & y & u & v & M & \end{array} \\ \left[\begin{array}{ccccc|c} 1 & 1 & 1 & 0 & 0 & 5 \\ 2 & \underline{-3} & 0 & 1 & 0 & -12 \\ \hline -40 & -30 & 0 & 0 & 1 & 0 \end{array}\right] \end{array}$$

$$\begin{array}{c} \\ u \\ y \\ M \end{array}\begin{array}{c} \begin{array}{cccccc} x & y & u & v & M & \end{array} \\ \left[\begin{array}{ccccc|c} \frac{5}{3} & 0 & 1 & \frac{1}{3} & 0 & 1 \\ -\frac{2}{3} & 1 & 0 & -\frac{1}{3} & 0 & 4 \\ \hline -60 & 0 & 0 & -10 & 1 & 120 \end{array}\right] \end{array}$$

$$\begin{array}{c} \\ x \\ y \\ M \end{array}\begin{array}{c} \begin{array}{cccccc} x & y & u & v & M & \end{array} \\ \left[\begin{array}{ccccc|c} 1 & 0 & \frac{3}{5} & \frac{1}{5} & 0 & \frac{3}{5} \\ 0 & 1 & \frac{2}{5} & -\frac{1}{5} & 0 & \frac{22}{5} \\ \hline 0 & 0 & 36 & 2 & 1 & 156 \end{array}\right] \end{array}$$

$x = \frac{3}{5}, y = \frac{22}{5}; M = 156$

11.

	x	y	u	v	M	
u	−1	−1	1	0	0	−3
v	−2	0	0	1	0	−5
M	3	1	0	0	1	0

	x	y	u	v	M	
y	1	1	−1	0	0	3
v	−2	0	0	1	0	−5
M	2	0	1	0	1	−3

	x	y	u	v	M	
y	0	1	−1	$\frac{1}{2}$	0	$\frac{1}{2}$
x	1	0	0	$-\frac{1}{2}$	0	$\frac{5}{2}$
M	0	0	1	1	1	−8

$x=\frac{5}{2}$, $y=\frac{1}{2}$ Since $M=-8$, the minimum is 8. Pivoting about the −2 at the start leads to the same final matrix.

13.

	x	y	u	v	w	t	M	
u	−2	−1	1	0	0	0	0	−11
v	1	1	0	1	0	0	0	10
w	$\frac{1}{3}$	1	0	0	1	0	0	6
t	$-\frac{1}{4}$	−1	0	0	0	1	0	−4
M	13	4	0	0	0	0	1	0

	x	y	u	v	w	t	M	
u	$-\frac{7}{4}$	0	1	0	0	−1	0	−7
v	$\frac{3}{4}$	0	0	1	0	1	0	6
w	$\frac{1}{12}$	0	0	0	1	1	0	2
y	$\frac{1}{4}$	1	0	0	0	−1	0	4
M	12	0	0	0	0	4	1	−16

	x	y	u	v	w	t	M	
u	$-\frac{5}{3}$	0	1	0	1	0	0	−5
v	$\frac{2}{3}$	0	0	1	−1	0	0	4
t	$\frac{1}{12}$	0	0	0	1	1	0	2
y	$\frac{1}{3}$	1	0	0	1	0	0	6
M	$\frac{35}{3}$	0	0	0	−4	0	1	−24

	x	y	u	v	w	t	M	
x	1	0	$-\frac{3}{5}$	0	$-\frac{3}{5}$	0	0	3
v	0	0	$\frac{2}{5}$	1	$-\frac{3}{5}$	0	0	2
t	0	0	$\frac{1}{20}$	0	$\frac{21}{20}$	1	0	$\frac{7}{4}$
y	0	1	$\frac{1}{5}$	0	$\frac{6}{5}$	0	0	5
M	0	0	7	0	3	0	1	−59

$x=3$, $y=5$; since $M=-59$, the minimum is 59. Other choices of pivot entries lead to the same final matrix.

15.

	x	y	u	v	w	t	M	
u	−2	−5	1	0	0	0	0	−30
v	3	−5	0	1	0	0	0	−5
w	8	3	0	0	1	0	0	101
t	−9	7	0	0	0	1	0	42
M	2	7	0	0	0	0	1	0

	x	y	u	v	w	t	M	
u	−5	0	1	−1	0	0	0	−25
y	$-\frac{3}{5}$	1	0	$-\frac{1}{5}$	0	0	0	1
w	$\frac{49}{5}$	0	0	$\frac{3}{5}$	1	0	0	98
t	$-\frac{24}{5}$	0	0	$\frac{7}{5}$	0	1	0	35
M	$\frac{31}{5}$	0	0	$\frac{7}{5}$	0	0	1	−7

	x	y	u	v	w	t	M	
x	1	0	$-\frac{1}{5}$	$\frac{1}{5}$	0	0	0	5
y	0	1	$-\frac{3}{25}$	$-\frac{2}{25}$	0	0	0	4
w	0	0	$\frac{49}{25}$	$-\frac{34}{25}$	1	0	0	49
t	0	0	$-\frac{24}{25}$	$\frac{59}{25}$	0	1	0	59
M	0	0	$\frac{31}{25}$	$\frac{4}{25}$	0	0	1	−38

$x=5$, $y=4$; since $M=-38$, the minimum is 38.

Copyright © 2018 Pearson Education, Inc.

17. Minimize $3a + 1.5b$ subject to the constraints:

$$\begin{cases} 30a+10b \ge 60 \\ 10a+10b \ge 40 \\ 20a+60b \ge 120 \\ a \ge 0,\ b \ge 0 \end{cases}$$

$$\begin{array}{c} \\ u \\ v \\ w \\ M \end{array}
\begin{array}{cccccc|c} a & b & u & v & w & M & \\ \hline \underline{-30} & -10 & 1 & 0 & 0 & 0 & -60 \\ -10 & -10 & 0 & 1 & 0 & 0 & -40 \\ -20 & -60 & 0 & 0 & 1 & 0 & -120 \\ \hline 3 & \frac{3}{2} & 0 & 0 & 0 & 1 & 0 \end{array}$$

$$\begin{array}{c} \\ a \\ v \\ w \\ M \end{array}
\begin{array}{cccccc|c} a & b & u & v & w & M & \\ \hline 1 & \frac{1}{3} & -\frac{1}{30} & 0 & 0 & 0 & 2 \\ 0 & -\frac{20}{3} & -\frac{1}{3} & 1 & 0 & 0 & -20 \\ 0 & \underline{-\frac{160}{3}} & -\frac{2}{3} & 0 & 1 & 0 & -80 \\ \hline 0 & \frac{1}{2} & \frac{1}{10} & 0 & 0 & 1 & -6 \end{array}$$

$$\begin{array}{c} \\ a \\ v \\ b \\ M \end{array}
\begin{array}{cccccc|c} a & b & u & v & w & M & \\ \hline 1 & 0 & -\frac{3}{80} & 0 & \frac{1}{160} & 0 & \frac{3}{2} \\ 0 & 0 & -\frac{1}{4} & 1 & \underline{-\frac{1}{8}} & 0 & -10 \\ 0 & 1 & \frac{1}{80} & 0 & -\frac{3}{160} & 0 & \frac{3}{2} \\ \hline 0 & 0 & \frac{3}{32} & 0 & \frac{3}{320} & 1 & -\frac{27}{4} \end{array}$$

$$\begin{array}{c} \\ a \\ w \\ b \\ M \end{array}
\begin{array}{cccccc|c} a & b & u & v & w & M & \\ \hline 1 & 0 & -\frac{1}{20} & \frac{1}{20} & 0 & 0 & 1 \\ 0 & 0 & 2 & -8 & 1 & 0 & 80 \\ 0 & 1 & \frac{1}{20} & -\frac{3}{20} & 0 & 0 & 3 \\ \hline 0 & 0 & \frac{3}{40} & \frac{3}{40} & 0 & 1 & -\frac{15}{2} \end{array}$$

1 serving of food A, 3 servings of food B

19. Maximize $30a + 50b + 60c$ subject to the constraints:

$$\begin{cases} a+b+c \le 600 \\ a \ge 100 \\ b \ge 50 \\ b+c \ge 200 \\ a \ge 0,\ b \ge 0,\ c \ge 0 \end{cases}$$

$$\begin{array}{c} \\ u \\ v \\ w \\ t \\ M \end{array}
\begin{array}{cccccccc|c} a & b & c & u & v & w & t & M & \\ \hline 1 & 1 & 1 & 1 & 0 & 0 & 0 & 0 & 600 \\ \underline{-1} & 0 & 0 & 0 & 1 & 0 & 0 & 0 & -100 \\ 0 & -1 & 0 & 0 & 0 & 1 & 0 & 0 & -50 \\ 0 & -1 & -1 & 0 & 0 & 0 & 1 & 0 & -200 \\ \hline -30 & -50 & -60 & 0 & 0 & 0 & 0 & 1 & 0 \end{array}$$

$$\begin{array}{c} \\ u \\ a \\ w \\ t \\ M \end{array}
\begin{array}{cccccccc|c} a & b & c & u & v & w & t & M & \\ \hline 0 & 1 & 1 & 1 & 1 & 0 & 0 & 0 & 500 \\ 1 & 0 & 0 & 0 & -1 & 0 & 0 & 0 & 100 \\ 0 & \underline{-1} & 0 & 0 & 0 & 1 & 0 & 0 & -50 \\ 0 & -1 & -1 & 0 & 0 & 0 & 1 & 0 & -200 \\ \hline 0 & -50 & -60 & 0 & -30 & 0 & 0 & 1 & 3000 \end{array}$$

Copyright © 2018 Pearson Education, Inc.

	a	b	c	u	v	w	t	M	
u	0	0	1	1	1	1	0	0	450
a	1	0	0	0	−1	0	0	0	100
b	0	1	0	0	0	−1	0	0	50
t	0	0	−1	0	0	−1	1	0	−150
M	0	0	−60	0	−30	−50	0	1	5500

	a	b	c	u	v	w	t	M	
u	0	0	0	1	1	0	1	0	300
a	1	0	0	0	−1	0	0	0	100
b	0	1	0	0	0	−1	0	0	50
c	0	0	1	0	0	1	−1	0	150
M	0	0	0	0	−30	10	−60	1	14,500

	a	b	c	u	v	w	t	M	
t	0	0	0	1	1	0	1	0	300
a	1	0	0	0	−1	0	0	0	100
b	0	1	0	0	0	−1	0	0	50
c	0	0	1	1	1	1	0	0	450
M	0	0	0	60	30	10	0	1	32,500

Stock 100 of brand A, 50 of brand B, and 450 of brand C.

21. Let x = number of computers shipped from Chicago to Detroit. Let y = number of computers shipped from Chicago to Fletcher. Then (40 – x) equals the number of computers shipped from Boston to Detroit and (30 – y) represents the number of computers shipped from Boston to Fletcher.

Minimize:

$125(40-x)+100x+180(30-y)+160y$

or $10400-25x-20y$

Subject to:

$$\begin{cases} x+y\le 80 \\ x+y\ge 20 \\ x\le 40 \\ y\le 30 \\ x\ge 0, y\ge 0 \end{cases}$$

	x	y	u	v	w	t	M	
u	1	1	1	0	0	0	0	80
v	−1	−1	0	1	0	0	0	−20
w	1	0	0	0	1	0	0	40
t	0	1	0	0	0	1	0	30
M	−25	−20	0	0	0	0	1	−10400

	x	y	u	v	w	t	M	
u	0	0	1	1	0	0	0	60
x	1	1	0	−1	0	0	0	20
w	0	−1	0	1	1	0	0	20
t	0	1	0	0	0	1	0	30
M	0	5	0	−25	0	0	1	−9900

	x	y	u	v	w	t	M	
u	0	1	1	0	−1	0	0	40
x	1	0	0	0	1	0	0	40
v	0	−1	0	1	1	0	0	20
t	0	1	0	0	0	1	0	30
M	0	−20	0	0	25	0	1	−9400

Copyright © 2018 Pearson Education, Inc.

$$\begin{array}{c|ccccccc|c} & x & y & u & v & w & t & M & \\ \hline u & 0 & 0 & 1 & 0 & -1 & -1 & 0 & 10 \\ x & 1 & 0 & 0 & 0 & 1 & 0 & 0 & 40 \\ v & 0 & 0 & 0 & 1 & 1 & 1 & 0 & 50 \\ y & 0 & 1 & 0 & 0 & 0 & 1 & 0 & 30 \\ \hline M & 0 & 0 & 0 & 0 & 25 & 20 & 1 & -8800 \end{array}$$

Therefore, the minimum cost is \$8800 when you ship all the computers from Chicago.

23.

$$\begin{array}{c|ccccc|c} & x & y & u & v & M & \\ \hline u & \underline{4} & 1 & 1 & 0 & 0 & 5 \\ v & -1 & -3 & 0 & 1 & 0 & -4 \\ \hline M & -1 & 2 & 0 & 0 & 1 & 0 \end{array}$$

$$\begin{array}{c|ccccc|c} & x & y & u & v & M & \\ \hline x & 1 & \frac{1}{4} & \frac{1}{4} & 0 & 0 & \frac{5}{4} \\ v & 0 & \underline{-\frac{11}{4}} & \frac{1}{4} & 1 & 0 & -\frac{11}{4} \\ \hline M & 0 & \frac{9}{4} & \frac{1}{4} & 0 & 1 & \frac{5}{4} \end{array}$$

$$\begin{array}{c|ccccc|c} & x & y & u & v & M & \\ \hline x & 1 & 0 & \frac{3}{11} & \frac{1}{11} & 0 & 1 \\ y & 0 & 1 & -\frac{1}{11} & -\frac{4}{11} & 0 & 1 \\ \hline M & 0 & 0 & \frac{5}{11} & \frac{9}{11} & 1 & -1 \end{array}$$

$x = 1, y = 1; M = -1$

Exercises 4.4

1. x (chef's knives) goes from 140 to $140 + 54\left(-\frac{50}{27}\right) = 40$, y (pocket knives) goes from 80 to $80 + 54\left(\frac{70}{27}\right) = 220$, and the profit goes from \$820 to $820 + 54\left(\frac{200}{27}\right) = \$1220$.

3. The slack variable involved is w. x (sets to Rockville) goes from 20 to $20 + (50 - 45)(1) = 25$, y (sets to Annapolis) goes from 25 to $25 + (50 - 45)(0) = 25$, and the cost goes from \$260 to $-[-260 + (50 - 45)(2)] = \$250$.

5. $100 + h\left(\frac{1}{4}\right) \geq 0$, $25 + h\left(-\frac{3}{16}\right) \geq 0$, and $100 + h\left(-\frac{1}{4}\right) \geq 0$, so $h \geq -400$, $h \leq \frac{400}{3}$ and $h \leq 400$ or $-400 \leq h \leq \frac{400}{3}$.

7. $\begin{bmatrix} 9 & 1 & 1 \\ 4 & 8 & -3 \end{bmatrix}$

9. $\begin{bmatrix} 7 \\ 6 \\ 5 \\ 1 \end{bmatrix}$

11. Yes

13. Minimize $[7\ 5\ 4]\begin{bmatrix} x \\ y \\ z \end{bmatrix}$ subject to the constraints $\begin{bmatrix} 3 & 8 & 9 \\ 1 & 2 & 5 \\ 4 & 1 & 7 \end{bmatrix}\begin{bmatrix} x \\ y \\ z \end{bmatrix} \geq \begin{bmatrix} 75 \\ 80 \\ 67 \end{bmatrix}$ and $\begin{bmatrix} x \\ y \\ z \end{bmatrix} \geq \begin{bmatrix} 0 \\ 0 \\ 0 \end{bmatrix}$.

15. Maximize $[3\ 5]\begin{bmatrix} x \\ y \end{bmatrix}$ subject to the constraints $\begin{bmatrix} 3 & 6 \\ 7 & 5 \\ 4 & 3 \end{bmatrix}\begin{bmatrix} x \\ y \end{bmatrix} \leq \begin{bmatrix} 90 \\ 138 \\ 120 \end{bmatrix}$ and $\begin{bmatrix} x \\ y \end{bmatrix} \geq \begin{bmatrix} 0 \\ 0 \end{bmatrix}$.

17. Minimize $2x + 3y$ subject to the constraints $\begin{cases} 7x + 4y \geq 33 \\ 5x + 8y \geq 44 \\ x + 3y \geq 55 \\ x \geq 0,\ y \geq 0 \end{cases}$.

Copyright © 2018 Pearson Education, Inc.

19. Constraints

Cell	Name	Final Value	Shadow Price	Constraint R.H. Side	Allowable Increase	Allowable Decrease
B1		45	–2	45	10	20
B2		45	0	15	30	1E + 30
B3		20	0	30	1E + 30	10
B4		25	–1	25	20	10
B5		20	0	0	20	1E + 30
B6		25	0	0	25	1E + 30

The first line of the table corresponds to the constraint for the number of sets shipped from College Park. The shadow price is –2, and the range feasibility is [45 – 20, 45 + 10] or [25, 55].

Exercises 4.5

1. Primal: Maximize $[4\ \ 2]\begin{bmatrix}x\\y\end{bmatrix}$ subject to

$$\begin{bmatrix}5 & 1\\3 & 2\end{bmatrix}\begin{bmatrix}x\\y\end{bmatrix}\le\begin{bmatrix}80\\76\end{bmatrix}.$$

Dual: Minimize $[80\ \ 76]\begin{bmatrix}u\\v\end{bmatrix}$ subject to

$$\begin{bmatrix}5 & 3\\1 & 2\end{bmatrix}\begin{bmatrix}u\\v\end{bmatrix}\ge\begin{bmatrix}4\\2\end{bmatrix}\text{ and }\begin{bmatrix}u\\v\end{bmatrix}\ge\begin{bmatrix}0\\0\end{bmatrix}.$$

Minimize $80u + 76v$ subject to the

$$\text{constraints }\begin{cases}5u+3v\ge 4\\ u+2v\ge 2\\ u\ge 0,\ v\ge 0\end{cases}.$$

3. Primal: Minimize $[10\ \ 12]\begin{bmatrix}x\\y\end{bmatrix}$ subject to

$$\begin{bmatrix}1 & 2\\-1 & 1\\2 & 3\end{bmatrix}\begin{bmatrix}x\\y\end{bmatrix}\ge\begin{bmatrix}1\\2\\1\end{bmatrix}\text{ and }\begin{bmatrix}x\\y\end{bmatrix}\ge\begin{bmatrix}0\\0\end{bmatrix}.$$

Dual: Maximize $[1\ \ 2\ \ 1]\begin{bmatrix}u\\v\\w\end{bmatrix}$ subject to

$$\begin{bmatrix}1 & -1 & 2\\2 & 1 & 3\end{bmatrix}\begin{bmatrix}u\\v\\w\end{bmatrix}\le\begin{bmatrix}10\\12\end{bmatrix}\text{ and }\begin{bmatrix}u\\v\\w\end{bmatrix}\ge\begin{bmatrix}0\\0\\0\end{bmatrix}.$$

Maximize $u + 2v + w$ subject to the

$$\text{constraints }\begin{cases}u-v+2w\le 10\\ 2u+v+3w\le 12\\ u\ge 0,\ v\ge 0,\ w\ge 0\end{cases}.$$

5. Primal: Minimize $\begin{bmatrix}3 & 5 & 1\end{bmatrix}\begin{bmatrix}x\\y\\z\end{bmatrix}$ subject

to $\begin{bmatrix}-2 & 4 & 6\\8 & 1 & 9\end{bmatrix}\begin{bmatrix}x\\y\\z\end{bmatrix}\ge\begin{bmatrix}-7\\10\end{bmatrix}$ and $\begin{bmatrix}x\\y\\z\end{bmatrix}\ge\begin{bmatrix}0\\0\\0\end{bmatrix}$.

Dual: Maximize $[-7\ \ 10]\begin{bmatrix}u\\v\end{bmatrix}$ subject to

$$\begin{bmatrix}-2 & 8\\4 & 1\\6 & 9\end{bmatrix}\begin{bmatrix}u\\v\end{bmatrix}\le\begin{bmatrix}3\\5\\1\end{bmatrix}\text{ and }\begin{bmatrix}u\\v\end{bmatrix}\ge\begin{bmatrix}0\\0\end{bmatrix}.$$

Maximize $-7u + 10v$ subject to the

$$\text{constraints }\begin{cases}-2u+8v\le 3\\ 4u+v\le 5\\ 6u+9v\le 1\\ u\ge 0,\ v\ge 0\end{cases}$$

7. $x = 12, y = 20, M = 88, u = \frac{2}{7}, v = \frac{6}{7},$ $M = 88$

9. $x = 0, y = 2, M = 24; u = 0, v = 12, w = 0,$ $M = 24$

11. Maximize $3u + 5v$ subject to the

$$\text{constraints }\begin{cases}u+2v\le 3\\ u\le 1\\ u\ge 0,\ v\ge 0\end{cases}.$$

Solve the dual.

 Copyright © 2018 Pearson Education, Inc.

$$\begin{array}{c} \\ x \\ y \\ M \end{array}\begin{array}{c} \begin{array}{ccccc} u & v & x & y & M \end{array} \\ \left[\begin{array}{ccccc|c} 1 & \underline{2} & 1 & 0 & 0 & 3 \\ 1 & 0 & 0 & 1 & 0 & 1 \\ \hline -3 & -5 & 0 & 0 & 1 & 0 \end{array}\right] \end{array}$$

$$\begin{array}{c} \\ v \\ y \\ M \end{array}\begin{array}{c} \begin{array}{ccccc} u & v & x & y & M \end{array} \\ \left[\begin{array}{ccccc|c} \frac{1}{2} & 1 & \frac{1}{2} & 0 & 0 & \frac{3}{2} \\ \underline{1} & 0 & 0 & 1 & 0 & 1 \\ \hline -\frac{1}{2} & 0 & \frac{5}{2} & 0 & 1 & \frac{15}{2} \end{array}\right] \end{array}$$

$$\begin{array}{c} \\ v \\ u \\ M \end{array}\begin{array}{c} \begin{array}{ccccc} u & v & x & y & M \end{array} \\ \left[\begin{array}{ccccc|c} 0 & 1 & \frac{1}{2} & -\frac{1}{2} & 0 & 1 \\ 1 & 0 & 0 & 1 & 0 & 1 \\ \hline 0 & 0 & \frac{5}{2} & \frac{1}{2} & 1 & 8 \end{array}\right] \end{array}$$

$x = \frac{5}{2}, y = \frac{1}{2}$, minimum = 8; $u = 1$, $v = 1$, maximum = 8

13. Minimize $6u + 9v + 12w$ subject to the constraints $\begin{cases} u + 3v \geq 10 \\ -2u + w \geq 12 \\ v + 3w \geq 10 \\ u \geq 0,\ v \geq 0,\ w \geq 0 \end{cases}$.

Solve the primal.

$$\begin{array}{c} \\ u \\ v \\ w \\ M \end{array}\begin{array}{c} \begin{array}{ccccccc} x & y & z & u & v & w & M \end{array} \\ \left[\begin{array}{ccccccc|c} 1 & -2 & 0 & 1 & 0 & 0 & 0 & 6 \\ 3 & 0 & 1 & 0 & 1 & 0 & 0 & 9 \\ 0 & \underline{1} & 3 & 0 & 0 & 1 & 0 & 12 \\ \hline -10 & -12 & -10 & 0 & 0 & 0 & 1 & 0 \end{array}\right] \end{array}$$

$$\begin{array}{c} \\ u \\ v \\ y \\ M \end{array}\begin{array}{c} \begin{array}{ccccccc} x & y & z & u & v & w & M \end{array} \\ \left[\begin{array}{ccccccc|c} 1 & 0 & 6 & 1 & 0 & 2 & 0 & 30 \\ \underline{3} & 0 & 1 & 0 & 1 & 0 & 0 & 9 \\ 0 & 1 & 3 & 0 & 0 & 1 & 0 & 12 \\ \hline -10 & 0 & 26 & 0 & 0 & 12 & 1 & 144 \end{array}\right] \end{array}$$

$$\begin{array}{c} \\ u \\ x \\ y \\ M \end{array}\begin{array}{c} \begin{array}{ccccccc} x & y & z & u & v & w & M \end{array} \\ \left[\begin{array}{ccccccc|c} 0 & 0 & \frac{17}{3} & 1 & -\frac{1}{3} & 2 & 0 & 27 \\ 1 & 0 & \frac{1}{3} & 0 & \frac{1}{3} & 0 & 0 & 3 \\ 0 & 1 & 3 & 0 & 0 & 1 & 0 & 12 \\ \hline 0 & 0 & \frac{88}{3} & 0 & \frac{10}{3} & 12 & 1 & 174 \end{array}\right] \end{array}$$

$x = 3, y = 12, z = 0$, maximum = 174; $u = 0$, $v = \frac{10}{3}$, $w = 12$, minimum = 174

15. Suppose we can hire workers out at a profit of u dollars per hour, sell the steel at a profit of v dollars per unit, and sell the wood at a profit of w dollars per unit. To find the minimum profit at which that should be done, minimize $90u + 138v + 120w$ subject to the constraints

$$\begin{cases} 3u + 7v + 4w \geq 3 \\ 6u + 5v + 3w \geq 5 \\ u \geq 0,\ v \geq 0,\ w \geq 0 \end{cases}.$$

17. Suppose we can buy anthracite at u dollars per ton, ordinary coal at v dollars per ton, and bituminous coal at w dollars per ton. To find the maximum cost at which this should be done, maximize $80u + 60v + 75w$ subject to the constraints

$$\begin{cases} 4u + 4v + 7w \leq 150 \\ 10u + 5v + 5w \leq 200. \\ u \geq 0,\ v \geq 0,\ w \geq 0 \end{cases}$$

19. The new primal problem is to maximize $30x + 50y + pz$, where p is the profit per Bowie knife and z is the number of Bowie knives produced, subject to the constraints

$$\begin{cases} 3x + 6y + 4z \leq 90 \\ 7x + 5y + 6z \leq 138 \\ 4x + 3y + 2z \leq 120 \\ x \geq 0,\ y \geq 0,\ z \geq 0 \end{cases}.$$

The dual is to minimize $90u + 138v + 120w$ subject to the constraints

$$\begin{cases} 3u + 7v + 4w \geq 3 \\ 6u + 5v + 3w \geq 5 \\ 4u + 6v + 2w \geq p \\ u \geq 0,\ v \geq 0,\ w \geq 0 \end{cases}.$$

Copyright © 2018 Pearson Education, Inc.

The original solution, with $u=\frac{200}{27}$, $v=\frac{10}{9}$, $w=0$, will still be optimal if

$$40\left(\frac{200}{27}\right)+60\left(\frac{10}{9}\right)+2(0)=36.30\geq p.$$

The minimum profit per Bowie knife that needs to be realized to warrant adding Bowie knives to the product line is \$36.30.

21. Dual: Maximize $5u + 8v$ subject to the constraints $\begin{cases} u+2v\leq 16 \\ 3u+4v\leq 42. \\ u\geq 0,\ v\geq 0 \end{cases}$

$$\begin{array}{c} \\ x \\ y \\ M \end{array}\begin{array}{c} \begin{array}{cccccc} u & v & x & y & M & \end{array} \\ \left[\begin{array}{ccccc|c} 1 & \underline{2} & 1 & 0 & 0 & 16 \\ 3 & 4 & 0 & 1 & 0 & 42 \\ \hline -5 & -8 & 0 & 0 & 1 & 0 \end{array}\right] \end{array}$$

$$\begin{array}{c} \\ v \\ y \\ M \end{array}\begin{array}{c} \begin{array}{cccccc} u & v & x & y & M & \end{array} \\ \left[\begin{array}{ccccc|c} \frac{1}{2} & 1 & \frac{1}{2} & 0 & 0 & 8 \\ \underline{1} & 0 & -2 & 1 & 0 & 10 \\ \hline -1 & 0 & 4 & 0 & 1 & 64 \end{array}\right] \end{array}$$

$$\begin{array}{c} \\ v \\ u \\ M \end{array}\begin{array}{c} \begin{array}{cccccc} u & v & x & y & M & \end{array} \\ \left[\begin{array}{ccccc|c} 0 & 1 & \frac{3}{2} & -\frac{1}{2} & 0 & 3 \\ \underline{1} & 0 & -2 & 1 & 0 & 10 \\ \hline 0 & 0 & 2 & 1 & 1 & 74 \end{array}\right] \end{array}$$

$x=2$, $y=1$, $M=74$

Chapter 4 Review Exercises

1.

$$\begin{array}{c} \\ u \\ v \\ M \end{array}\begin{array}{c} \begin{array}{cccccc} x & y & u & v & M & \end{array} \\ \left[\begin{array}{ccccc|c} 2 & 1 & 1 & 0 & 0 & 7 \\ -1 & \underline{1} & 0 & 1 & 0 & 1 \\ \hline -3 & -4 & 0 & 0 & 1 & 0 \end{array}\right] \end{array}$$

$$\begin{array}{c} \\ u \\ y \\ M \end{array}\begin{array}{c} \begin{array}{cccccc} x & y & u & v & M & \end{array} \\ \left[\begin{array}{ccccc|c} \underline{3} & 0 & 1 & -1 & 0 & 6 \\ -1 & 1 & 0 & 1 & 0 & 1 \\ \hline -7 & 0 & 0 & 4 & 1 & 4 \end{array}\right] \end{array}$$

$$\begin{array}{c} \\ x \\ y \\ M \end{array}\begin{array}{c} \begin{array}{cccccc} x & y & u & v & M & \end{array} \\ \left[\begin{array}{ccccc|c} 1 & 0 & \frac{1}{3} & -\frac{1}{3} & 0 & 2 \\ 0 & 1 & \frac{1}{3} & \frac{2}{3} & 0 & 3 \\ \hline 0 & 0 & \frac{7}{3} & \frac{5}{3} & 1 & 18 \end{array}\right] \end{array}$$

$x=2$, $y=3$, $M=18$

2.

$$\begin{array}{c} \\ u \\ v \\ M \end{array}\begin{array}{c} \begin{array}{cccccc} x & y & u & v & M & \end{array} \\ \left[\begin{array}{ccccc|c} 1 & \underline{1} & 1 & 0 & 0 & 7 \\ 4 & 3 & 0 & 1 & 0 & 24 \\ \hline -2 & -5 & 0 & 0 & 1 & 0 \end{array}\right] \end{array}$$

$$\begin{array}{c} \\ y \\ v \\ M \end{array}\begin{array}{c} \begin{array}{cccccc} x & y & u & v & M & \end{array} \\ \left[\begin{array}{ccccc|c} 1 & 1 & 1 & 0 & 0 & 7 \\ 1 & 0 & -3 & 1 & 0 & 3 \\ \hline 3 & 0 & 5 & 0 & 1 & 35 \end{array}\right] \end{array}$$

$x=0$, $y=7$, $M=35$

3.

$$\begin{array}{c} \\ u \\ v \\ w \\ M \end{array}\begin{array}{c} \begin{array}{ccccccc} x & y & u & v & w & M & \end{array} \\ \left[\begin{array}{cccccc|c} 1 & \underline{2} & 1 & 0 & 0 & 0 & 14 \\ 1 & 1 & 0 & 1 & 0 & 0 & 9 \\ 3 & 2 & 0 & 0 & 1 & 0 & 24 \\ \hline -2 & -3 & 0 & 0 & 0 & 1 & 0 \end{array}\right] \end{array}$$

$$\begin{array}{c} \\ y \\ v \\ w \\ M \end{array}\begin{array}{c} \begin{array}{ccccccc} x & y & u & v & w & M & \end{array} \\ \left[\begin{array}{cccccc|c} \frac{1}{2} & 1 & \frac{1}{2} & 0 & 0 & 0 & 7 \\ \underline{\frac{1}{2}} & 0 & -\frac{1}{2} & 1 & 0 & 0 & 2 \\ 2 & 0 & -1 & 0 & 1 & 0 & 10 \\ \hline -\frac{1}{2} & 0 & \frac{3}{2} & 0 & 0 & 1 & 21 \end{array}\right] \end{array}$$

Copyright © 2018 Pearson Education, Inc.

$$\begin{array}{c|cccccc|c} & x & y & u & v & w & M & \\ \hline y & 0 & 1 & 1 & -1 & 0 & 0 & 5 \\ x & 1 & 0 & -1 & 2 & 0 & 0 & 4 \\ w & 0 & 0 & 1 & -4 & 1 & 0 & 2 \\ \hline M & 0 & 0 & 1 & 1 & 0 & 1 & 23 \end{array}$$

$x = 4, y = 5, M = 23$

4.

$$\begin{array}{c|cccccc|c} & x & y & u & v & w & M & \\ \hline u & 1 & 2 & 1 & 0 & 0 & 0 & 10 \\ v & 4 & 3 & 0 & 1 & 0 & 0 & 30 \\ w & -2 & \underline{1} & 0 & 0 & 1 & 0 & 0 \\ \hline M & -3 & -7 & 0 & 0 & 0 & 1 & 0 \end{array}$$

$$\begin{array}{c|cccccc|c} & x & y & u & v & w & M & \\ \hline u & \underline{5} & 0 & 1 & 0 & -2 & 0 & 10 \\ v & 10 & 0 & 0 & 1 & -3 & 0 & 30 \\ y & -2 & 1 & 0 & 0 & 1 & 0 & 0 \\ \hline M & -17 & 0 & 0 & 0 & 7 & 1 & 0 \end{array}$$

$$\begin{array}{c|cccccc|c} & x & y & u & v & w & M & \\ \hline x & 1 & 0 & \frac{1}{5} & 0 & -\frac{2}{5} & 0 & 2 \\ v & 0 & 0 & -2 & 1 & 1 & 0 & 10 \\ y & 0 & 1 & \frac{2}{5} & 0 & \frac{1}{5} & 0 & 4 \\ \hline M & 0 & 0 & \frac{17}{5} & 0 & \frac{1}{5} & 1 & 34 \end{array}$$

$x = 2, y = 4, M = 34$

5.

$$\begin{array}{c|ccccc|c} & x & y & u & v & M & \\ \hline u & \underline{-7} & -5 & 1 & 0 & 0 & -40 \\ v & -1 & -4 & 0 & 1 & 0 & -9 \\ \hline M & 1 & 1 & 0 & 0 & 1 & 0 \end{array}$$

$$\begin{array}{c|ccccc|c} & x & y & u & v & M & \\ \hline x & 1 & \frac{5}{7} & -\frac{1}{7} & 0 & 0 & \frac{40}{7} \\ v & 0 & \underline{-\frac{23}{7}} & -\frac{1}{7} & 1 & 0 & -\frac{23}{7} \\ \hline M & 0 & \frac{2}{7} & \frac{1}{7} & 0 & 1 & -\frac{40}{7} \end{array}$$

$$\begin{array}{c|ccccc|c} & x & y & u & v & M & \\ \hline x & 1 & 0 & -\frac{4}{23} & \frac{5}{23} & 0 & 5 \\ y & 0 & 1 & \frac{1}{23} & -\frac{7}{23} & 0 & 1 \\ \hline M & 0 & 0 & \frac{3}{23} & \frac{2}{23} & 1 & -6 \end{array}$$

$x = 5, y = 1, M = -6$; the minimum is 6.

6.

$$\begin{array}{c|ccccc|c} & x & y & u & v & M & \\ \hline u & \underline{-1} & -1 & 1 & 0 & 0 & -6 \\ v & -1 & -2 & 0 & 1 & 0 & 0 \\ \hline M & 3 & 2 & 0 & 0 & 1 & 0 \end{array}$$

$$\begin{array}{c|ccccc|c} & x & y & u & v & M & \\ \hline x & 1 & \underline{1} & -1 & 0 & 0 & 6 \\ v & 0 & -1 & -1 & 1 & 0 & 6 \\ \hline M & 0 & -1 & 3 & 0 & 1 & -18 \end{array}$$

$$\begin{array}{c|ccccc|c} & x & y & u & v & M & \\ \hline y & 1 & 1 & -1 & 0 & 0 & 6 \\ v & 1 & 0 & -2 & 1 & 0 & 12 \\ \hline M & 1 & 0 & 2 & 0 & 1 & -12 \end{array}$$

$x = 0, y = 6, M = -12$; the minimum is 12.

7.

$$\begin{array}{c|cccccc|c} & x & y & u & v & w & M & \\ \hline u & -1 & -4 & 1 & 0 & 0 & 0 & -8 \\ v & -1 & -1 & 0 & 1 & 0 & 0 & -5 \\ w & \underline{-2} & -1 & 0 & 0 & 1 & 0 & -7 \\ \hline M & 20 & 30 & 0 & 0 & 0 & 1 & 0 \end{array}$$

$$\begin{array}{c|cccccc|c} & x & y & u & v & w & M & \\ \hline u & 0 & -\frac{7}{2} & 1 & 0 & -\frac{1}{2} & 0 & -\frac{9}{2} \\ v & 0 & -\frac{1}{2} & 0 & 1 & \underline{-\frac{1}{2}} & 0 & -\frac{3}{2} \\ x & 1 & \frac{1}{2} & 0 & 0 & -\frac{1}{2} & 0 & \frac{7}{2} \\ \hline M & 0 & 20 & 0 & 0 & 10 & 1 & -70 \end{array}$$

$$\begin{array}{c|cccccc|c} & x & y & u & v & w & M & \\ \hline u & 0 & \underline{-3} & 1 & -1 & 0 & 0 & -3 \\ w & 0 & 1 & 0 & -2 & 1 & 0 & 3 \\ x & 1 & 1 & 0 & -1 & 0 & 0 & 5 \\ \hline M & 0 & 10 & 0 & 20 & 0 & 1 & -100 \end{array}$$

Copyright © 2018 Pearson Education, Inc.

	x	y	u	v	w	M	
y	0	1	$-\frac{1}{3}$	$\frac{1}{3}$	0	0	1
w	0	0	$\frac{1}{3}$	$-\frac{7}{3}$	1	0	2
x	1	0	$\frac{1}{3}$	$-\frac{4}{3}$	0	0	4
M	0	0	$\frac{10}{3}$	$\frac{50}{3}$	0	1	−110

$x = 4$, $y = 1$, $M = -110$; the minimum is 110.

8.

	x	y	u	v	w	M	
u	−2	−1	1	0	0	0	−10
v	−3	−2	0	1	0	0	−18
w	−1	−2	0	0	1	0	−10
M	5	7	0	0	0	1	0

	x	y	u	v	w	M	
u	$-\frac{3}{2}$	0	1	0	$-\frac{1}{2}$	0	−5
v	−2	0	0	1	−1	0	−8
y	$\frac{1}{2}$	1	0	0	$-\frac{1}{2}$	0	5
M	$\frac{3}{2}$	0	0	0	$\frac{7}{2}$	1	−35

	x	y	u	v	w	M	
u	$-\frac{1}{2}$	0	1	$-\frac{1}{2}$	0	0	−1
w	2	0	0	−1	1	0	8
y	$\frac{3}{2}$	1	0	$-\frac{1}{2}$	0	0	9
M	$-\frac{11}{2}$	0	0	$\frac{7}{2}$	0	1	−63

	x	y	u	v	w	M	
x	1	0	−2	1	0	0	2
w	0	0	4	−3	1	0	4
y	0	1	3	−2	0	0	6
M	0	0	−11	9	0	1	−52

	x	y	u	v	w	M	
x	1	0	0	$-\frac{1}{2}$	$\frac{1}{2}$	0	4
u	0	0	1	$-\frac{3}{4}$	$\frac{1}{4}$	0	1
y	0	1	0	$\frac{1}{4}$	$-\frac{3}{4}$	0	3
M	0	0	0	$\frac{3}{4}$	$\frac{11}{4}$	1	−41

$x = 4$, $y = 3$, $M = -41$; the minimum is 41.

9.

	x	y	z	t	u	v	w	M	
t	1	0	0	1	0	0	0	0	4
u	0	1	0	0	1	0	0	0	6
v	0	0	1	0	0	1	0	0	8
w	4	3	2	0	0	0	1	0	38
M	−36	−48	−70	0	0	0	0	1	0

	x	y	z	t	u	v	w	M	
t	1	0	0	1	0	0	0	0	4
u	0	1	0	0	1	0	0	0	6
z	0	0	1	0	0	1	0	0	8
w	4	3	0	0	0	−2	1	0	22
M	−36	−48	0	0	0	70	0	1	560

	x	y	z	t	u	v	w	M	
t	1	0	0	1	0	0	0	0	4
y	0	1	0	0	1	0	0	0	6
z	0	0	1	0	0	1	0	0	8
w	4	0	0	0	−3	−2	1	0	4
M	−36	0	0	0	48	70	0	1	848

	x	y	z	t	u	v	w	M	
t	0	0	0	1	$\frac{3}{4}$	$\frac{1}{2}$	$-\frac{1}{4}$	0	3
y	0	1	0	0	1	0	0	0	6
z	0	0	1	0	0	1	0	0	8
x	1	0	0	0	$-\frac{3}{4}$	$-\frac{1}{2}$	$\frac{1}{4}$	0	1
M	0	0	0	0	21	52	9	1	884

$x = 1$, $y = 6$, $z = 8$, $M = 884$

Copyright © 2018 Pearson Education, Inc.

10.

$$\begin{array}{c} \\ t \\ u \\ v \\ M \end{array}\begin{array}{cccccccc|c} x & y & z & w & t & u & v & M & \\ 6 & 9 & 12 & 15 & 1 & 0 & 0 & 0 & 672 \\ 1 & -1 & \underline{2} & 2 & 0 & 1 & 0 & 0 & 92 \\ 5 & 10 & -5 & 4 & 0 & 0 & 1 & 0 & 280 \\ \hline -3 & -4 & -5 & -4 & 0 & 0 & 0 & 1 & 0 \end{array}$$

$$\begin{array}{c} \\ t \\ z \\ v \\ M \end{array}\begin{array}{cccccccc|c} x & y & z & w & t & u & v & M & \\ 0 & \underline{15} & 0 & 3 & 1 & -6 & 0 & 0 & 120 \\ \frac{1}{2} & -\frac{1}{2} & 1 & 1 & 0 & \frac{1}{2} & 0 & 0 & 46 \\ \frac{15}{2} & \frac{15}{2} & 0 & 9 & 0 & \frac{5}{2} & 1 & 0 & 510 \\ \hline -\frac{1}{2} & -\frac{13}{2} & 0 & 1 & 0 & \frac{5}{2} & 0 & 1 & 230 \end{array}$$

$$\begin{array}{c} \\ y \\ z \\ v \\ M \end{array}\begin{array}{cccccccc|c} x & y & z & w & t & u & v & M & \\ 0 & 1 & 0 & \frac{1}{5} & \frac{1}{15} & -\frac{2}{5} & 0 & 0 & 8 \\ \frac{1}{2} & 0 & 1 & \frac{11}{10} & \frac{1}{30} & \frac{3}{10} & 0 & 0 & 50 \\ \underline{\frac{15}{2}} & 0 & 0 & \frac{15}{2} & -\frac{1}{2} & \frac{11}{2} & 1 & 0 & 450 \\ \hline -\frac{1}{2} & 0 & 0 & \frac{23}{10} & \frac{13}{30} & -\frac{1}{10} & 0 & 1 & 282 \end{array}$$

$$\begin{array}{c} \\ y \\ z \\ x \\ M \end{array}\begin{array}{cccccccc|c} x & y & z & w & t & u & v & M & \\ 0 & 1 & 0 & \frac{1}{5} & \frac{1}{15} & -\frac{2}{5} & 0 & 0 & 8 \\ 0 & 0 & 1 & \frac{3}{5} & \frac{1}{15} & -\frac{1}{15} & -\frac{1}{15} & 0 & 20 \\ 1 & 0 & 0 & 1 & -\frac{1}{15} & \frac{11}{15} & \frac{2}{15} & 0 & 60 \\ \hline 0 & 0 & 0 & \frac{14}{5} & \frac{2}{5} & \frac{4}{15} & \frac{1}{15} & 1 & 312 \end{array}$$

$x = 60, y = 8, z = 20, w = 0, M = 312$

11. Minimize $14u + 9v + 24w$ subject to the constraints $\begin{cases} u+v+3w \ge 2 \\ 2u+v+2w \ge 3 \\ u \ge 0,\ v \ge 0,\ w \ge 0 \end{cases}$.

12. Maximize $8u + 5v + 7w$ subject to the constraints $\begin{cases} u+v+2w \le 20 \\ 4u+v+w \le 30 \\ u \ge 0,\ v \ge 0,\ w \ge 0 \end{cases}$.

13. Primal: $x = 4, y = 5$, maximum = 23; Dual: $u = 1, v = 1, w = 0$, minimum = 23

14. Primal: $x = 4, y = 1$, minimum = 110; Dual: $u = \frac{10}{3}, v = \frac{50}{3}, w = 0$, maximum = 110

15. $A = \begin{bmatrix} 1 & 2 \\ 1 & 1 \\ 3 & 2 \end{bmatrix}, B = \begin{bmatrix} 14 \\ 9 \\ 24 \end{bmatrix},$

$C = \begin{bmatrix} 2 & 3 \end{bmatrix}, X = \begin{bmatrix} x \\ y \end{bmatrix}$

Primal: Maximize CX subject to $AX \le B$, $X \ge 0$.

Dual: $U = \begin{bmatrix} u \\ v \\ w \end{bmatrix}$

Minimize $B^T U$ subject to $A^T U \ge C^T$, $U \ge 0$.

16. $A = \begin{bmatrix} 1 & 4 \\ 1 & 1 \\ 2 & 1 \end{bmatrix}, B = \begin{bmatrix} 8 \\ 5 \\ 7 \end{bmatrix},$

$C = \begin{bmatrix} 20 & 30 \end{bmatrix}, X = \begin{bmatrix} x \\ y \end{bmatrix}$

Primal: Minimize CX subject to $AX \ge B$, $X \ge 0$.

Dual: $U = \begin{bmatrix} u \\ v \\ w \end{bmatrix}$

Maximize $B^T U$ subject to $A^T U \le C^T, U \ge 0$

Copyright © 2018 Pearson Education, Inc.

17. Let x be the number of type A lens, y be the number of type B lens, and z be the number of type C lens produced. Maximize $12x+10y+8z$ subject to the constraints:

$$\begin{cases} 4x+2y+2z \le 360 \\ 2x+6y+4z \le 600 \\ 4x+2y+4z \le 480 \\ x \ge 0, y \ge 0, z \ge 0 \end{cases}$$

$$\begin{array}{c} \\ u \\ v \\ w \\ M \end{array} \begin{array}{c} \begin{array}{ccccccc|c} x & y & z & u & v & w & M & \end{array} \\ \left[\begin{array}{ccccccc|c} \underline{4} & 2 & 2 & 1 & 0 & 0 & 0 & 360 \\ 2 & 6 & 4 & 0 & 1 & 0 & 0 & 600 \\ 4 & 2 & 4 & 0 & 0 & 1 & 0 & 480 \\ \hline -12 & -10 & -8 & 0 & 0 & 0 & 1 & 0 \end{array}\right] \end{array}$$

$$\begin{array}{c} \\ x \\ v \\ w \\ M \end{array} \begin{array}{c} \begin{array}{ccccccc|c} x & y & z & u & v & w & M & \end{array} \\ \left[\begin{array}{ccccccc|c} 1 & \frac{1}{2} & \frac{1}{2} & \frac{1}{4} & 0 & 0 & 0 & 90 \\ 0 & 5 & 3 & -\frac{1}{2} & 0 & 0 & 0 & 420 \\ 0 & 0 & 2 & -1 & 0 & 1 & 0 & 120 \\ \hline 0 & -4 & -2 & 3 & 0 & 0 & 1 & 1080 \end{array}\right] \end{array}$$

$$\begin{array}{c} \\ x \\ y \\ w \\ M \end{array} \begin{array}{c} \begin{array}{ccccccc|c} x & y & z & u & v & w & M & \end{array} \\ \left[\begin{array}{ccccccc|c} 1 & 0 & \frac{1}{5} & \frac{3}{10} & 0 & 0 & 0 & 48 \\ 0 & 1 & \frac{3}{5} & -\frac{1}{10} & 0 & 0 & 0 & 84 \\ 0 & 0 & 2 & -1 & 0 & 1 & 0 & 120 \\ \hline 0 & 0 & \frac{2}{5} & \frac{13}{5} & 0 & 0 & 0 & 1416 \end{array}\right] \end{array}$$

Since the last row has all positive values, the manufacturer should produce 48 type A lens, 84 type B lens, and no type C lens, for a maximum profit of $1,416.

18. Let x be the number of pounds of oranges, y be the number of pounds of cherries, and z be the number of pounds of blueberries used. Maximize $-230x-260y-250z$ subject to the constraints:

$$\begin{cases} x+y+z \le 75 \\ 3x+7y+5z \le 251 \\ -4x-5y-3z \le -300 \\ x \ge 0, y \ge 0, z \ge 0 \end{cases}$$

$$\begin{array}{c} \\ u \\ v \\ w \\ M \end{array} \begin{array}{c} \begin{array}{ccccccc|c} x & y & z & u & v & w & M & \end{array} \\ \left[\begin{array}{ccccccc|c} 1 & 1 & 1 & 1 & 0 & 0 & 0 & 75 \\ 3 & 7 & 5 & 0 & 1 & 0 & 0 & 251 \\ -4 & -\underline{5} & -3 & 0 & 0 & 1 & 0 & -300 \\ \hline 230 & 260 & 250 & 0 & 0 & 0 & 1 & 0 \end{array}\right] \end{array}$$

Copyright © 2018 Pearson Education, Inc.

$$\begin{array}{c} \\ u \\ v \\ y \\ M \end{array}\begin{array}{c}\begin{array}{ccccccc|c} x & y & z & u & v & w & M & \end{array}\\ \left[\begin{array}{ccccccc|c} \frac{1}{5} & 0 & \frac{2}{5} & 1 & 0 & \frac{1}{5} & 0 & 15 \\ -\frac{13}{5} & 0 & \frac{4}{5} & 0 & 0 & \frac{7}{5} & 0 & -169 \\ \frac{4}{5} & 1 & \frac{3}{5} & 0 & 0 & -\frac{1}{5} & 0 & 60 \\ \hline 22 & 0 & 94 & 0 & 0 & 52 & 1 & -15600 \end{array}\right]\end{array}$$

$$\begin{array}{c} \\ u \\ x \\ y \\ M \end{array}\begin{array}{c}\begin{array}{ccccccc|c} x & y & z & u & v & w & M & \end{array}\\ \left[\begin{array}{ccccccc|c} 0 & 0 & \frac{6}{13} & 0 & 0 & \frac{4}{13} & 0 & 2 \\ 1 & 0 & -\frac{4}{13} & 0 & 0 & -\frac{7}{13} & 0 & 65 \\ 0 & 1 & \frac{11}{13} & 0 & 0 & \frac{3}{13} & 0 & 8 \\ \hline 0 & 0 & \frac{1310}{13} & 0 & 0 & \frac{830}{13} & 0 & -17,030 \end{array}\right]\end{array}$$

Since the last row has all positive values, the camp counselor should use 65 pounds of oranges, 8 pounds of cherries, and no blueberries, for a minimum calorie count of 17,030 (the opposite of the value because we changed it from a minimization problem to a maximization problem).

19. a. Let a be the number of attack sticks and b the number of defense sticks. Maximize $16a + 20b$ subject to the

constraints $\begin{cases} 2a+b \le 120 \\ a+3b \le 150 \\ 2a+2b \le 140 \\ a \ge 0,\ b \ge 0 \end{cases}$.

$$\begin{array}{c} \\ u \\ v \\ w \\ M \end{array}\begin{array}{c}\begin{array}{cccccc|c} a & b & u & v & w & M & \end{array}\\ \left[\begin{array}{cccccc|c} 2 & 1 & 1 & 0 & 0 & 0 & 120 \\ 1 & \underline{3} & 0 & 1 & 0 & 0 & 150 \\ 2 & 2 & 0 & 0 & 1 & 0 & 140 \\ \hline -16 & -20 & 0 & 0 & 0 & 1 & 0 \end{array}\right]\end{array}$$

$$\begin{array}{c} \\ u \\ b \\ w \\ M \end{array}\begin{array}{c}\begin{array}{cccccc|c} a & b & u & v & w & M & \end{array}\\ \left[\begin{array}{cccccc|c} \frac{5}{3} & 0 & 1 & -\frac{1}{3} & 0 & 0 & 70 \\ \frac{1}{3} & 1 & 0 & \frac{1}{3} & 0 & 0 & 50 \\ \frac{4}{3} & 0 & 0 & -\frac{2}{3} & 1 & 0 & 40 \\ \hline -\frac{28}{3} & 0 & 0 & \frac{20}{3} & 0 & 1 & 1000 \end{array}\right]\end{array}$$

$$\begin{array}{c} \\ u \\ b \\ a \\ M \end{array}\begin{array}{c}\begin{array}{cccccc|c} a & b & u & v & w & M & \end{array}\\ \left[\begin{array}{cccccc|c} 0 & 0 & 1 & \frac{1}{2} & -\frac{5}{4} & 0 & 20 \\ 0 & 1 & 0 & \frac{1}{2} & -\frac{1}{4} & 0 & 40 \\ 1 & 0 & 0 & -\frac{1}{2} & \frac{3}{4} & 0 & 30 \\ \hline 0 & 0 & 0 & 2 & 7 & 1 & 1280 \end{array}\right]\end{array}$$

30 attack sticks, 40 defense sticks

b. The new problem is to maximize $16a + 20b + pc$, where c is the number of tennis rackets and p is the profit on each racket, subject to the

constraints: $\begin{cases} 2a+b+c \le 120 \\ a+3b+4c \le 150 \\ 2a+2b+2c \le 140 \\ c \ge 0,\ b \ge 0,\ c \ge 0 \end{cases}$.

The dual problem is to minimize $120u+150v+140w$ subject to the

constraints: $\begin{cases} 2u+v+2w \ge 16 \\ u+3v+2w \ge 20 \\ u+4v+2w \ge p \\ u \ge 0,\ v \ge 0,\ w \ge 0 \end{cases}$.

The original solution, with $u = 0$, $v = 2$, $w = 7$ will be optimal if $0+4(2)+2(7) \ge p$. Since $0+4(2)+2(7) = \$22$, that is the profit per tennis racket that needs to be realized to justify the diversification.

Copyright © 2018 Pearson Education, Inc.

20. The new problem is to maximize $70a+210b+140c+pd$, where d is the number of brand D stereo systems and p is the profit per system, subject to the constraints:

$$\begin{cases} a+b+c+d \le 100 \\ 5a+4b+4c+3d \le 480 \\ 40a+20b+30c+30d \le 3200 \\ a \ge 0,\ b \ge 0,\ c \ge 0,\ d \ge 0 \end{cases}.$$

The dual problem is to minimize $100u+480v+3200w$, subject to the constraints:

$$\begin{cases} u+5v+40w \ge 70 \\ u+4v+20w \ge 210 \\ u+4v+30w \ge 140 \\ u+3v+30w \ge p \\ u \ge 0,\ v \ge 0,\ w \ge 0 \end{cases}.$$

The original solution, with $u = 210$, $v = 0$, $w = 0$, will still be optimal if $210 + 3(0) + 30(0) \ge p$. Since $210 + 3(0) + 30(0) = \$210$, that is the required profit per brand D system.
The original solution was to sell 100 units of brand B at a profit of \$210 each. Brand D units have to be at least this profitable. (Differences in storage space and commission turned out not to matter.)

 Copyright © 2018 Pearson Education, Inc.

Chapter 5

Exercises 5.1

1. **a.** $S' = \{5, 6, 7\}$

   b. $S \cup T = \{1, 2, 3, 4, 5, 7\}$

   c. $S \cap T = \{1, 3\}$

   d. $S' \cap T = \{5, 7\}$

3. **a.** $R \cup S = \{a, b, c, \text{e, i, o, u}\}$

   b. $R \cap S = \{\text{a}\}$

   c. $S \cap T = \varnothing$

   d. $S' \cap R = \{b, c\}$

5. $\varnothing$, $\{1\}$, $\{2\}$, $\{1, 2\}$

7. **a.** $F \cap B =$ {all freshman college students who like basketball}

   b. $B' =$ {all college students who do not like basketball}

   c. $F' \cap B' =$ {all college students who are neither freshman nor like basketball}

   d. $F \cup B =$ {all college students who are either freshman or like basketball}

9. **a.** $S =$ {1999, 2003, 2006, 2010, 2013}

   b. $T =$ {1996, 1997, 1998, 1999, 2003, 2009, 2013}

   c. $S \cap T =$ {1999, 2003, 2013}

   d. $S \cup T =$ {1996, 1997, 1998, 1999, 2003, 2006, 2009, 2010, 2013}

   e. $S' \cap T =$ {1996, 1997, 1998, 2009}

   f. $S \cap T' =$ { 2006, 2010}

11. From 1996 to 2015, there were only two years in which the Standard and Poor's Index increased by 2% or more during the first five days and not increase by 16% or more for the entire year.

13. **a.** $R \cup S = \{a, b, c, d\}$
    $(R \cup S)' = \{\text{e}, f\}$

    b. $R \cup S \cup T = \{a, b, c, \text{d}, e, f\}$

    c. $R \cap S = \{a, \text{b}\}$
    $R \cap S \cap T = (R \cap S) \cap T = \varnothing$

    d. $T' = \{a, b, c, d\}$
    $R \cap S \cap T' = (R \cap S) \cap T' = \{a, \text{b}\}$

    e. $R' = \{d, e, f\}$; $S \cap T = \varnothing$
    $R' \cap S \cap T = R' \cap (S \cap T) = \varnothing$

    f. $S \cup T = \{a, \text{b, d}, e, f\}$

    g. $R \cup S = \{a, b, c, d\}$;
    $R \cup T = \{a, b, c, e, f\}$
    $(R \cup S) \cap (R \cup T) = \{a, b, c\}$

    h. $R \cap S = \{a, \text{b}\}$; $R \cap T = \varnothing$
    $(R \cap S) \cup (R \cap T) = \{a, b\}$

    i. $R' = \{d, e, f\}; T' = \{a, b, c, d\}$
    $R' \cap T' = \{d\}$

15. $(S')' = S$

17. $S \cup S' = U$

19. $T \cap S \cap T' = S \cap (T \cap T') = S \cap \varnothing = \varnothing$

21. {divisions that had increases in labor costs or total revenue} $= L \cup T$

23. {divisions that made a profit despite an increase in labor costs} $= L \cap P$

25. {profitable divisions with increases in labor costs and total revenue} $= P \cap L \cap T$

27. {applicants who have not received speeding tickets} $= S'$

29. {applicants who have received speeding tickets, caused accidents, or were arrested for drunk driving}
    $= S \cup A \cup D$

Copyright © 2018 Pearson Education, Inc.

31. {applicants who have not both caused accidents and received speeding tickets but who have been arrested for drunk driving}
$= (A \cap S)' \cap D$

33. $A \cap D$ = {students at Mount College who are younger than 35}

35. $A \cap B$ = {people who are both student and teachers at Mount College}

37. $A \cup C' = A \cup D$ = {people at Mount College who are students or are at most 35}

39. $D' = C$ = {people at Mount College who are at least 35}

41. {people who don't like vanilla ice cream} = V'

43. {people who like vanilla but not chocolate or strawberry ice cream} = $V \cap (C \cup S)'$

45. {people who like neither chocolate nor vanilla ice cream} = $(V \cup C)'$

47. **a.** $R = \{B, C, D, E\}$

b. $S = \{C, D, E, F\}$

c. $T = \{A, D, E, F\}$

d. $R' = \{A, F\}$
$R' \cup S = \{A, C, D, E, F\}$

e. $R' \cap T = \{A, F\}$

f. $R \cap S = \{C, D, E\}$
$R \cap S \cap T = (R \cap S) \cap T = \{D, E\}$

49. There are eight different ways. They are no toppings; peppers; onions; mushrooms; peppers and onions; peppers and mushrooms; onions and mushrooms; all three toppings.

51. Any subset of T with 2 as an element is an example. Possible answer: {2}

53. If S is a subset of T, then $S \cup T = T$.

55. True; 5 is an element of the set {3, 5, 7}.

57. True; {b} is a subset of the set {b, c}.

59. False; 0 is not an element of the empty set $\varnothing$.

61. True; any set is a subset of itself.

Exercises 5.2

1. $n(S \cup T) = n(S) + n(T) - n(S \cap T)$
$= 4 + 4 - 2 = 6$

3. $n(S \cup T) = n(S) + n(T) - n(S \cap T)$
$15 = 6 + 9 - n(S \cap T)$
$n(S \cap T) = 6 + 9 - 15 = 0$

5. $n(S \cup T) = n(S) + n(T) - n(S \cap T)$
$10 = n(S) + 7 - 5$
$n(S) = 10 - 7 + 5 = 8$

7. S is a subset of T.

9. Let P = {adults in South America fluent in Portuguese} and
S = {adults in South America fluent in Spanish}.
Then $P \cup S$ = {adults in South America fluent in Portuguese or Spanish} and
$P \cap S$ = {adults in South America fluent in Portuguese and Spanish}.
$n(P) = 170$, $n(S) = 155$,
$n(P \cup S) = 314$ (numbers in millions)
$n(P \cup S) = n(P) + n(S) - n(P \cap S)$
$314 = 170 + 155 - n(P \cap S)$
$n(P \cap S) = 170 + 155 - 314 = 11$
11 million are fluent in both languages.

11. Let U = {all letters of the alphabet},
let V = {letters with vertical symmetry}, and
H = {letters with horizontal symmetry}.
Then $V \cup H$ = {letters with vertical or horizontal symmetry} and
$V \cap H$ = {letters with both vertical and horizontal symmetry}.
$n(V) = 11$, $n(H) = 9$, $n(V \cap H) = 4$
$n(V \cup H) = n(V) + n(H) - n(V \cap H)$
$n(V \cup H) = 11 + 9 - 4 = 16$
$n((V \cup H)') = n(U) - n(V \cup H) = 26 - 16 = 10$
There are 10 letters with no symmetry.

 Copyright © 2018 Pearson Education, Inc.

13. Let A = {cars with a navigation system} and P = {cars with push-button start}.
Then $A \cup P$ = {cars with a navigation system or push-button start} and
$A \cap P$ = {cars with both a navigation system and push-button start},
$n(A) = 325$, $n(P) = 216$, $n(A \cap P) = 89$

$$\begin{aligned} n(A \cup P) &= n(A) + n(P) - n(A \cap P) \\ &= 325 + 216 - 89 \\ &= 452 \end{aligned}$$

452 cars were manufactured with at least one of the two options.

15. Consists of points not in T but in S.

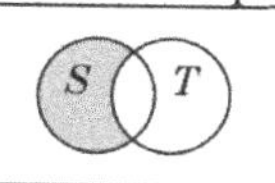

17. Consists of points in T or not in S.

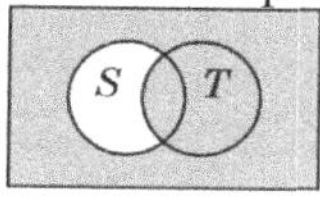

19. $(S \cap T')' = S' \cup T$
Consists of points in T or not in S.

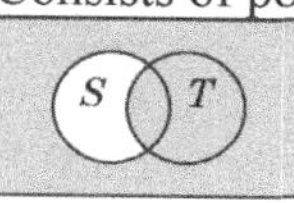

21. Consists of points in S but not in T or points in T but not in S.

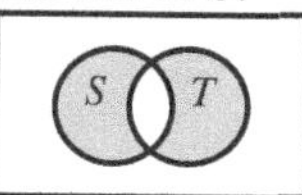

23. $S \cup (S \cap T) = S$
Consists of points in S.

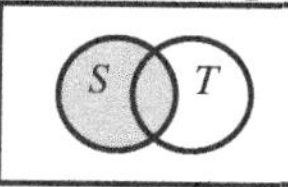

25. $S \cup S' = U$
Consists of all points.

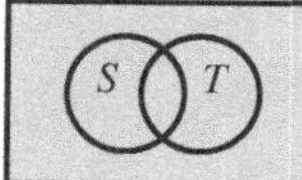

27. Consists of points in R and S but not in T.

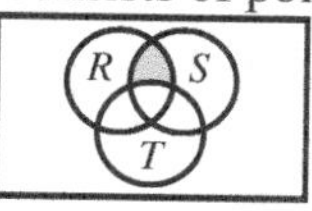

29. Consists of points in R or points in both S and T.

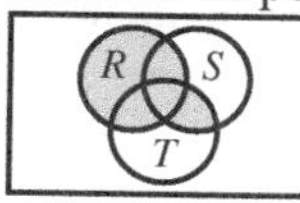

31. Consists of points in R but not in S or points in both R and T.

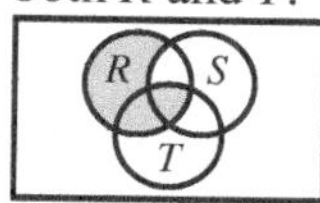

33. Consists of points in both R and T.

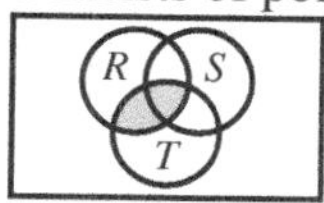

35. Consists of points not in R, S, and T.

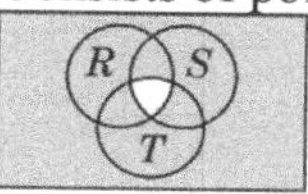

37. Consists of points in R and T or points in S but not T.

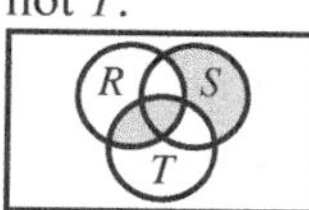

39. $S' \cup (S \cap T)' = S' \cup S' \cup T' = S' \cup T'$

41. $(S' \cup T)' = S \cap T'$

43. $$\begin{aligned} T \cup (S \cap T)' &= T \cup S' \cup T' \\ &= (T \cup T') \cup S' \\ &= U \end{aligned}$$

45. S'

47. $R \cap T$

49. $R' \cap S \cap T$

51.

$T \cup (R \cap S')$

Copyright © 2018 Pearson Education, Inc.

53. First draw a Venn diagram for $(R \cap S') \cup (S \cap T') \cup (T \cap R')$.

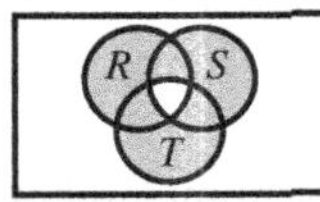

The set consists of the complement.
$(R \cap S \cap T) \cup (R' \cap S' \cap T')$

55. Everyone who is not a citizen or is both over the age of 18 and employed

57. Everyone over the age of 18 who is unemployed

59. Noncitizens who are 5 years of age or older

Exercises 5.3

1. $5 + 6 = 11$

3. $6 + 5 + 15 + 20 = 46$

5. 11

7. $19 + 10 + 5 + 6 + 15 + 20 = 75$

9. $10 + 5 + 15 = 30$

11. $n(S \cap T') = n(S) - n(S \cap T) = 12 - 5 = 7$
$n(S' \cap T) = n(T) - n(S \cap T) = 7 - 5 = 2$
$n(S \cup T) = n(S) + n(T) - n(S \cap T)$
$= 7 + 2 + 5 = 14$
$n(S' \cap T') = n(U) - n(S \cup T) = 17 - 14 = 3$

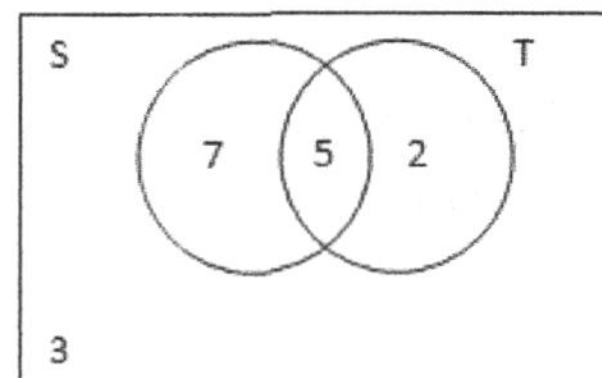

 Copyright © 2018 Pearson Education, Inc.

13. $n(S\cap T)=n(S)+n(T)-n(S\cup T)$
$=12+14-18=8$
$n(S\cap T')=n(S)-n(S\cap T)=12-8=4$
$n(S'\cap T)=n(T)-n(S\cap T)=14-8=6$
$n(S'\cap T')=n(U)-n(S\cup T)=20-18=2$

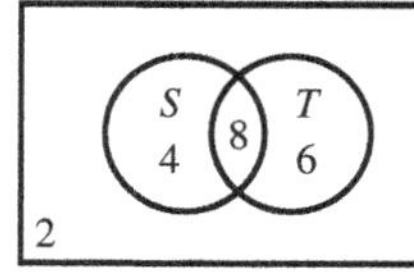

15. $n(S\cup T)=n(U)-n(S'\cap T')\ =75-40=35$
$n(S\cap T)=n(S)+n(T)-n(S\cup T)\ =15+25-35=5$
$n(S\cap T')=n(S)-n(S\cap T)=15-5=10$
$n(S'\cap T)=n(T)-n(S\cap T)=25-5=20$

S
T
10
5
20
40

17. $n(S\cap T)=n(S)+n(T)-n(S\cup T)\ =3+4-6=1$
$n(U)=n(S'\cup T')+n(S\cap T)=9+1=10$
$n(S\cap T')=n(S)-n(S\cap T)=3-1=2$
$n(S'\cap T)=n(T)-n(S\cap T)=4-1=3$
$n(S'\cap T')=n(U)-n(S\cup T)=10-6=4$

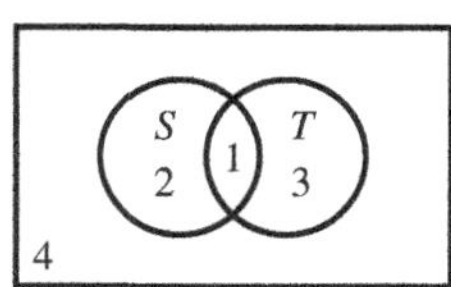

19. $n(R\cap S\cap T')=n(R\cap S)-n(R\cap S\cap T)\ =5-2=3$
$n(R\cap S'\cap T)=n(R\cap T)-n(R\cap S\cap T)\ =7-2=5$
$n(R'\cap S\cap T)=n(S\cap T)-n(R\cap S\cap T)\ =3-2=1$
$n(R\cap S'\cap T')\ =n(R)-n(R\cap S\cap T')-n(R\cap S'\cap T)\ -n(R\cap S\cap T)\ =12-3-5-2=2$
$n(R'\cap S\cap T')=n(S)-n(R\cap S\cap T')-n(R'\cap S\cap T)-n(R\cap S\cap T)=12-3-1-2=6$
$n(R'\cap S'\cap T)\ =n(T)-n(R\cap S'\cap T)-n(R'\cap S\cap T)\ -n(R\cap S\cap T)\ =9-5-1-2=1$
$n(R\cup S\cup T)\ =3+5+1+2+6+1+2=20$

Copyright © 2018 Pearson Education, Inc.

$n(R'\cap S'\cap T') = n(U) - n(R\cup S\cup T) = 28 - 20 = 8$

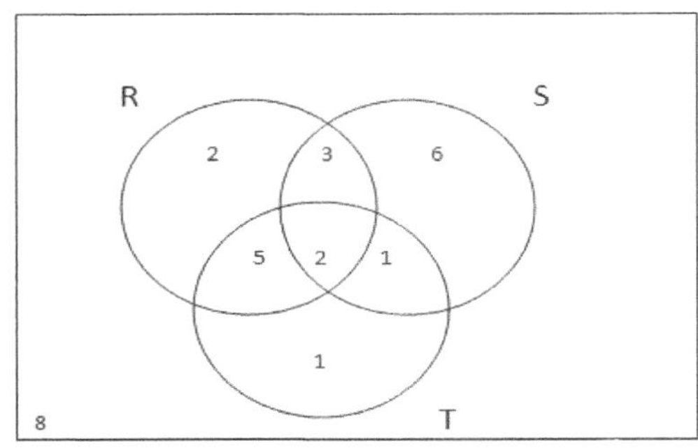

21. $n(R) = n(R\cup S) + n(R\cap S) - n(S) = 21 + 7 - 14 = 14$

$n(U) = n(R) + n(R') = 22 + 14 = 36$

$n(R\cap S\cap T') = n(R\cap S) - n(R\cap S\cap T) = 7 - 5 = 2$

$n(R\cap S'\cap T) = n(R\cap T) - n(R\cap S\cap T) = 11 - 5 = 6$

$n(R'\cap S\cap T) = n(S\cap T) - n(R\cap S\cap T) = 9 - 5 = 4$

$n(R\cap S'\cap T') = n(R) - n(R\cap S\cap T') - n(R\cap S'\cap T) - n(R\cap S\cap T) = 14 - 2 - 6 - 5 = 1$

$n(R'\cap S\cap T') = n(S) - n(R\cap S\cap T') - n(R'\cap S\cap T) - n(R\cap S\cap T) = 14 - 2 - 4 - 5 = 3$

$n(R'\cap S'\cap T) = n(T) - n(R\cap S'\cap T) - n(R'\cap S\cap T) - n(R\cap S\cap T) = 22 - 6 - 4 - 5 = 7$

$n(R\cup S\cup T) = 1 + 3 + 7 + 2 + 6 + 4 + 5 = 28$

$n(R'\cap S'\cap T') = n(U) - n(R\cup S\cup T) = 36 - 28 = 8$

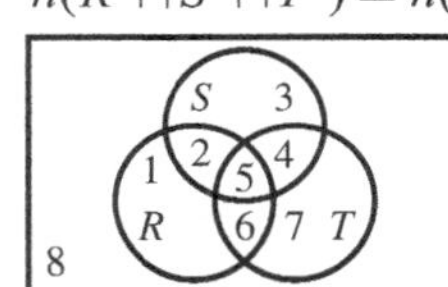

23. Let U = {high school students surveyed}, R = {students who like rock music}, and H = {students who like hip-hop music}.

$n(U) = 70$; $n(R) = 35$; $n(H) = 15$; $n(R\cap H) = 5$

$n(R\cup H) = n(R) + n(H) - n(R\cap H) = 35 + 15 - 5 = 45$

$n((R\cup H)') = n(U) - n(R\cup H) = 70 - 45 = 25$

25 students do not like either rock or hip-hop music.

25. Let U = {lines}, V = {lines with verbs}, A = {lines with adjectives}

$n(U) = 14$; $n(V) = 11$; $n(A) = 9$;

$n(V\cap A) = 7$

$n(V\cap A') = n(V) - n(V\cap A) = 11 - 7 = 4$

Four lines have a verb with no adjective.

$n(V'\cap A) = n(A) - n(V\cap A) = 9 - 7 = 2$

Two lines have an adjective but no verb.

$n(V\cup A) = n(V) + n(A) - n(V\cap A) = 11 + 9 - 7 = 13$

$n((V\cup A)') = n(U) - n(V\cap A) = 14 - 13 = 1$

One line has neither an adjective nor a verb.

 Copyright © 2018 Pearson Education, Inc.

For Exercises 26–30, let U = {students who took the exam}, F = {students who correctly answered the first question}, S = {students who correctly answered the second question}. Then $n(U) = 150$, $n(F) = 90$, $n(S) = 71$, $n(F \cap S) = 66$. Draw and complete the Venn diagram as follows.

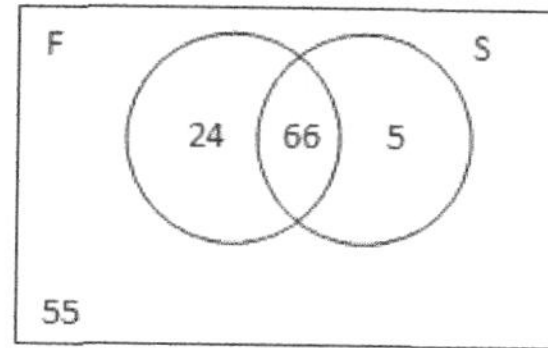

27. $n((F \cup S)') = 55$

29. $n(S \cap F') = 5$

31. Let U = {students in finite math}, M = {male students}, B = {students who are business majors}, and F = {first-year students}. $n(U) = 35$; $n(M) = 22$; $n(B) = 19$; $n(F) = 27$; $n(M \cap B) = 14$; $n(M \cap F) = 17$; $n(B \cap F) = 15$; $n(M \cap B \cap F) = 11$

a. Draw a Venn diagram as shown.

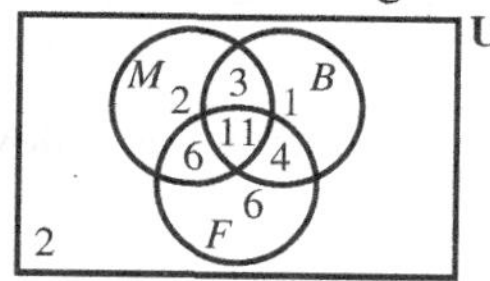

b. $n(F' \cap M' \cap B') = 2$
There are two upperclass women nonbusiness majors.

c. $n(M' \cap B) = 1 + 4 = 5$
There are five women business majors.

33. $n(R \cap (W \cup B)') = 28$

35. $n((R \cup W \cup B)') = 2$

37. $n(R \cap W \cap B') = 51$

For Exercises 39–44, let U = {people surveyed}, I = {people who learned from the Internet}, T = {people who learned from television}, N = {people who learned from newspapers}. Then $n(U) = 400$, $n(I) = 180$, $n(T) = 190$, $n(N) = 190$, $n(I \cap T) = 80$, $n(I \cap N) = 90$, $n(T \cap N) = 50$, $n(I \cap T \cap N) = 30$. Draw and complete the Venn diagram as follows.

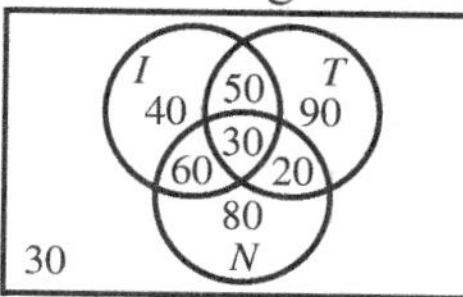

39. $n((I \cap N') \cup (I' \cap N)) = 40 + 50 + 80 + 20$
$= 190$

41. $n((I \cup T) \cap N') = 40 + 90 + 50 = 180$

43. $n((I \cap T' \cap N') \cup (I' \cap T \cap N') \cup (I' \cap T' \cap N))$
$= 40 + 90 + 80$
$= 210$

45. Draw and complete the Venn diagram as follows:

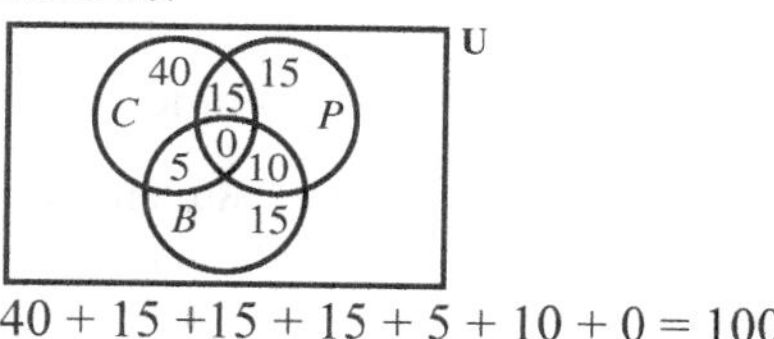

40 + 15 +15 + 15 + 5 + 10 + 0 = 100

Copyright © 2018 Pearson Education, Inc.

For Exercises 46–50, $n(U) = 4000$, $n(F) = 2000$, $n(S) = 3000$, $n(L) = 500$, $n(F \cap S) = 1500$, $n(F \cap L) = 300$, $n(S \cap L) = 200$, $n(F \cap S \cap L) = 50$. Draw and complete the Venn diagram as follows.

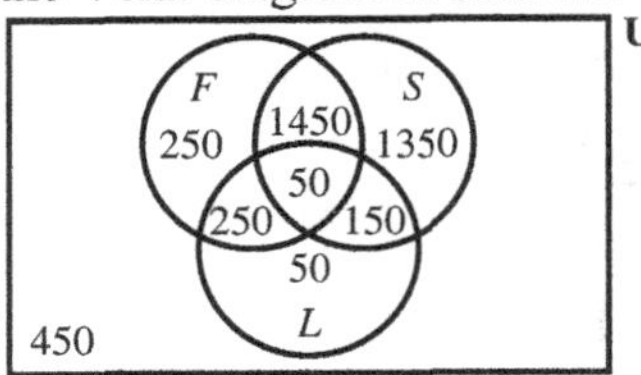

47. $(L \cup F \cup S)' = 450$

49. $L \cup S \cup F' = 4000 - 250 = 3750$

51. Let U = {college students surveyed}, F = {first-year students}, D = {voted Democratic}, $n(U) = 100$; $n(F) = 50$; $n(D) = 55$

$n(F' \cap D') = n((F \cup D)') = 25$

$n(F \cup D) = n(U) - n((F \cup D)') = 100 - 25 = 75$

$$\begin{aligned} n(F \cap D) &= n(F) + n(D) - n(F \cup D) \\ &= 50 + 55 - 75 \\ &= 30 \end{aligned}$$

30 freshmen voted Democratic.

53. Let U = {students}, D = {students who passed the diagnostic test}, C = {students who passed the course}.

$n(U) = 30$; $n(D) = 21$; $n(C) = 23$; $n(D \cap C') = 2$

$n(D \cap C) = n(D) - n(D \cap C') = 21 - 2 = 19$

$n(D' \cap C) = n(C) - n(D \cap C) = 23 - 19 = 4$

Four students passed the course even though they failed the diagnostic test.

For Exercises 55–60, let U = {students}, S = {seniors}, B = {biology majors}. Then $n(U) = 61$, $n(S \cap B) = 6$, $n(S' \cap B) = 17$, and $n(S' \cap B') = 12$. Therefore $n(S \cap B') = n(U) - n(B) = n(U) - n(S \cap B) - n(S' \cap B) - n(S' \cap B') = 61 - 6 - 17 - 12 = 26$

Draw and complete the Venn diagram as follows.

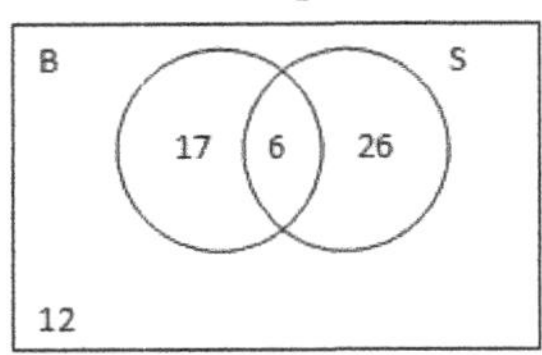

55. $17 + 6 + 26 = 49$

57. $17 + 12 = 29$

59. 26

For Exercises 61–68, let U = {surveyed students}, R = {students who like rock}, C = {students who like country}, J = {students who like rap}. Then $n(U) = 190$, $n(R) = 114$, $n(C) = 50$, $n(R \cap J) = 15$, $n(C \cap J) = 11$, $n(R' \cap C' \cap J) = 20$, $n(R \cap C' \cap J) = 10$, $n(R \cap C \cap J') = 9$ and $n((R \cup C \cup J)') = 20$. Draw and complete the Venn diagram as follows.

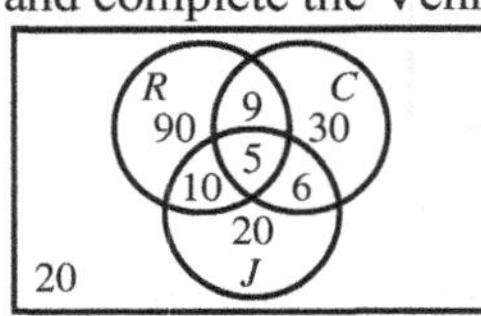

61. $n(R \cap C' \cap J') = 90$

63. $n(R' \cap C \cap J) = 6$

65. $n((R \cap C' \cap J') \cup (R' \cap C \cap J') \cup (R' \cap C' \cap J)) = 90 + 30 + 20 = 140$

67.
$$\begin{aligned} n((R \cap C) \cup (R \cap J) \cup (C \cap J)) &= 9 + 10 + 6 + 5 \\ &= 30 \end{aligned}$$

Copyright © 2018 Pearson Education, Inc.

69. Let U = {executives}, C = {executives who visited *CNN Money*}, B = {executives who visited *Bloomberg*}, W = {executives who visited *The Wall Street Journal*}.
$n(U) = 160$, $n(C) = 70$, $n(B) = 60$, $n(W) = 55$, $n(C \cap B) = 20$, $n(B \cap W) = 20$, $n(M \cap T \cap N) = 5$
Draw and complete the Venn diagram as follows.

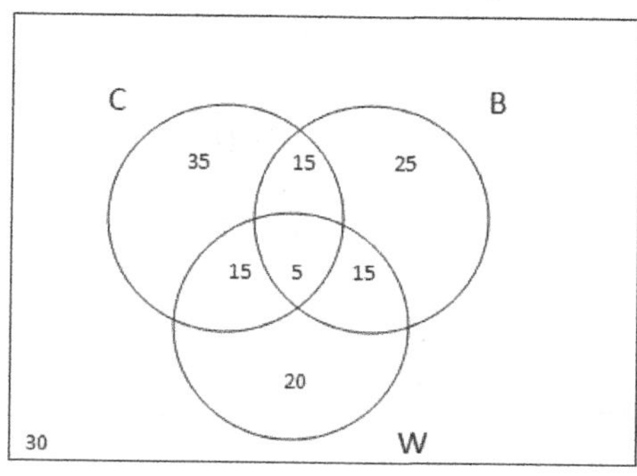

$n((C \cup B \cup W)') = 30$

71. Let U = {students}, P = {students who play piano}, V = {students who play violin} and C = {students who play clarinet}. Then let $x = n(P \cap V \cap C)$ and complete the Venn diagram as follows.

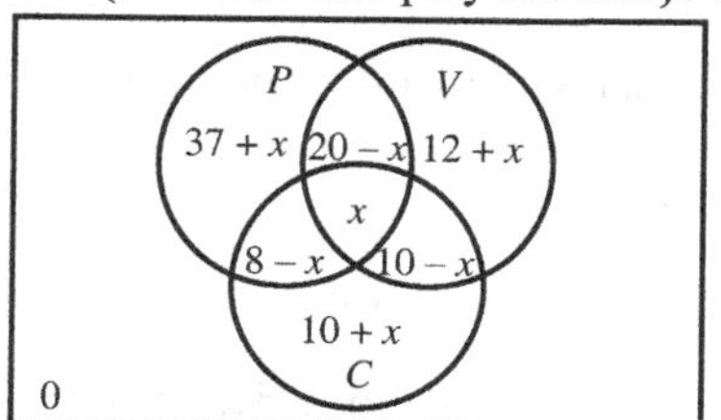

$37 + x + 12 + x + 10 + x + (20 - x) + (8 - x) + (10 - x) + x = 97 + x = 100$, so $x = 3$.

Exercises 5.4

1. $4 \cdot 2 = 8$ routes

3. $3 \cdot 2 = 6$ routes

5. $44 \cdot 43 \cdot 42 = 79{,}464$ possibilities

7. $20 \cdot 19 \cdot 18 = 6840$ possibilities

9. 30 because $30 \cdot 29 = 870$

11. **a.** $8 \cdot 7 \cdot 6 \cdot 5 \cdot 4 \cdot 3 \cdot 2 \cdot 1 = 40{,}320$ ways

b. $5 \cdot 4 \cdot 3 \cdot 2 \cdot 1 \cdot 3 \cdot 2 \cdot 1 = 720$ ways

13. $4 \cdot 3 \cdot 2 \cdot 1 = 24$ words

15. $2 \cdot 3 = 6$ outfits

17. $3 \cdot 12 \cdot 10 \cdot 10 \cdot 10 \cdot 10 = 360{,}000$ serial numbers

19. $10^9 - 1 = 999{,}999{,}999$ social security numbers

21. $8 \cdot 2 \cdot 10 = 160$ area codes

23. $9 \cdot 10 \cdot 10 \cdot 1 \cdot 1 = 900$ 5-digit palindromes

25. $26 \cdot 26 \cdot 1 \cdot 1 = 676$ 4-letter palindromes

27. $15 \cdot 15 = 225$ matchups

29. $3200 \cdot 2 \cdot 24 \cdot 52 = 7{,}987{,}200$ deals per year

31. $26 \cdot 26 \cdot 26 = 17{,}576$ sets of unique initials. Since there are 20,000 students, at least two students have the same set of initials

33. $7 \cdot 5 = 35$ different possible halftime scores

35. $5 \cdot 4 = 20$ different mismatched sets

37. $2^6 = 64$ possible sequences

39. $2^5 = 32$ possible ways

41. $4^{10} = 1{,}048{,}576$ possible ways

43. $10^5 = 100{,}000$ possible zip codes

Copyright © 2018 Pearson Education, Inc.

45. $8\cdot 7\cdot 6\cdot 5\cdot 4\cdot 3\cdot 2\cdot 1 = 40{,}320$ ways

$40320\cdot 15 = 604{,}800$ seconds

$\frac{604800}{60} = 10{,}080$ minutes

$\frac{10080}{60} = 168$ hours

$\frac{168}{24} = 7$ days

47. $6\cdot 7\cdot 4 = 168$ days or 24 weeks

49. $5\cdot 11\cdot (7\cdot 2+1)\cdot 10 = 8250$ different ways

51. $2^4 = 16$ possible ways

53. $2\cdot 38\cdot 38 = 2888$ different outcomes

55. **a.** $9\cdot 8\cdot 7\cdot 6\cdot 5\cdot 4\cdot 3\cdot 2\cdot 1 = 362{,}880$

b. $8\cdot 7\cdot 6\cdot 5\cdot 4\cdot 3\cdot 2\cdot 1\cdot 1 = 40{,}320$

c. $1\cdot 6\cdot 5\cdot 4\cdot 3\cdot 2\cdot 1\cdot 1\cdot 1 = 720$

57. $\frac{10\cdot 9}{2} + 10\cdot 10 = 145$ handshakes

59. $4\cdot 3\cdot 3\cdot 3\cdot 3\cdot 3 = 972$ ways

61. $7\cdot 4\cdot 2^6 = 1792$ different ballots

$8\cdot 5\cdot 3^6 = 29{,}160$ different ballots

63. $2^4 = 16$ possible ways

Exercises 5.5

1. $P(4, 2) = 4\cdot 3 = 12$

3. $P(6, 3) = 6\cdot 5\cdot 4 = 120$

5. $C(10, 3) = \frac{P(10, 3)}{3!} = \frac{10\cdot 9\cdot 8}{3\cdot 2\cdot 1} = 120$

7. $C(5, 4) = \frac{P(5, 4)}{4!} = \frac{5\cdot 4\cdot 3\cdot 2}{4\cdot 3\cdot 2\cdot 1} = 5$

9. $P(7, 1) = 7$

11. $P(n, 1) = n$

13. $C(4, 4) = \frac{P(4, 4)}{4!} = \frac{4\cdot 3\cdot 2\cdot 1}{4\cdot 3\cdot 2\cdot 1} = 1$

15.
$$\begin{aligned} C(n, n-2) &= \frac{P(n, n-2)}{(n-2)!} \\ &= \frac{n\cdot (n-1)\text{L } 4\cdot 3}{(n-2)\cdot (n-3)\text{L } 2\cdot 1} \\ &= \frac{n\cdot (n-1)}{2\cdot 1} \\ &= \frac{n(n-1)}{2} \end{aligned}$$

17. $6! = 6\cdot 5\cdot 4\cdot 3\cdot 2\cdot 1 = 720$

19. $\frac{9!}{7!} = \frac{9\cdot 8\cdot 7\cdot 6\cdot 5\cdot 4\cdot 3\cdot 2\cdot 1}{7\cdot 6\cdot 5\cdot 4\cdot 3\cdot 2\cdot 1} = 9\cdot 8 = 72$

21. Permutation; order matters

23. Combination; order does not matter

25. Neither

27. $4! = 4\cdot 3\cdot 2\cdot 1 = 24$ ways

29. $C(9,7) = \frac{P(9,7)}{7!} = \frac{9\cdot 8\cdot 7\cdot 6\cdot 5\cdot 4\cdot 3}{7\cdot 6\cdot 5\cdot 4\cdot 3\cdot 2\cdot 1} = 36$ selections

31. $C(8,4) = \frac{P(8,4)}{4!} = \frac{8\cdot 7\cdot 6\cdot 5}{4\cdot 3\cdot 2\cdot 1} = 70$ ways

33.
$$\begin{aligned} P(65,5) &= 65\cdot 64\cdot 63\cdot 62\cdot 61 \\ &= 991{,}186{,}560 \text{ ways} \end{aligned}$$

35.
$$\begin{aligned} C(10,5) &= \frac{P(10,5)}{5!} \\ &= \frac{10\cdot 9\cdot 8\cdot 7\cdot 6}{5\cdot 4\cdot 3\cdot 2\cdot 1} \\ &= 252 \text{ ways} \end{aligned}$$

Copyright © 2018 Pearson Education, Inc.

37. $C(100,3) = \frac{P(100,3)}{3!} = \frac{100\cdot 99\cdot 98}{3\cdot 2\cdot 1} = 161,700$ possible samples

$C(7,3) = \frac{P(7,3)}{3!} = \frac{7\cdot 6\cdot 5}{3\cdot 2\cdot 1} = 35$ defective samples

39. $P(150,3) = 150\cdot 149\cdot 148 = 3,307,800$ ways

41. $C(52,5) = \frac{P(52,5)}{5!} = \frac{52\cdot 51\cdot 50\cdot 49\cdot 48}{5\cdot 4\cdot 3\cdot 2\cdot 1} = 2,598,960$ hands

43. $C(13, 5) = \frac{P(13, 5)}{5!} = \frac{13\cdot 12\cdot 11\cdot 10\cdot 9}{5\cdot 4\cdot 3\cdot 2\cdot 1} = 1287$ hands

45. $5! = 5\cdot 4\cdot 3\cdot 2\cdot 1 = 120$ ways

47. **a.** $C(10,4) = \frac{P(10,4)}{4!} = \frac{10\cdot 9\cdot 8\cdot 7}{4\cdot 3\cdot 2\cdot 1} = 210$ ways

b. $C(10,6) = \frac{P(10,6)}{6!} = \frac{10\cdot 9\cdot 8\cdot 7\cdot 6\cdot 5}{6\cdot 5\cdot 4\cdot 3\cdot 2\cdot 1} = 210$ ways

c. They are the same because taking four sweaters is the same as leaving 6 sweaters.

49. $C(8, 2) = \frac{P(8, 2)}{2!} = \frac{8\cdot 7}{2\cdot 1} = 28$ games

51. $26C(69, 5) = \frac{26P(69, 5)}{5!} = \frac{26\cdot 69\cdot 68\cdot 67\cdot 66\cdot 65}{5\cdot 4\cdot 3\cdot 2\cdot 1} = 292,201,338$ outcomes

53. $\frac{C(59,6)}{C(49,6)} = \frac{45,057,474}{13,983,816} \approx 3.22$ Choice (b)

55. Moe: $C(10, 3) = \frac{P(10, 3)}{3!} = \frac{10\cdot 9\cdot 8}{3\cdot 2\cdot 1} = 120$ choices

Joe: $C(7,4) = \frac{P(7,4)}{4!} = \frac{7\cdot 6\cdot 5\cdot 4}{4\cdot 3\cdot 2\cdot 1} = 35$ choices

Thus Joe is correct.

57. $4!\, P(4,3)\cdot P(5,3)\cdot P(6,3)\cdot P(7,3)$

$24\cdot 24\cdot 60\cdot 120\cdot 210 = 870,912,000$ pictures

59. $3!\ 3!\ 3!\ 3!$

$6\cdot 6\cdot 6\cdot 6 = 1296$ ways

61. Through trial are error, you will find that 10 people were at the party.

63. $C(15,3) + (15\cdot 14) + 15 = 455 + 210 + 15 = 680$ side dish options

65. $720 - 3! - 5!$

$720 - 6 - 120 = 594$

67. **a.** $C(45,5) = 1,221,759$ possible lottery tickets

b. $C(100,4) = 3,921,225$ possible lottery tickets

c. The first lottery has a better chance of winning

69. Yes; the number of ways to shuffle a deck of cards is $52! \approx 8\times 10^{67}$.

Copyright © 2018 Pearson Education, Inc.

Exercises 5.6

1. **a.** $2^8 = 256$ outcomes

   b. $C(8, 4) = \dfrac{8\cdot 7\cdot 6\cdot 5}{4\cdot 3\cdot 2\cdot 1} = 70$ outcomes

3. a. $C(7,5)+C(7,6)+C(7,7) = 21+7+1 = 29$ outcomes

   b. $2^7 - 29 = 128 - 29 = 99$ outcomes

5. **a.** $2C(6,3) = \dfrac{2\cdot 6\cdot 5\cdot 4}{3\cdot 2\cdot 1} = 40$ ways

   b. $2C(5,3) = \dfrac{2\cdot 5\cdot 4\cdot 3}{3\cdot 2\cdot 1} = 20$ ways

7. $C(11,5)\cdot C(6,5)\cdot C(1,1) = 462\cdot 6\cdot 1 = 2772$ ways

9. $C(8,5) = \dfrac{8\cdot 7\cdot 6\cdot 5\cdot 4}{5\cdot 4\cdot 3\cdot 2\cdot 1} = 56$ ways

11. $C(7,2) = \dfrac{7\cdot 6}{2\cdot 1} = 21$ ways

13. $C(5,2)\cdot C(4,2) = 10\cdot 6 = 60$ ways

15. $C(6,2) = \dfrac{6\cdot 5}{2\cdot 1} = 15$ ways

17. **c.** The two points where the combinations stop are the two points that make up the intersection B. Therefore, the sum of these two points will be the single combination.

    d. $C(8,3)+C(8,4) = C(9,4)$
    $56+70 = 126$

19. $C(8,3) = 56$ outcomes

21. $C(10,6) - C(7,4) = 210 - 35 = 175$ ways

23. **a.** $C(12,5) = 792$ samples

    b. $C(7,5) = 21$ samples

    c. $C(7,2)\cdot C(5,3) = 21\cdot 10 = 210$ samples

    d. $C(7,4)\cdot C(5,1) + 21 = 35\cdot 5 + 21 = 196$ samples

25. **a.** $C(10,3) = 120$ ways

    b. $C(8,3) = 56$ ways

    c. $120 - 56 = 64$

27. $C(4,2)\cdot C(6,2) = 6\cdot 15 = 90$ ways

29. $C(4,3)\cdot C(4,2) = 4\cdot 6 = 24$ ways

31. $13C(4,3)\cdot 12C(4,2) = 13\cdot 4\cdot 12\cdot 6 = 3744$ ways

33. $C(7,5)\cdot 5!\ \cdot 21\cdot 20 = 21\cdot 120\cdot 21\cdot 20 = 1{,}058{,}400$ ways

35. $C(10,5)\cdot P(21,5) = 252\cdot 2{,}441{,}880 = 615{,}353{,}760$ ways

37. $6!\ \cdot 7\cdot 3! = 720\cdot 7\cdot 6 = 30{,}240$ ways

39. $C(9,5) = 126$ ways

41. $C(26,22)\cdot C(10,7) = 14{,}950\cdot 120 = 1{,}794{,}000$ ways

43. $C(12,6) = 924$ ways

45. $\dfrac{C(100,50)}{2^{100}} \approx 0.07959 = 7.96\%$

47. $\dfrac{C(50,10)\cdot C(50,10)}{C(100,20)} \approx 0.1969 = 19.7\%$

Exercises 5.7

1. $\dbinom{18}{16} = C(18,\ 16) = C(18,\ 2) = \dfrac{18\cdot 17}{2\cdot 1} = 153$

Copyright © 2018 Pearson Education, Inc.

3. $\binom{6}{2} = C(6, 2) = \frac{6 \cdot 5}{2 \cdot 1} = 15$

5. $\binom{8}{1} = C(8, 1) = \frac{8}{1} = 8$

7. $\binom{7}{0} = C(7, 0) = 1$

9. $\binom{8}{8} = C(8, 8) = 1$

11. $\binom{n}{n-1} = C(n, n-1) = C(n, 1) = \frac{n}{1} = n$

13. $0! = 1$

15. $n \cdot (n-1)! = n!$

17. $\binom{6}{0} + \binom{6}{1} + \binom{6}{2} + \binom{6}{3} + \binom{6}{4} + \binom{6}{5} + \binom{6}{6}$
$= 2^6$
$= 64$

19. The number of terms in a binomial expansion is the exponent plus 1, so there are 20 terms.

21. $\binom{10}{0}x^{10} + \binom{10}{1}x^9 y + \binom{10}{2}x^8 y^2$
$= x^{10} + 10x^9 y + 45x^8 y^2$

23. $\binom{15}{13}x^2 y^{13} + \binom{15}{14}xy^{14} + \binom{15}{15}y^{15}$
$= 105x^2 y^{13} + 15xy^{14} + y^{15}$

25. $\binom{20}{10}x^{10} y^{10} = 184,756x^{10} y^{10}$

27. $\binom{4}{2} = C(4,2) = 6$

29. $\binom{11}{7} = 330$

31. $\binom{9}{0}x^9 + \binom{9}{1}x^8(2y) + \binom{9}{2}x^7(2y)^2$
$= x^9 + 9x^8(2y) + 36x^7(4y^2)$
$= x^9 + 18x^8 y + 144x^7 y^2$

33. $\binom{12}{6}x^6(-3y)^6 = 924x^6(729y^6)$
$= 673,596x^6 y^6$

35. $\binom{7}{3}x^4(-3y)^3 = 35x^4(-27y^3) = -945x^4 y^3$

37. $2^8 = 256$ subsets

39. $2^4 = 16$ tips

41. $2^5 = 32$ options (The number of subsets of any size taken from a set of five elements.)

43. $2^8 - 1 = 255$ ways

45. $2 \cdot 3 \cdot 2^{13} = 49,152$ types

47. $2^7 - C(7, 6) - C(7, 7) = 128 - 7 - 1$
$= 120$ ways

49. $2^8 - C(8, 0) - C(8, 1) = 256 - 1 - 8$
$= 247$ ways

51.

$\left(2^9 - C(9,1) - C(9,0)\right)\left(2^{10} - C(10,1) - C(10,0)\right)$
$= (512 - 9 - 1)(1024 - 10 - 1)$
$= (502)(1013)$
$= 508,526$ ways

53. No, the exponents on the variables add to 7, but the combination portion of the terms shows that they should add to 8.

55. $\binom{5}{0} + \binom{5}{2} + \binom{5}{4} = \binom{5}{1} + \binom{5}{3} + \binom{5}{5}$
$1 + 10 + 5 = 5 + 10 + 1$
$16 = 16$

Copyright © 2018 Pearson Education, Inc.

57. $2^{10} - 2 = 1022$

Exercises 5.8

1. $\frac{5!}{3!1!1!} = 20$

3. $\frac{6!}{2!1!2!1!} = 180$

5. $\frac{7!}{3!2!2!} = 210$

7. $\frac{12!}{4!4!4!} = 34,650$

9. $\frac{12!}{5!3!2!2!} = 166,320$

11. $\frac{1}{5!} \cdot \frac{15!}{(3!)^5} = 1,401,400$

13. $\frac{1}{3!} \cdot \frac{18!}{(6!)^3} = 2,858,856$

15. $\binom{20}{7,\ 5,\ 8} = \frac{20!}{7!5!8!} = 99,768,240$ reports

17. $\binom{8}{2,1,4,1} = \frac{8!}{2!1!4!1!} = 840$ words

19. $\binom{9}{3,2,4} = \frac{9!}{3!2!4!} = 1260$ ways

21. $\frac{1}{5!} \cdot \frac{20!}{(4!)^5} = 2,546,168,625$ ways

23. $\binom{30}{10,\ 2,\ 18} = \frac{30!}{10!2!18!}$
$= 5,708,552,850$ ways

25. $\binom{4}{1,\ 1,\ 2} = \frac{4!}{1!1!2!} = 12$ ways

27. $\frac{1}{7!} \cdot \frac{14!}{(2!)^7} = 135,135$ ways

29. The number of ways ten students are to be divided into two five member teams for a basketball game is $\frac{10!}{(2)!(5!)^2} = 126.$

31. $\binom{n}{1,1,\cdots,1} = \frac{n!}{1!1!\cdots 1!} = n!$

33. $\binom{38}{10,\ 12,\ 10,\ 6}$
$= \frac{38!}{10!12!10!6!}$
$= 115,166,175,166,136,334,240$ ways

35. $\binom{52}{13,13,13,13} = \frac{52!}{13!13!13!13!}$
$\approx 5.4 \times 10^{28}$

Therefore, there are more than one octillion.

Chapter 5 Review Exercises

1. $\varnothing, \{a\}, \{b\}, \{a, b\}$

2. $(S \cup T')' = S' \cap T$

3. $C(16,\ 2) = \frac{16!}{2!14!} = 120$ possibilities

4. $2 \cdot 5! = 240$ ways

5.

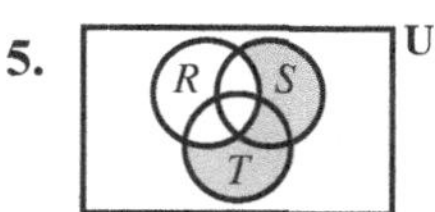

6. $\binom{12}{0}x^{12} + \binom{12}{1}x^{11}(-2y) + \binom{12}{2}x^{10}(-2y)^2$
$= x^{12} - 24x^{11}y + 264x^{10}y^2$

7. $C(8,\ 3) \cdot C(6,\ 2) = \frac{8!}{3!5!} \cdot \frac{6!}{2!4!} = 56 \cdot 15 = 840$

 Copyright © 2018 Pearson Education, Inc.

8. Let U = {people given pills}, P = {people who received placebos}, I = {people who showed improvement}.

$n(U) = 60$; $n(P) = 15$; $n(I) = 40$; $n(P' \cap I) = 30$

Draw and complete a Venn diagram as shown.

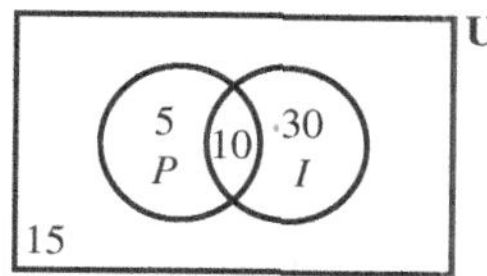

$n(P' \cap I') = 15$

Fifteen of the people who received the drug showed no improvement.

9. $7 \cdot 5 = 35$ combinations

10. $\begin{pmatrix} 12 \\ 2,\ 4,\ 6 \end{pmatrix} = \dfrac{12!}{2!4!6!} = 13{,}860$

11. Let U = {applicants}, F = {applicants who speak French}, S = {applicants who speak Spanish}, and G = {applicants who speak German}.

$n(U) = 115$; $n(F) = 70$; $n(S) = 65$; $n(G) = 65$; $n(F \cap S) = 45$; $n(S \cap G) = 35$; $n(F \cap G) = 40$;

$n(F \cap S \cap G) = 35$. Draw and complete a Venn diagram.

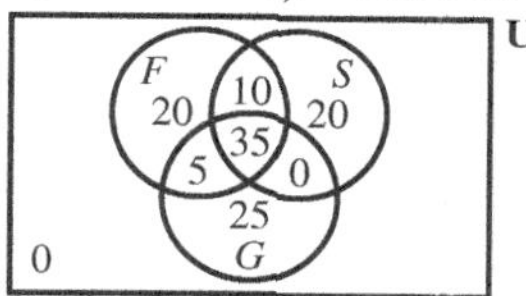

$n((F \cup S \cup G)') = 0$

None of the people speak none of the three languages.

12. $\begin{pmatrix} 17 \\ 15 \end{pmatrix} = \begin{pmatrix} 17 \\ 2 \end{pmatrix} = \dfrac{17 \cdot 16}{2 \cdot 1} = 136$

For Exercises 13–20, let U = {members of the Earth Club}, W = {members who thought the priority is clean water}, A = {members who thought the priority is clean air}, R = {members who thought the priority is recycling}. Then $n(U) = 100$, $n(W) = 45$, $n(A) = 30$, $n(R) = 42$, $n(W \cap A) = 13$, $n(A \cap R) = 20$, $n(W \cap R) = 16$, $n(W \cap A \cap R) = 9$. Draw and complete the Venn diagram as follows.

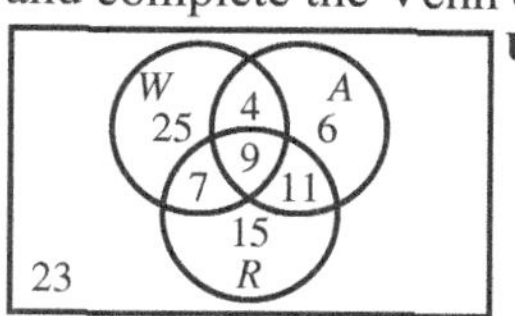

13. $n(A \cap W' \cap R') = 6$

14. $n((W \cap A') \cup (W' \cap A)) = (25 + 7) + (6 + 11) = 32 + 17 = 49$

Copyright © 2018 Pearson Education, Inc.

15. $n((W \cup R) \cap A') = 25 + 15 + 7 = 47$

16. $n(A \cap R \cap W') = 11$

17. $n((W \cap A' \cap R') \cup (W' \cap A \cap R') \cup (W' \cap A' \cap R)) = 25 + 6 + 15 = 46$

18. $n(R') = 23 + 25 + 6 + 4 = 58$

19. $n(R \cap A') = 15 + 7 = 22$

20. $n((W \cup A \cup R)') = 23$

21. $C(9, 4) = C(9, 5) = 126$

22. $2^{20} = 1{,}048{,}576$ ways

23. Let S = {students who ski}, H = {students who play ice hockey}. Then

$$\begin{aligned} n(S \cup H) &= n(S) + n(H) - n(S \cap H) \\ &= 400 + 300 - 150 \\ &= 550. \end{aligned}$$

24. $6 \cdot 10 \cdot 8 = 480$ meals

25. $\begin{pmatrix} 5 \\ 1,\ 3,\ 1 \end{pmatrix} = \frac{5!}{1!3!1!} = 20$ ways

26. The first digit can be anything but 0, the hundreds digit must be 3, and the last digit must be even.

$9 \cdot 1 \cdot 5 \cdot 10^4 = 450{,}000$

27. $9^2 \cdot 10^8 = 8{,}100{,}000{,}000$

28. $P(7,3) = 7 \cdot 6 \cdot 5 = 210$ ways

29. Strings of length 8 formed from the symbols *a*, *b*, *c*, *d*, *e*:

$5^8 = 390{,}625$ strings

Strings of length 8 formed from the symbols *a*, *b*, *c*, *d*:

$4^8 = 65{,}536$ strings

Strings with at least one *e*:

390,625 – 65,536 = 325,089 strings

30. $C(12, 5) = \frac{12!}{5!7!} = 792$

31. $C(30, 14) = \frac{30!}{14!16!}$
$= 145{,}422{,}675$ groups

32. U = {households}
F = {households that get *Fancy Diet Magazine*}
C = {households that get *Clean Living Journal*}

$$\begin{aligned} n(F \cup C) &= n(F) + n(C) - n(F \cap C) \\ &= 4000 + 10{,}000 - 1500 \\ &= 12{,}500 \end{aligned}$$

$$\begin{aligned} n((F \cup C)') &= n(U) - n(F \cup C) \\ &= 40{,}000 - 12{,}500 \\ &= 27{,}500 \end{aligned}$$

27,500 households get neither.

33. $3^{10} = 59{,}049$ paths

34. $C(60, 10) = \frac{60!}{10!50!} = 75{,}394{,}027{,}566$ ways

35. $5^{10} = 9{,}765{,}625$ tests

36. $21^6 = 85{,}766{,}121$ strings

37. $C(10, 4) = \frac{10!}{4!6!} = 210$ ways

38. $\frac{1}{3!} \cdot \frac{21!}{(7!)^3} = 66{,}512{,}160$ ways

39. $14! = 87{,}178{,}291{,}200$ ways

40. $\frac{1}{4!} \cdot \frac{20!}{(5!)^4} = 488{,}864{,}376$ ways

41. $3 \cdot 5 \cdot 4 = 60$ ways

42. $3 \cdot C(10, 2) = 135$

Copyright © 2018 Pearson Education, Inc.

43. A diagonal corresponds to a pair of vertices, except that adjacent pairs must be excluded. Hence there are

$$C(n, 2) - n = \frac{n(n-1)}{2} - n = \frac{n(n-3)}{2} \text{ diagonals.}$$

44. $8 \cdot 6 = 48$

45. $5! \cdot 4! \cdot 3! \cdot 2! \cdot 1! = 120 \cdot 24 \cdot 6 \cdot 2 \cdot 1 = 34{,}560$

46. $P(12, 5) = 12 \cdot 11 \cdot 10 \cdot 9 \cdot 8 = 95{,}040$

47. There are four suits. Once a suit is chosen, select 5 of the 13 cards.
Poker hands with cards of the same suit:
$4 \cdot C(13, 5) = 4 \cdot 1287 = 5148$ hands

48. $C(4, 3) \cdot C(48, 2) = 4 \cdot 1128 = 4512$ hands

49. Exactly the first two digits alike: $9 \cdot 1 \cdot 9 = 81$
First and last digit alike: $9 \cdot 9 \cdot 1 = 81$
Last two digits alike: $9 \cdot 9 \cdot 1 = 81$
Numbers with exactly two digits alike:
$81 + 81 + 81 = 243$

50. $9 \cdot 9 \cdot 8 = 648$ numbers

51. $3 \cdot 3 = 9$ pairings

52. $6 \cdot 3 \cdot 2 \cdot 2 \cdot 1 \cdot 1 = 72$ ways

53.
$$\begin{aligned} 24 \cdot 23 + 24 \cdot 23 \cdot 22 &= 552 + 12{,}144 \\ &= 12{,}696 \text{ names} \end{aligned}$$

54. $3 \cdot 4! \cdot 2 = 3 \cdot 24 \cdot 2 = 144$ ways

55. Any two lines will intersect, so the number of intersections is $C(10, 2) = 45$.

56. First teacher: $\frac{1}{4!} \cdot \frac{24!}{(6!)^4} = 96{,}197{,}645{,}544$ ways

Second teacher:
$\frac{1}{6!} \cdot \frac{24!}{(4!)^6} = 4{,}509{,}264{,}634{,}875$ ways
The second teacher has more options.

57. $\binom{10}{3,\ 4,\ 3} = \frac{10!}{3!4!3!} = 4200$

58. Let n be the number of books.
$n! = 120$, so $n = 5$. There are five books.

59. $7 \cdot 6 \cdot 5 \cdot 1 \cdot 4 \cdot 3 \cdot 2 \cdot 1 \cdot 1 = 5040$ orders

60. **a.** $C(6, 2) \cdot C(6, 3) = 15 \cdot 20 = 300$

b. Select five out of twelve states. Then determine the junior or senior senator from each state.
$C(12, 5) \cdot 2^5 = 25{,}344$

61. $C(7, 2) + C(7, 1) + C(7, 0) = 29$ ways

62. There are two choices for every row, so the number of paths from the A at the top to the final A is $2^6 = 64$ ways.

63. There are $2 \cdot 26^2$ 3-letter call letters and $2 \cdot 26^3$ 4-letter call letters, so $2 \cdot 26^2 + 2 \cdot 26^3 = 36{,}504$ call letters are possible.

64. Find n such that $n! = 479{,}001{,}600$.
Testing numbers on computer, $n = 12$.

65. $\binom{25}{10,\ 9,\ 6} = 16{,}360{,}143{,}800$ ways

66. First choose the three letters. There are $C(26, 3)$ ways to do this. Then choose the three numbers. There are $C(10, 3)$ ways to do this. Then order the six. There are 6! ways to do this.
The number of license plates is
$C(26, 3) \cdot C(10, 3) \cdot 6! = 224{,}640{,}000$.

Conceptual Exercises

67. If $A \cap B = \varnothing$ then A and B have no elements in common.

68. If $n(A \cup B) = n(A) + n(B)$, then A and B are disjoint sets and have no elements in common.

69. The intersection of sets S and T will be the same as set T when all the elements of T are also in set S, in symbols, when $T \subseteq S$.

70. The union of sets S and T will be the same as set T when all the elements of S are also in set T, in symbols, when $S \subseteq T$.

71. True;
$$n(S \cup T) = n(S) + n(T) - n(S \cap T)$$
$$n(S \cup T) + n(S \cap T) = n(S) + n(T)$$

72. True; the empty set is a subset of every set.

Copyright © 2018 Pearson Education, Inc.

73. $n(n-1)! = n(n-1)(n-2)(n-3)...1 = n!$
For example,
$5 \cdot 4! = 5(4 \cdot 3 \cdot 2 \cdot 1) = 5 \cdot 4 \cdot 3 \cdot 2 \cdot 1 = 5!$.

74. Let $n = 1$. We have $1(1-1)! = 1(0)! = 1! = 1$ So, since 1 times any number is the number, we have $0! = 1! = 1$.

75. A permutation is an arrangement of items in which order is important. A combination is a subset of objects taken from a larger set in which the order of the objects does not make a difference.

76. The number of committees of size 6 is the same as the number of committees of size 4. For each committee of size 6, there is a committee consisting of the four people who were not chosen. For example, use letters A, B,C, D, E, F, G, H, I, J to represent the people. For the committee $\{A, B, C, D, E, F\}$ the complementary committee is $\{G, H, I, J\}$.

77. $C(10, 3) = C(10, 7)$. The number of subsets of size 3 taken from a 10 member set is the same as the number of subsets of size 7. For example, consider the set $N = \{1, 2, 3, 4, 5, 6, 7, 8, 9, 10\}$. For the subset $A = \{1, 2, 3\}$, there is the corresponding subset $B = \{4, 5, 6, 7, 8, 9, 10\}$.

78. A committee of size 5 either includes or excludes John Doe. If the committee includes him, there are $C(10, 4)$ ways to choose the other 4 members. If the committee excludes him, there are $C(10, 5)$ ways to choose the 5 members. Hence $C(10, 4) + C(10, 5) = C(11, 5)$.

 Copyright © 2018 Pearson Education, Inc.

Chapter 6

Exercises 6.1

1. a. The set of all possible pairs: {RS, RT, RU, RV, ST, SU, SV, TU, TV, UV}

b. The set of pairs containing R: {RS, RT, RU, RV}

c. The set of pairs containing neither R nor S: {TU, TV, UV}

3. a. {HH, HT, TH, TT}

b. {HH, HT}

5. a. {(I, red), (I, white), (II, red), (II, white)}

b. All combinations with I: {(I, red), (I, white)}

7. a. $S = \{\text{all positive numbers of minutes}\}$

b. $E \cap F =$ "more than 5 but less than 8 minutes"
$E \cap G = \varnothing$ (There's no time longer than 5 minutes but less than 4 minutes.)
$E' =$ "5 minutes or less"
$F' =$ "8 minutes or more"
$E' \cap F = E' =$ "5 minutes or less"
$E' \cap F \cap G = G =$ "less than 4 minutes"
$E \cup F = S$

9. a. Eight possible combinations: {(Fr, Lib), (Fr, Con), (So, Lib), (So, Con), (Jr, Lib), (Jr, Con), (Sr, Lib), (Sr, Con)}

b. All combinations with Con: {(Fr, Con), (So, Con), (Jr, Con), (Sr, Con)}

c. {(Jr, Lib)}

d. All combinations with neither Fr nor Con: {(So, Lib), (Jr, Lib), (Sr, Lib)}

11. a. No; $E \cap F = \{2\}$

b. Yes; $F \cap G = \varnothing$

13. All combinations of members of S: $\varnothing, \{a\}, \{b\}, \{c\}, \{a, b\}, \{a, c\}, \{b, c\}, S$

15. Yes; $(E \cup F) \cap (E' \cap F') = \{1, 2, 3\} \cap \{4\} = \varnothing$

17. a. {0, 1, 2, 3, 4, 5, 6, 7, 8, 9, 10}

b. More than half heads: {6, 7, 8, 9, 10}

19. a. No; there are blue-eyed people at least 18 years old.

b. Yes; a brown-eyed person younger than 18 doesn't have blue eyes.

c. Yes; a brown-eyed person younger than 18

21. {0, 1, 2, 3, 4, 5, 6, 7, 8}

23. A possible outcome is (7, 4). There are $9 \cdot 9 = 81$ outcomes in the sample space.

25. A possible outcome is {2, 6, 9, 10}. There are $C(14, 4) - 1 = 1000$ possible combinations, so $\frac{250}{1000} = 0.25 = 25\%$ were assigned to the Minnesota Timberwolves.

27. A sample space for the choice of murderers would be {Colonel Mustard, Miss Scarlet, Professor Plum, Mrs. White, Mr. Green, Mrs. Peacock}

A sample space for the entire solution would be formed by placing each suspect, with each murder weapon, in each room.

a. 6 suspects × 6 weapons × 9 rooms = 324 outcomes

b. $E \cap F =$ "The murder occurred in the library with a gun."

c. $E \cup F =$ "Either the murder occurred in the library, or it was done with a gun."

Copyright © 2018 Pearson Education, Inc.

Exercises 6.2

1. Judgemental; the probability is an opinion.

3. Logical; the probability is based on theory.

5. The probability distribution is as follows:

Number of Heads	Probability
0	$\frac{1}{4}$
1	$\frac{2}{4}=\frac{1}{2}$
2	$\frac{1}{4}$

7. $\frac{1}{38}+\frac{1}{38}=\frac{2}{38}=\frac{1}{19}$

9. a. $\frac{191}{4487}\approx 0.04257$

b. $\frac{191+81}{4487}=\frac{272}{4487}\approx 0.06062$

c. $\frac{4487-272}{4487}=\frac{4215}{4487}\approx 0.9394$

11. a. $\frac{6}{26}=\frac{3}{13}\approx 0.2308$

b. $\frac{5}{26}\approx 0.1923$

c. $\frac{6+5-2}{26}=\frac{9}{26}\approx 0.3462$

13. a. E = "the numbers add up to 9" = {(3, 6), (4, 5), (5, 4), (6, 3)}

$\Pr(E)=\frac{4}{36}=\frac{1}{9}\approx 0.1111$

b. $\Pr(\text{sum is }2)=\Pr((1,1))=\frac{1}{36}$;

$\Pr(\text{sum is }3)=\Pr((1,2))+\Pr((2,1))=\frac{2}{36}$

$\Pr(\text{sum is }4)=\Pr((1,3))+\Pr((2,2))+\Pr((3,1))=\frac{3}{36}$

The probability that the sum is less than 5 is $\frac{1}{36}+\frac{2}{36}+\frac{3}{36}=\frac{1}{6}\approx 0.1667$.

15.

Kind of High School	Probability
Public	$\frac{115620}{141000}=0.820$
Private	$\frac{24252}{141000}=0.172$
Home School	$\frac{1128}{141000}=0.008$

17. $\Pr(B, C, \text{ or } D)=0.34+0.21+0.09=0.64$

19. a. $\Pr(E)=0.1+0.5=0.6$;
$\Pr(F)=0.5+0.2=0.7$

b. $\Pr(E')=1-0.6=0.4$

c. $\Pr(E\cap F)=0.5$

d. $\Pr(E\cup F)=0.6+0.7-0.5=0.8$

21. a.

Number of Colleges Applied to	Probability
1	0.10
2	$0.17-0.10=0.07$
3	$0.27-0.17=0.10$
4	$0.40-0.27=0.13$
≥ 5	$1-0.40=0.60$

Copyright © 2018 Pearson Education, Inc.

b. $0.10+0.13+0.60=0.83$

23. a. No; The probabilities add to more than 1.

b. No; One of the probabilities is a negative value.

25. $1-\left(\frac{2}{3}+\frac{1}{4}\right)=\frac{1}{12}$

27. Pr(Liberal) = .28; Pr(middle) = $2x$; Pr(Conseravtive) = x

$$0.28+2x+x=1$$
$$3x=0.72$$
$$x=0.24$$

29. The probability is 1 because the sum of a pair of dice must be odd or even.

31. $\Pr(E\cup F)=\Pr(E)+\Pr(F)$
$=0.4+0.5$
$=0.9$

33. $\Pr(S)=0.05+0.40=0.45$

35. $\Pr(T \text{ only})=0.25$

37.

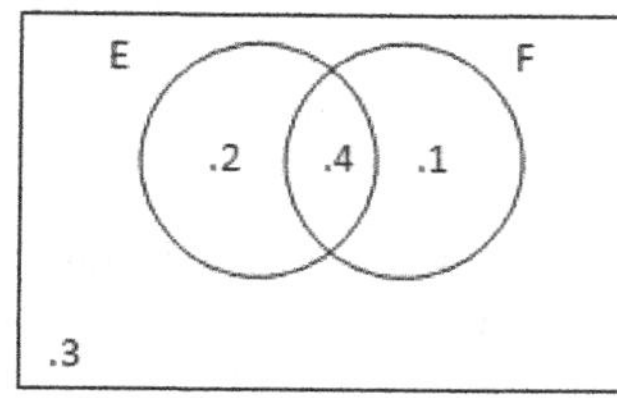

a. $\Pr(E\cup F)=0.2+0.4+0.1=0.7$

b. $\Pr(E\cap F')=0.2$

39. $\Pr(H\cap P)=\Pr(H)+\Pr(P)-\Pr(H\cup P)$
$=0.7+0.8-0.9$
$=0.6$

41. $10 \text{ to } 1=\frac{10}{10+1}=\frac{10}{11}$

43. $.2=\frac{1}{5}\Rightarrow 1 \text{ to } (5-1)=1 \text{ to } 4$

45. $.3125=\frac{5}{16}\Rightarrow 5 \text{ to } (16-5)=5 \text{ to } 11$

47. $2 \text{ to } 9=\frac{2}{2+9}=\frac{2}{11}$

49. a. Sparks: $5 \text{ to } 3=\frac{3}{5+3}=\frac{3}{8}$

Meteors: $3 \text{ to } 1=\frac{1}{3+1}=\frac{1}{4}$

Asteroids: $3 \text{ to } 2=\frac{2}{3+2}=\frac{2}{5}$

Suns: $4 \text{ to } 1=\frac{1}{4+1}=\frac{1}{5}$

b. $\frac{3}{8}+\frac{1}{4}+\frac{2}{5}+\frac{1}{5}=\frac{15}{40}+\frac{10}{40}+\frac{16}{40}+\frac{8}{40}$
$=\frac{49}{40}$

c. Bookies have to make a living. The payoffs are a little lower than they should be; thus allowing the bookie to make a profit.

51. There are more members (13) than Zodiac signs (12) so two or more members will always have the same Zodiac sign; thus the probability is 1.

Exercises 6.3

1. a. $\frac{9}{17}\approx 0.5294$

b. $\frac{8}{17}\approx 0.4706$

c. $\Pr(\{3, 6, 9, 12, 15\})=\frac{5}{17}\approx 0.2941$

d. $\Pr(\{1,3,5,6,7,9,11,12,13,15,17\})=\frac{11}{17}$
≈ 0.6471

Copyright © 2018 Pearson Education, Inc.

3. a. $\frac{C(5, 2)}{C(11, 2)} = \frac{10}{55} = \frac{2}{11} \approx 0.1818$

b. $1 - \frac{C(5, 2)}{C(11, 2)} = 1 - \frac{10}{55} = \frac{9}{11} \approx 0.8182$

5. a. $\frac{C(6,4) + C(7,4)}{C(13,4)} = \frac{15 + 35}{715}$

$= \frac{50}{715} = \frac{10}{143} \approx 0.0699$

b. $\frac{C(6,4) + C(6,3)C(7,1)}{C(13,4)} = \frac{15 + 20 \cdot 7}{715}$

$= \frac{15 + 140}{715}$

$= \frac{155}{715}$

$= \frac{31}{143} \approx 0.2168$

7. $1 - \frac{C(5, 3)}{C(7, 3)} = 1 - \frac{10}{35}$

$= \frac{5}{7} \approx 0.7143$

9. $1 - \frac{C(9, 3)}{C(13, 3)} = 1 - \frac{84}{286} = \frac{101}{143} \approx 0.7063$

11. $1 - \frac{C(4, 3)}{C(10, 3)} = 1 - \frac{4}{120} = \frac{29}{30} \approx 0.9667$

13. $\frac{C(10, 7)}{C(22, 7)} = \frac{5}{7106} \approx 0.0007$

15. Ways for no girls to be chosen: $C(12, 7)$
Ways for exactly 1 girl to be chosen:
$C(12, 6) \times C(10, 1)$

$1 - \frac{C(12, 7) + C(12, 6) \times C(10, 1)}{C(22, 7)}$

$= 1 - \frac{792 + 924 \times 10}{170{,}544} = \frac{16}{17} \approx 0.9412$

17. $1 - \frac{7 \cdot 6 \cdot 5}{7 \cdot 7 \cdot 7} = 1 - \frac{210}{343} = \frac{133}{343} \approx 0.3878$

19. $1 - \frac{30 \times 29 \times 28 \times 27}{30^4} = \frac{47}{250} = 0.188$

21. $1 - \frac{P(20,8)}{20^8} \approx 0.8016$

23. Pr(at least one birthday on June 13)

$= 1 - \left(\frac{364}{365}\right)^{25} \approx 0.06629$

Because in Table 1 no particular date is being matched. Any two (or more) identical birthdays count as a success.

25. $\frac{6 \cdot 5}{6^2} = \frac{30}{36} = \frac{5}{6} \approx 0.8333$

27. $\frac{3^4}{6^4} = \frac{81}{1296} = \frac{1}{16} \approx 0.0625$

29. $\frac{C(10, 4)}{2^{10}} = \frac{210}{1024} = \frac{105}{512} \approx 0.2051$

31. $1 - \frac{7 \times 6 \times 5 \times 4}{7^4} = \frac{223}{343} \approx 0.6501$

33. The tourist must travel 8 blocks of which 3 are south. Thus he has $C(8, 3) = 56$ ways to get to B from A.

a. To get from A to B through C there are $C(3, 1) \cdot C(5, 2) = 30$ ways.
The probability is $\frac{30}{56} = \frac{15}{28} \approx 0.5357$.

b. To get from A to B through D there are $C(5, 1) \cdot C(3, 2) = 15$ ways.
The probability is $\frac{15}{56} \approx 0.2679$.

c. To get from A to B through C and D there are $C(3, 1) \cdot C(2, 0) \cdot C(3, 2) = 9$ ways.
The probability is $\frac{9}{56} \approx 0.1607$.

d. The number of ways to get from A to B through C or D is $30 + 15 - 9 = 36$.
The probability is $\frac{36}{56} = \frac{9}{14} \approx 0.6429$.

Copyright © 2018 Pearson Education, Inc.

35. $$1-\frac{4\cdot4\cdot4}{5\cdot5\cdot5}=1-\frac{64}{125}=\frac{61}{125}\approx 0.488$$

37. $$1-\frac{12\cdot11\cdot10}{15\cdot14\cdot13}=1-\frac{1320}{2730}=\frac{1410}{2730}\approx 0.5165$$

His chances are increased.

39. $$\frac{5\cdot4\cdot3}{5^3}=\frac{60}{125}=\frac{12}{25}=0.48$$

41. There are 5! = 120 ways to arrange the family. First determine where the parents will stand. There are two ways with the man at one end. If the man doesn't stand at one end, there are $3\cdot2=6$ ways for the couple to stand together. Next determine the order of the children. There are 3! = 6 ways to order the children. Thus there are $(2+6)\cdot6=48$ ways to stand with the parents together.

The probability is $\frac{48}{120}=\frac{2}{5}=0.4.$

43. There are $13\cdot12=156$ ways to choose the two denominations, $C(4, 3)=4$ ways to choose the suits for the three of a kind and $C(4, 2)=6$ ways to choose the suits for the pair; thus there are $156\cdot4\cdot6=3744$ possible full house hands, so the probability is $\frac{3744}{C(52, 5)}\approx 0.0014.$

45. There are $C(13, 2)=78$ choices for the denominations of the two pairs, and $C(4, 2)=6$ choices each for their suits; the remaining card has $52-8=44$ choices, for a total of $78\cdot6^2\cdot44=123{,}552$ possible two pair hands, so the probability is $\frac{123{,}552}{C(52, 5)}\approx 0.0475.$

47. a. There are 4 ways to select the suit of the 4 card group, $C(13, 4)$ ways to select their denominations, and $C(13, 3)^3$ ways to select 3 cards each from the remaining 3 suits. The probability of a 4-3-3-3 bridge hand is

$$\frac{4\cdot C(13, 4)\cdot C(13, 3)^3}{C(52, 13)}\approx 0.1054.$$

b. There are $C(4, 2)=6$ ways to choose the suits of the 4-card groups and 2 ways to choose the suit of the 3-card group (then the suit of the 2-card group is uniquely determined). The denominations can then be chosen in $C(13, 4)^2\cdot C(13, 3)\cdot C(13, 2)$ ways, so the probability of a 4-4-3-2 bridge hand is

$$\frac{6\cdot2\cdot C(13, 4)^2\cdot C(13, 3)\cdot C(13, 2)}{C(52, 13)}\approx 0.2155.$$

49. $$\frac{2}{C(40, 6)}=\frac{1}{1{,}919{,}190}=0.0000005211$$

51. $$\frac{C(5,3)\cdot C(34,2)}{C(39, 5)}=\frac{10\cdot561}{575{,}757}=\frac{5610}{575{,}757}\approx 0.0097$$

53. $$\frac{25-20}{80}=\frac{5}{80}=\frac{1}{16}=0.0625=6.25\%$$
$$\frac{25-20}{25}=\frac{5}{25}=\frac{1}{5}=0.2=20\%$$

16 people will be needed because it helps 1 out of every 16 people.

55. $$\frac{20-15}{100}=\frac{5}{100}=\frac{1}{20}=0.05=5\%$$
$$\frac{20-15}{20}=\frac{5}{20}=\frac{1}{4}=0.25=25\%$$

20 people will be needed because it helps 1 out of every 20 people.

57. Let x represent the number of people in the control group that developed the condition.

$$\frac{x-12}{x}=\frac{0.25}{1}$$
$$0.25x=x-12$$
$$-0.75x=-12$$
$$x=16$$

59. $$1-\frac{P(100,15)}{100^{15}}\approx 0.6687$$

Copyright © 2018 Pearson Education, Inc.

61. The probability that no two people choose the same card is $\frac{P(52,\ n)}{52^n}$, so the probability that at least two people pick the same card is
$P_n = 1 - \frac{P(52,\ n)}{52^n}$.
For $n = 5$, $P_5 \approx 0.1797$.
$P_8 \approx 0.4324$ and $P_9 \approx 0.5197$.
P_n increases as n increases, so $n = 9$ is the smallest value of n for which $P_n > 0.5$.

63. The probability that one or more people in a group of size n were born on a given specific day is $P_n = 1 - \left(\frac{364}{365}\right)^n$.
For $n = 100$, $P_n \approx 0.240$.
$P_{252} \approx 0.4991$ and $P_{253} \approx 0.5005$, so $n = 253$ is the smallest value of n for which $P_n > 0.5$.

65. When $k = 6$, $1 - \frac{C(n-k+1, k)}{C(n, k)}$ drops below 50% starting when $n = 48$.

Exercises 6.4

1. a. $\Pr(E) = 0.3 + 0.2 = 0.5$

b. $\Pr(F) = 0.2 + 0.4 = 0.6$

c. $\Pr\left(E \middle| F\right) = \frac{0.2}{0.6} = 0.3333$

d. $\Pr\left(F \middle| E\right) = \frac{0.2}{0.5} = 0.4$

3. a. $\Pr\left(E \middle| F\right) = \frac{0.1}{0.4} = \frac{1}{4}$

b. $\Pr\left(F \middle| E\right) = \frac{0.1}{0.5} = \frac{1}{5}$

c. $\Pr\left(E \middle| F'\right) = \frac{0.4}{0.6} = \frac{2}{3}$

d. $\Pr\left(E' \middle| F'\right) = \frac{0.2}{0.6} = \frac{1}{3}$

5. a. $\Pr(E \cap F) = \frac{1}{3} + \frac{5}{12} - \frac{2}{3} = \frac{1}{12}$

b. $\Pr\left(E \middle| F\right) = \frac{\frac{1}{12}}{\frac{5}{12}} = \frac{1}{5}$

c. $\Pr\left(F \middle| E\right) = \frac{\frac{1}{12}}{\frac{1}{3}} = \frac{1}{4}$

7. a.

$$\Pr(F \mid E) = \frac{\Pr(E \cap F)}{\Pr(E)}$$

$$0.25 = \frac{\Pr(E \cap F)}{0.4}$$

$$\Pr(E \cap F) = 0.1$$

b. $\Pr(E \cup F) = 0.4 + 0.3 - 0.1 = 0.6$

c. $\Pr\left(E \middle| F\right) = \frac{0.1}{0.3} = \frac{1}{3}$

d. $\Pr(E' \cap F) = 0.3 - 0.1 = 0.2$

9. $\Pr(8 \mid \text{not } 7) = \frac{\Pr(8 \cap \text{not } 7)}{\Pr(\text{not } 7)}$

$\Pr(8 \mid \text{not } 7) = \frac{\frac{5}{36}}{\frac{30}{36}}$

$\Pr(8 \mid \text{not } 7) = \frac{5}{30} = \frac{1}{6}$

11. 0; because exactly one coin shows heads therefore there are two tails.

13. $\frac{\text{[number of outcomes that four are white]}}{\text{[number of outcomes that at least 1 is white]}}$

$$\frac{C(7,4)}{C(12,4) - C(5,4)} = \frac{35}{495 - 5} = \frac{35}{490} = \frac{1}{14} \approx 0.0714$$

15. $\Pr(\text{both girls}|\text{first girl}) = \frac{1}{2}$

 Copyright © 2018 Pearson Education, Inc.

17. $\text{Pr(grad}|\text{more \$45,000)}=\dfrac{\text{Pr(grad and >45000)}}{\text{Pr(> 45000)}}$

$=\dfrac{0.10}{0.25}$

$=\dfrac{2}{5}=0.4$

19. **a.** $\text{Pr(Masters)}=\dfrac{851}{2898}\approx 0.2937$

b. $\text{Pr(Male)}=\dfrac{1201}{2898}\approx 0.4144$

c. $\text{Pr(Female|Masters)}=\dfrac{522}{851}\approx 0.6134$

d. $\text{Pr(Doctors|Female)}=\dfrac{93}{1697}\approx 0.0548$

21. **a.** $\text{Pr(Officer)}=\dfrac{228.6}{1291.8}\approx 0.1770$

b. $\text{Pr(Marine)}=\dfrac{183.2}{1291.8}\approx 0.1418$

c. $\text{Pr(Officer and Marine)}=\dfrac{20.7}{1291.8}\approx 0.0160$

d. $\text{Pr(Officer|Marine)}=\dfrac{20.7}{183.2}\approx 0.1130$

e. $\text{Pr(Marine|Officer)}=\dfrac{20.7}{228.6}\approx 0.0906$

23. $\text{Pr(\$5|\$5)}=\dfrac{\frac{1}{3}}{\frac{1}{2}}=\dfrac{2}{3}$

25. $\dfrac{4}{52}\cdot\dfrac{3}{51}=\dfrac{12}{2652}=\dfrac{1}{221}\approx 0.004525$

27. $\dfrac{1}{2}$; because the flip of the coin the fifth time is independent of the first four times.

29. $\text{Pr(Kasich}|\text{Fem.)}=\dfrac{\text{Pr(Kasich and Fem.)}}{\text{Pr(Fem.)}}$

$0.09=\dfrac{\text{Pr(Kasich and Fem.)}}{0.48}$

$\text{Pr(Kasich and Fem.)}=0.0432$

31. $\text{Pr(Win with 2 point shot)}=0.48\cdot 0.5=0.24$

$\text{Pr(Win with 3 point shot)}=0.29$

Therefore, there is a better chance of winning if you take the three – point shot.

33. $\Pr(E\cap F)=0.4+0.5-0.7=0.2$

$\Pr(E\mid F)=\dfrac{0.2}{0.5}=0.4=\Pr(E)$

$\Pr(F\mid E)=\dfrac{0.2}{0.4}=0.5=\Pr(F)$

Therefore, the two events are independent.

35. $\Pr(E\cap F)=(0.5)(0.6)=0.3$

$\Pr(E\cup F)=0.5+0.6-0.3=0.8$

37. Since the events are independent, $\Pr(F\mid E)=\Pr(F)=0.6$

39. Since the events are independent,

$\Pr(F)=\dfrac{\Pr(E\cap F)}{1-\Pr(E')}=\dfrac{0.1}{1-0.6}=\dfrac{0.1}{0.4}=0.25$

41. Since the events are independent,

$\Pr[(A\cap B\cap C)']=1-(0.4)(0.1)(0.2)$

$=1-0.008$

$=0.992$

43. No; because the selection of the first ball affects the selection of the second ball.

45. Yes; because $\Pr(E\cap F)=\Pr(E)\cdot\Pr(F)$.

47. No; because $\Pr(E\cap F)\neq\Pr(E)\cdot\Pr(F)$.

49. No; because $\Pr(E\cap F)\neq\Pr(E)\cdot\Pr(F)$.

Copyright © 2018 Pearson Education, Inc.

51. a. Pr(pass all) = (0.80)(0.75)(0.60) = 0.36

b. Pr(pass first 2) = (0.80)(0.75)(0.40) = 0.24

Pr(pass 1st and 3rd) = (0.80)(0.25)(0.60) = 0.12

Pr(pass last 2) = (0.2)(0.75)(0.60) = 0.09

Pr(pass ≥ 2) = 0.36 + 0.24 + 0.12 + 0.09 = 0.81

53. $(0.99)^5(0.98)^5(0.975)^3 \approx 0.7967$

55. $(1-0.7)^4 = (0.3)^4 = 0.0081$

57. a. $1-(0.7)^4 = 1-0.2401 = 0.7599$

b. $(0.7599)^{10} \approx 0.06420$

c. $1-(0.9358)^{20} \approx 0.7347$

59. Scoring 0: $1-0.6 = 0.4$

Scoring 1: $(0.6)(1-0.6) = 0.24$

Scoring 2: $(0.6)(0.6) = 0.36$

61. Answer will vary.

63. $(0.6)(0.4) = 0.24$

$(0.4)(0.6) = 0.24$

Answers will vary

65. Answers will vary.

67. Answers will vary.

69. 26; $1-\left(\frac{37}{38}\right)^{26} \approx 0.5001$

Exercises 6.5

1.

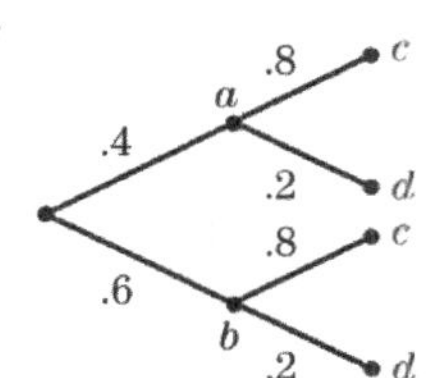

3.

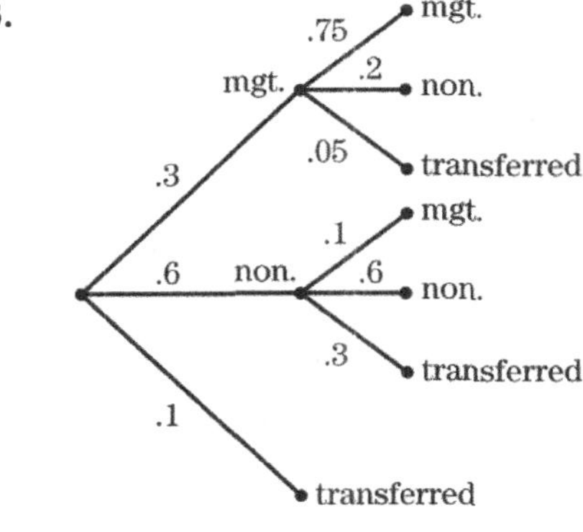

5. $0.30 \times 0.75 + 0.60 \times 0.10 = 0.285$

7. $0.10 + 0.30 \times 0.05 + 0.60 \times 0.30 = 0.295$

9. Pr(white, then red) + Pr(red, then red)

$= \frac{2}{3} \times \frac{1}{2} + \frac{1}{3} \times \frac{3}{4} = \frac{7}{12}$

11.

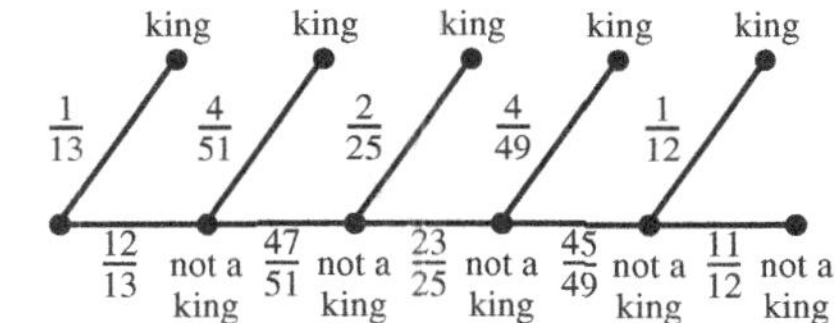

Pr(king on 1st draw) + Pr(king on 2nd draw)
+ Pr(king on 3rd draw)
= 1 – Pr(not a king on 3rd draw)

$= 1 - \frac{12}{13} \times \frac{47}{51} \times \frac{23}{25} = 1 - \frac{4324}{5525} = \frac{1201}{5525} \approx 0.22$

 Copyright © 2018 Pearson Education, Inc.

13.

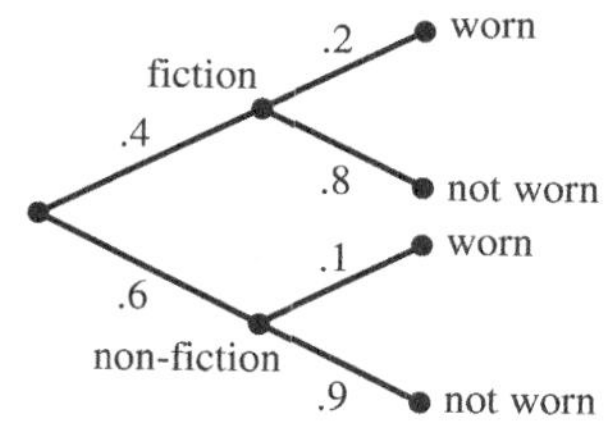

$0.40 \times 0.20 + 0.60 \times 0.10 = 0.14$

15. $\text{Pr(male|color-blind)} = \dfrac{\frac{1}{2}\times 0.08}{\frac{1}{2}\times 0.08 + \frac{1}{2}\times 0.005} = \dfrac{16}{17} \approx 0.9412$

17. $0.5 \times 0.9 + 0.5 \times 0.7 = 0.8$

19. $\Pr(\text{fake}|\text{HH}) = \dfrac{\frac{1}{4}\times 1}{\frac{3}{4}\times\frac{1}{4} + \frac{1}{4}\times 1} = \dfrac{4}{7}$

21. **a.** Pr(wins the point) $= 0.60 \times 0.75 + 0.40 \times 0.75 \times 0.50 = 0.60$

b. Pr(first serve good|wins service point) $= \dfrac{\Pr(\text{first serve good and wins service point})}{\Pr(\text{wins service point})} = \dfrac{0.60 \times 0.75}{0.60}$
$= 0.75$

23.

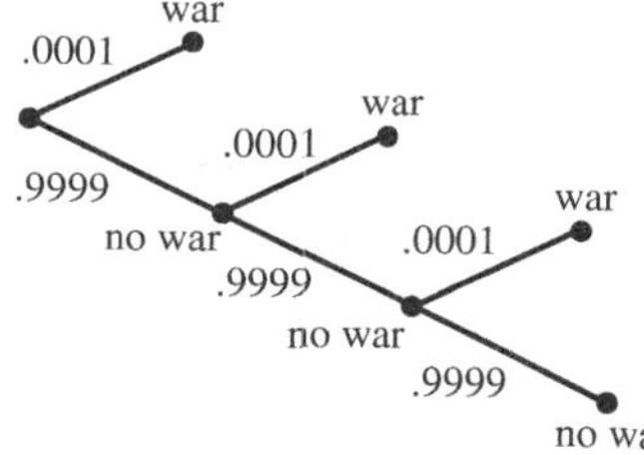

$0.0001 + 0.9999 \times 0.0001 + 0.9999^2 \times 0.0001$ or $1 - (0.9999)^3 \approx 0.00029997$

25.

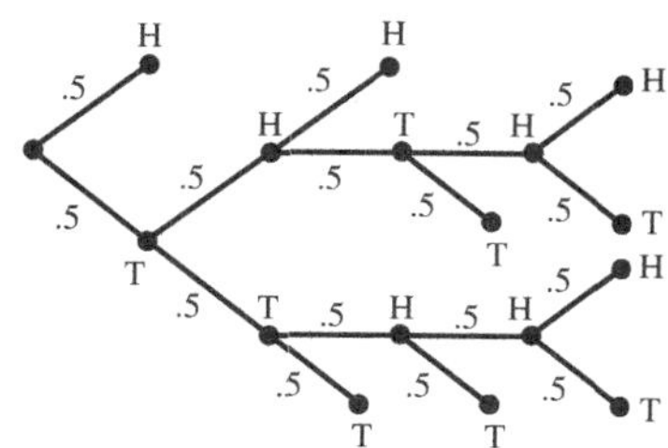

$.5 + (.5)^3 + 2(.5)^5 = \dfrac{11}{16}$

27. **a.** $\Pr(\text{white}) = \frac{1}{2} \times \frac{1}{2} = \frac{1}{4}$

$\Pr(\text{red}) = 1 - \frac{1}{4} = \frac{3}{4}$

b. $\Pr(\text{red}) = 0.6 \times 0.5 + 0.4 \times 1 = 0.7$

29. Pr(night|part-timer)

$$= \frac{0.60 \times 2}{0.40 \times 5 + .60 \times 2} = \frac{3}{8} = 0.375$$

31. $pr(W_2) = pr(R_1 \cap W_2) + pr(W_1 \cap W_2)$

$$= pr(R_1)pr(W_2|R_1) + pr(W_1)p(W_2|W_1)$$

$$= \left(\frac{5}{10}\right) \cdot \left(\frac{12}{13}\right) + \left(\frac{5}{10}\right) \cdot 1$$

$$= \frac{25}{26}$$

33. Pr(get same number of heads)
= Pr(2 heads) + Pr(1 head) + Pr(0 heads)

$$= \frac{1}{4} \cdot \frac{1}{4} + \frac{1}{2} \cdot \frac{1}{2} + \frac{1}{4} \cdot \frac{1}{4} = \frac{3}{8}$$

35. **a.** Since printer B produces 201 models 99 out of 100 weeks, Pr (printer B produces more models than printer A) = 0.99.

b. For printer A to produce more than printer B, printer B would have to break down. Since this only occurs 1% of the time, the probability that printer B produces more models than printer A is $0.99^{200} \approx 0.1340$.

37. a. $\Pr(\text{red die} > \text{blue die}) = \frac{1}{2} + \frac{1}{2} \cdot \frac{5}{6}$

$$= \frac{1}{2} + \frac{1}{12} = \frac{7}{12}$$

b. $\Pr(\text{blue die} > \text{green die}) = \frac{1}{6} + \frac{5}{6} \cdot \frac{1}{2}$

$$= \frac{1}{6} + \frac{5}{12} = \frac{7}{12}$$

c. $\Pr(\text{green die} > \text{red die}) = 0 + \frac{5}{6} \cdot \frac{5}{6}$

$$= 0 + \frac{25}{36} = \frac{25}{36}$$

d. Since the red die beats the blue die more than half the time and the blue die beats the green die more than half the time, the red die appears to be the strongest of the three dice and the green appears to be the weakest. However, the green die beats the red die more than half the time.

39. True; Sensitivity gives the percent of people that have a condition given the fact that they tested positive.

41. True; Based on the definition of specificity.

43. True; Sensitivity gives the percentage of people that have a given condition.

45. $\Pr(\text{Hep}|\text{Pos.}) = \frac{(0.0005)(0.95)}{(0.0005)(0.95) + (0.9995)(0.1)}$

$$= \frac{0.000475}{0.000475 + 0.09995}$$

$$= \frac{0.000475}{0.100425} \approx 0.00473$$

47. $\Pr(\text{Cond.}|\text{Positive}) = \frac{9}{19} \approx 0.474$

49. $\Pr(\text{Used}|\text{Positive}) = \frac{(0.05)(1)}{(0.05)(1) + (0.95)(.01)}$

$$= \frac{0.05}{0.05 + .0095}$$

$$= \frac{0.05}{0.0595} \approx 0.84 = 84\%$$

 Copyright © 2018 Pearson Education, Inc.

Exercises 6.6

1. $\Pr\left(\text{over } 60 \middle| \text{acc.}\right) = \dfrac{0.10\times 0.04}{(0.05\times 0.06)+(0.10\times 0.04)+(0.25\times 0.02)+(0.20\times 0.015)+(0.30\times 0.025)+(0.10\times 0.04)}$

$= \dfrac{0.004}{0.0265}$

$= \dfrac{8}{53}$

3. $\Pr\left(\text{sophomore} \middle| \text{A}\right) = \dfrac{0.30\times 0.4}{(0.10\times 0.2)+(0.30\times 0.4)+(0.40\times 0.3)+(0.20\times 0.1)} = \dfrac{0.12}{0.28} = \dfrac{3}{7}$

5. $\Pr\left(\geq \$75{,}000 \middle| 2 \text{ or more cars}\right) = \dfrac{0.05\times 0.9}{(0.10\times 0.2)+(0.20\times 0.5)+(0.35\times 0.6)+(0.30\times 0.75)+(0.05\times 0.9)}$

$= \dfrac{0.045}{0.6}$

$= \dfrac{3}{40}$

$= 0.075$

7. $\Pr\left(\text{passed exam} \middle| \text{A}\right) = \dfrac{0.80\times 0.40}{0.80\times 0.40 + 0.20\times 0.20}$

$= \dfrac{0.32}{0.36}$

$= \dfrac{8}{9}$

9. **a.** $(0.20\times 0.20)+(0.15\times 0.15)+(0.25\times 0.12)+(0.30\times 0.10)+(0.10\times 0.10) = 0.1325$

b. $\Pr\left(\text{division C} \middle| \text{bilingual}\right) = \dfrac{0.25\times 0.12}{0.1325} = \dfrac{0.03}{0.1325} = \dfrac{12}{53} \approx 0.23$

11. $\Pr\left(\text{cancer} \middle| \text{positive}\right) = \dfrac{\Pr(\text{cancer})\times \Pr\left(\text{positive} \middle| \text{cancer}\right)}{\Pr(\text{cancer})\times \Pr\left(\text{positive} \middle| \text{cancer}\right) + \Pr(\text{no cancer})\times \Pr\left(\text{positive} \middle| \text{no cancer}\right)}$

$= \dfrac{0.02\times 0.75}{(0.02\times 0.75)+(0.98\times 0.30)} = \dfrac{5}{103} \approx 0.049$

13. **a.** $1 - 0.99 = 0.01$

b. $\Pr\left(\text{pregnant} \middle| \text{positive}\right)$

$= \dfrac{\Pr(\text{pregnant})\times \Pr\left(\text{positive} \middle| \text{pregnant}\right)}{\Pr(\text{pregnant})\times \Pr\left(\text{positive} \middle| \text{pregnant}\right) + \Pr(\text{not pregnant})\times \Pr\left(\text{positive} \middle| \text{not pregnant}\right)}$

$= \dfrac{0.40\times 0.99}{(0.40\times 0.99)+(0.60\times 0.02)} = \dfrac{33}{34} \approx 0.971$

Copyright © 2018 Pearson Education, Inc.

15. $\Pr(\text{steroids}|\text{positive}) = \dfrac{\Pr(\text{steroids}) \times \Pr(\text{positive}|\text{steroids})}{\Pr(\text{steroids}) \times \Pr(\text{positive}|\text{steroids}) + \Pr(\text{no steroids}) \times \Pr(\text{positive}|\text{no steroids})}$

$$= \frac{0.10 \times 0.93}{(0.10 \times 0.93) + (0.90 \times 0.02)} = \frac{31}{37} \approx 0.838$$

17. $\Pr(\text{Male}|\text{Clinton}) = \dfrac{\Pr(\text{Male}) \times \Pr(\text{Clinton}|\text{Male})}{\Pr(\text{Male}) \times \Pr(\text{Clinton}|\text{Male}) + \Pr(\text{Female}) \times \Pr(\text{Clinton}|\text{Female})}$

$$= \frac{0.43 \times 0.57}{(0.43 \times 0.57) + (0.57 \times 0.70)} = \frac{0.2451}{0.2451 + 0.399} = \frac{43}{113} \approx 0.3805$$

19. a. $\Pr(\text{one is}) = \dfrac{13}{52} = \dfrac{1}{4}$

b. $\Pr(\text{none is}|\text{random one isn't})$

$$= \frac{\Pr(\text{none is}) \times (\text{random one isn't}|\text{none is})}{\Pr(\text{none is}) \times \Pr(\text{random one isn't}|\text{none is}) + \Pr(\text{one is}) \times \Pr(\text{random one isn't}|\text{one is})}$$

$$= \frac{\frac{3}{4} \times 1}{\left(\frac{3}{4} \times 1\right) + \left(\frac{1}{4} \times \frac{12}{13}\right)} = \frac{13}{17} \approx 0.765$$

c. $\Pr(\text{one is}|10 \text{ randoms aren't})$

$$= \frac{\Pr(\text{one is}) \times \Pr(10 \text{ randoms aren't}|\text{one is})}{\Pr(\text{one is}) \times \Pr(10 \text{ randoms aren't}|\text{one is}) + \Pr(\text{none is}) \times \Pr(10 \text{ randoms aren't}|\text{none is})}$$

$$= \frac{\frac{1}{4} \times \left(\frac{12}{13}\right)^{10}}{\left[\frac{1}{4} \times \left(\frac{12}{13}\right)^{10}\right] + \left(\frac{3}{4} \times 1\right)}$$

$\approx .130$

21. a. $\Pr(\text{Lakeside}|\text{winner})$

$$= \frac{\Pr(\text{Lakeside}) \times \Pr(\text{winner}|\text{Lakeside})}{\Pr(\text{Lakeside}) \times \Pr(\text{winner}|\text{Lakeside}) + \Pr(\text{Pylesville}) \times \Pr(\text{winner}|\text{Pylesville}) + \Pr(\text{Millerville}) \times \Pr(\text{winner}|\text{Millerville})}$$

$$= \frac{0.40 \times 0.05}{(0.40 \times 0.05) + (0.20 \times 0.02) + (0.40 \times 0.03)} = \frac{5}{9}$$

b. $\dfrac{0.20 \times 0.02}{(0.40 \times 0.05) \times (0.20 \times 0.02) + (0.40 \times 0.03)} = \dfrac{1}{9} \approx 11\%$

 Copyright © 2018 Pearson Education, Inc.

23.

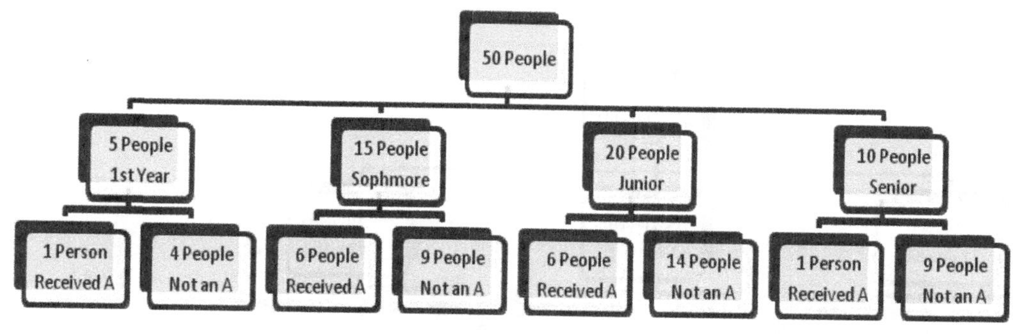

$$\Pr(\text{Sophmore} \mid \text{A}) = \frac{6}{1+6+6+1} = \frac{6}{14} = \frac{3}{7}$$

25.

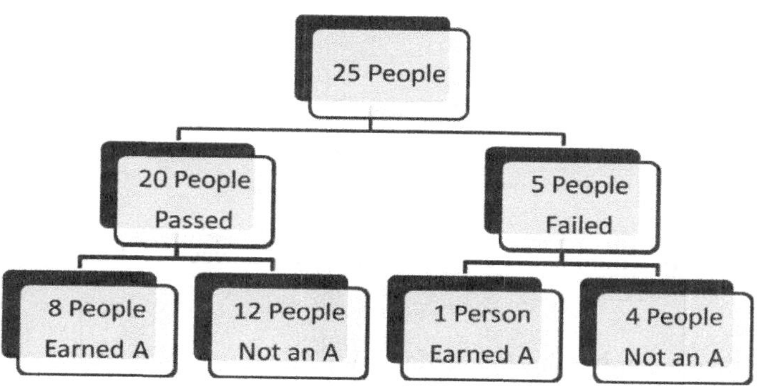

$$\Pr(\text{Passed} \mid \text{A}) = \frac{8}{8+1} = \frac{8}{9}$$

27.

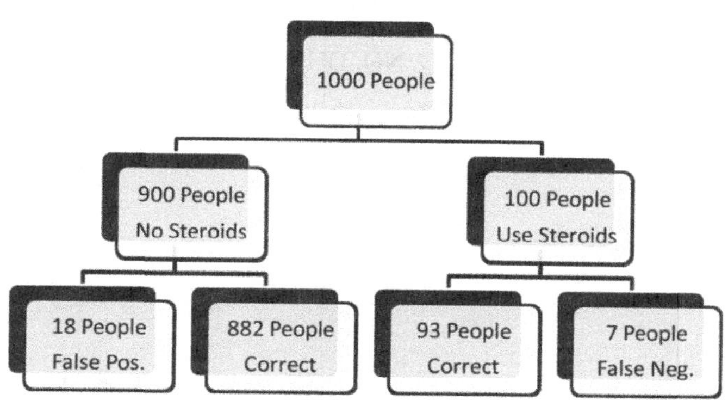

Copyright © 2018 Pearson Education, Inc.

$$\Pr(\text{Used} \mid \text{Positive}) = \frac{93}{93+18} = \frac{93}{111} = \frac{31}{37} \approx 0.838$$

29.

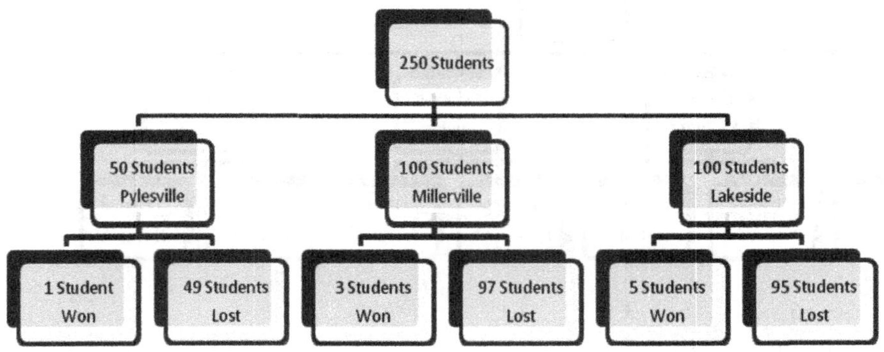

$$\Pr(\text{Lakeside} \mid \text{Winner}) = \frac{5}{1+3+5} = \frac{5}{9}$$

Exercises 6.7

1. Use seq(randInt(1,6),X,1,36,1)→L_1.

   Theoretical probabilities: $\frac{1}{6}$ for each fall.

3. Use seq(randInt(1,1000),X,1,10,1)→L_1 where a number from 1–840 represents a successful freethrow and 841–1000 represents a miss.

5. Use seq(randInt(1,4),X,1,10,1)→L_1, where a = 1, b = 2, c = 3, and d = 4.

7. Answers will vary (see Example 4).

9. Answers will vary.

Chapter 6 Review Exercises

1. **a.** The set of all possible pairs: {PN, PD, PQ, PH, ND, NQ, NH, DQ, DH, QH}

   b. The set of pairs containing an even number of cents: {PN, PQ, NQ, DH}

2. a. A male junior is elected

   b. A female junior is not elected

   c. A male or a junior is elected

3. $\Pr(E \cap F) = 0.4 + 0.3 - 0.5 = 0.2$

4. $\Pr(E \cup F) = 0.5 + 0.3 = 0.8$

 Copyright © 2018 Pearson Education, Inc.

5. $\dfrac{120-(\text{speak Chinese or Spanish})-(\text{speak French only})}{120}=\dfrac{120-(30+50-12)-(75-30-15+7)}{120}=\dfrac{1}{8}$

6. There are 15 – 5 = 10 who like only bicycling and 20 – 5 = 15 who like only jogging, so the probability is $\dfrac{10+15}{50}=0.5.$

7. $\Pr(\text{within 20 minutes})=\dfrac{13}{13+12}=\dfrac{13}{25};\ \Pr(\text{more than 20 minutes})=1-\dfrac{13}{25}=\dfrac{12}{25}$

8. $\dfrac{1}{1+3708}=\dfrac{1}{3709}$

9. $26\%=\dfrac{13}{50};$ $a = 13$ and $a + b = 50$, so $b = 37$. The odds the person selected is under 18 are 13 to 37; the odds the person selected is 18 or older are 37 to 13.

10. $25\%=\dfrac{1}{4};$ $a = 1$ and $a + b = 4$, so $b = 3$. The odds are 1 to 3.

11. $\dfrac{5}{10}\times\dfrac{4}{9}\times\dfrac{3}{8}=\dfrac{1}{12}$

12. There are $C(4, 2) = 6$ ways to choose a "pair" of socks from the drawer; of these, 2 have the same color, so the probability is $\dfrac{2}{6}=\dfrac{1}{3}.$

13.
$$\begin{aligned}1-\frac{C(5,1)\cdot C(95,3)+C(95,4)}{C(100,4)}&=1-\frac{5\cdot 138,415+3,183,545}{3,921,225}\\&=1-\frac{3,875,620}{3,921,225}\\&\approx 0.01163\end{aligned}$$

14. $\dfrac{C(5,1)\cdot C(4,2)}{C(9,3)}=\dfrac{5\cdot 6}{84}=\dfrac{30}{84}=\dfrac{5}{14}\approx 0.3571$

15. a. $\Pr(\text{prepared every question})=\dfrac{C(8,6)}{C(10,6)}=\dfrac{28}{210}=\dfrac{2}{15}$

b. $\Pr(\text{not prepared on test})=\dfrac{C(8,4)}{C(10,6)}=\dfrac{70}{210}=\dfrac{1}{3}$

16. a.
$$\begin{aligned}\Pr(\text{winning})&=\Pr(7)+\Pr(11)\\&=\frac{6}{36}+\frac{2}{36}\\&=\frac{8}{36}=\frac{2}{9}\end{aligned}$$

Copyright © 2018 Pearson Education, Inc.

$$\begin{aligned}\Pr(\text{losing}) &= \Pr(2)+\Pr(3)+\Pr(12)\\ &= \frac{1}{36}+\frac{2}{36}+\frac{1}{36}\\ &= \frac{4}{36}=\frac{1}{9}\end{aligned}$$

b. $\Pr(\text{winning}) = \Pr(6) = \frac{5}{36}$

$\Pr(\text{losing}) = \Pr(7) = \frac{6}{36} = \frac{1}{6}$

17. $1-\left(\frac{1}{2}\right)^5 = \frac{31}{32}$

18. Pr(no tails) + Pr(1 tail each) + Pr(2 tails each) + Pr(3 tails each) $= \left(\frac{1}{8}\right)^2+\left(\frac{3}{8}\right)^2+\left(\frac{3}{8}\right)^2+\left(\frac{1}{8}\right)^2 = \frac{5}{16}$

19. $\frac{2}{7}\times\frac{1}{6}=\frac{1}{21}$

20. $\Pr(\text{select 4 winning teams}) = \left(\frac{1}{17}\right)^4 = \frac{1}{83,521}$

$\text{Odds against} = 1-\frac{1}{83,521} : \frac{1}{83,521} = 83,520:1$

21. a. $\left(\frac{1}{36}\right)^3$

b. $\left(\frac{1}{10}\right)^4$

c. $\left(\frac{26}{36}\right)^3\left(\frac{1}{2}\right)^4 = \left(\frac{17,576}{46,656}\right)\left(\frac{1}{16}\right) = \frac{2197}{93,312}$

22. a. Pr(all three cards are aces)

$= pr(A_1 \cap A_2 \cap A_3) = \frac{4}{52}\cdot\frac{4}{52}\cdot\frac{4}{52} = \frac{1}{2197}$

b. Pr (at least one ace) = 1 – Pr(no aces)

$= 1-\left(\frac{48}{52}\right)^3 = 1-\left(\frac{12}{13}\right)^3 = \frac{469}{2197}$

Copyright © 2018 Pearson Education, Inc.

23. Record the six consecutive outcomes; there are 6^6 possibilities, of which 6! contain each number exactly once. Hence the probability is $\frac{6!}{6^6} = \frac{5}{324} \approx 0.0154.$

24. $\Pr(4 \text{ different numbers}) = \frac{6 \cdot 5 \cdot 4 \cdot 3}{6 \cdot 6 \cdot 6 \cdot 6} = \frac{5}{18},$ so the odds in favor of getting four different numbers are $\frac{5}{18} : 1 - \frac{5}{18} = 5 \text{ to } 13.$

25. $1 - \frac{365 \cdot 364 \cdot 363 \cdot 362 \cdot 361}{365^5} \approx 0.0271.$

26. $1 - \frac{6 \cdot 5 \cdot 4}{7 \cdot 7 \cdot 7} = 1 - \frac{120}{343} = \frac{223}{343} \approx 0.6501$

27. $\Pr(E \mid F) = \frac{.4 + .3 - .5}{.3} = \frac{.2}{.3} = \frac{2}{3}$

28. $\Pr(F) = \frac{\frac{1}{10}}{\frac{1}{7}} = \frac{7}{10}$

29. Pr(at least one tail appears in three coins given that at least one head appeared)
1) Look at the sample space: HHH, HHT, HTH, HTT, THH, THT, TTH, TTT.
2) The first seven outcomes have at least one head.
3) Pr (one or more tails in the first seven outcomes) = 6/7

30. $\Pr\left(\text{one } 3 \middle| \text{no doubles}\right) = \frac{5+5}{30} = \frac{1}{3}$

31. **a.** Pr(employed)=$\frac{147.48}{154.81} \approx 0.9527$

b. Pr(Male)=$\frac{80.74}{154.81} \approx 0.5215$

c. Pr(Female|employed)=$\frac{70.70}{147.48} \approx 0.4794$

d. Pr(employed|Female)=$\frac{70.70}{74.07} \approx 0.9545$

32. **a.** Pr(eng)=$\frac{15}{50} = \frac{3}{10}$

b. Pr(Eng|public)=$\frac{10}{25} = \frac{2}{5}$

c. Pr(Private|eng)=$\frac{5}{15} = \frac{1}{3}$

Copyright © 2018 Pearson Education, Inc.

d. $\text{Pr(Public|eng)}=\frac{10}{15}=\frac{2}{3}$

33. $(0.08)(0.5)=0.04$

34. $$\begin{aligned}\Pr(R_1\cap G_2\cap G_3\cap R_4)&=\frac{10}{30}\cdot\frac{20}{29}\cdot\frac{19}{28}\cdot\frac{9}{27}\\&=\frac{95}{1827}\approx 0.0520\end{aligned}$$

35. No; $\Pr(F|E)=\frac{1}{6}>\Pr(F)=\frac{5}{36}$

36. No; $\Pr(F|E)=\frac{3}{4}\neq\Pr(F)=\frac{1}{2}$

37. Yes

38. Yes

39. **a.** $\frac{1}{4}\times\frac{1}{3}=\frac{1}{12}$

b. $\frac{1}{4}+\frac{1}{3}-\frac{1}{12}=\frac{1}{2}$

40. a. $(0.4)(0.75)=0.3$

b. $0.4+0.75-0.3=0.85$

41. $\Pr(A\cup B)=\frac{1}{2};\ \Pr(A'\cap B)=\frac{1}{3}$

$\Pr(A)+\Pr(A'\cap B)=\Pr(A\cup B)$

$\Pr(A)=\frac{1}{2}-\frac{1}{3}=\frac{1}{6}$

42. $\Pr(A\text{ and }B)=(0.4)(0.3)=0.12$

$\Pr(A\text{ only})=0.3-0.12=0.18$

$\Pr(B\text{ only})=0.4-0.12=0.28$

$\Pr(\text{exactly 1})=0.18+0.28=0.46$

43. $\left(\frac{1}{3}\times\frac{2}{3}\right)+\left(\frac{2}{3}\times\frac{1}{3}\right)=\frac{4}{9}$

44. $\Pr(\text{3 is drawn on 1st or 2nd draw})=\frac{1}{3}+\left(\frac{1}{3}\right)\left(\frac{1}{2}\right)+\left(\frac{1}{3}\right)\left(\frac{1}{2}\right)=\frac{2}{3}$

Copyright © 2018 Pearson Education, Inc.

45. You should switch. If you stay with your original choice your probability of winning remains $\frac{1}{3}$, whereas if you switch you lose only if your original choice was correct, so your probability of winning is $\frac{2}{3}$.

46. $(0.60 \times 0.90) + (0.40 \times 0.05) = 0.56 = 56\%$

47. $\Pr(\text{both parents left-handed} | \text{child left-handed})$

$$= \frac{\Pr(\text{all three left-handed})}{\Pr(\text{child left-handed})}$$

$$= \frac{0.4 \times 0.25 \times 0.25}{(0.4 \times 0.25 \times 0.25) + (0.2 \times 0.25 \times 0.75) + (0.2 \times 0 .75 \times 0.25) + (0.1 \times 0.75 \times 0.75)}$$

$$= \frac{4}{25}$$

48. $$\Pr(\text{correct} | \text{rejected}) = \frac{\Pr(\text{correct}) \times \Pr(\text{rejected} | \text{correct})}{\Pr(\text{correct}) \times \Pr(\text{rejected} | \text{correct}) + \Pr(\text{incorrect}) \times \Pr(\text{rejected} | \text{incorrect})}$$

$$= \frac{0.80 \times 0.05}{(0.80 \times 0.05) + (0.20 \times 0.90)} = \frac{2}{11}$$

49. $$\Pr(C | \text{wrong}) = \frac{\Pr(C) \times \Pr(\text{wrong} | C)}{\Pr(C) \times \Pr(\text{wrong} | C) + \Pr(A) \times \Pr(\text{wrong} | A) + \Pr(B) \times \Pr(\text{wrong} | B)}$$

$$= \frac{0.20 \times 0.05}{(0.20 \times 0.05) + (0.40 \times 0.02) + (0.40 \times 0.03)}$$

$$= \frac{0.01}{0.03}$$

$$= \frac{1}{3}$$

50. If n is the number of dragons, then $\frac{\text{\# of heads on 1-headed dragons}}{\text{\# of heads}} = \frac{\frac{n}{3}}{\frac{n}{3} + 2 \cdot \frac{n}{3} + 3 \cdot \frac{n}{3}} = \frac{1}{6}$

51. If E and F are independent events, then the outcome of F does not affect the outcome of E and vice versa. If the outcome of F does not affect the outcome of E, then neither would the outcome of F'.
Example: Let E = event you get a six on a die and F = event you get a H on a coin toss. Now, E and F are independent. F' is the event you get a T on a coin toss. The events E and F' are independent, so whether you get a H or T on the coin toss does not affect the probability of the six on a die.

52. If you know $\Pr(E \cup F)$, then you can compute
$\Pr(E \cap F) = \Pr(E) + \Pr(F) - \Pr(E \cup F)$. Alternatively, if you know that E and F are independent events, then you can compute
$\Pr(E \cap F) = \Pr(E) \cdot \Pr(F)$.

Copyright © 2018 Pearson Education, Inc.

53. If two events are independent, then $\Pr(E \cap F) = \Pr(E) \cdot \Pr(F)$. Since the condition also states that they have nonzero probabilities, the right side of the equation has a nonzero value. Therefore, the probability of the intersection is nonzero, and events E and F must have at least one outcome in come, thus they are not mutually exclusive.

54. Suppose A and B are two mutually exclusive events with nonzero probabilities. Then $\Pr(A \cap B) = 0$ and $\Pr(A)\Pr(B) \neq 0$, so A and B are not independent.

55. True; The formula for a conditional probability is $\Pr(E \mid F) = \dfrac{\Pr(E \cap F)}{\Pr(F)}$.

56. No; because Pr(E) + Pr (F) > 1.

Copyright © 2018 Pearson Education, Inc.

Chapter 7

Exercises 7.1

1.

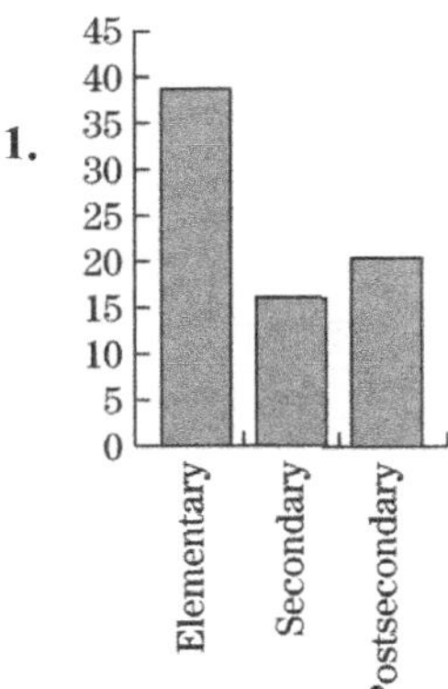

3.

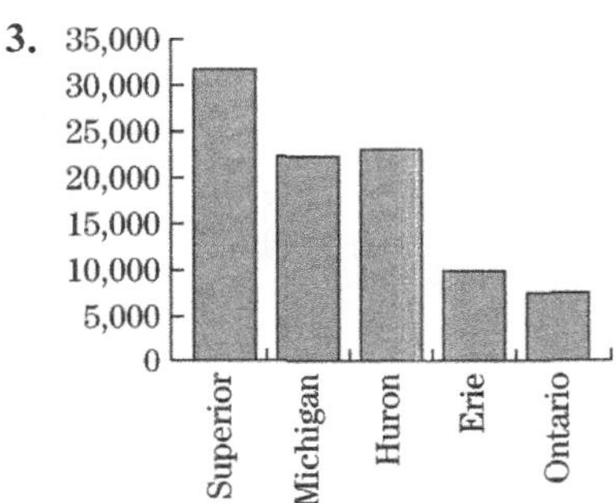

5.

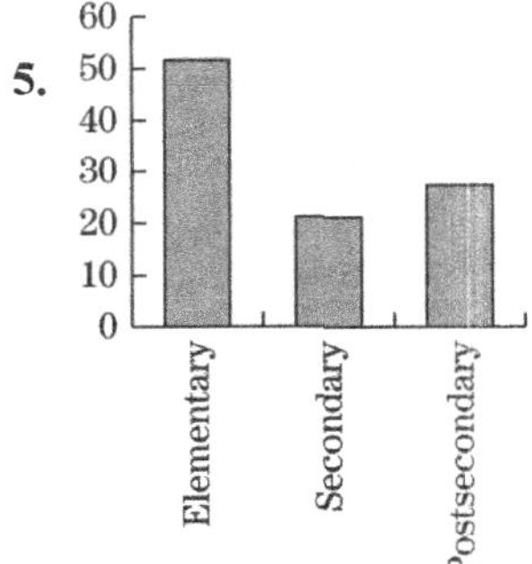

7. To find the central angle, multiply the percentage 0.10 by 360 to obtain 36°.

9.

Lake	*Percent*	360° × *Percent*
Superior	33.5	120.6°
Michigan	23.6	85.0°
Huron	24.4	87.8°
Erie	10.5	37.8°
Ontario	8.0	28.8°

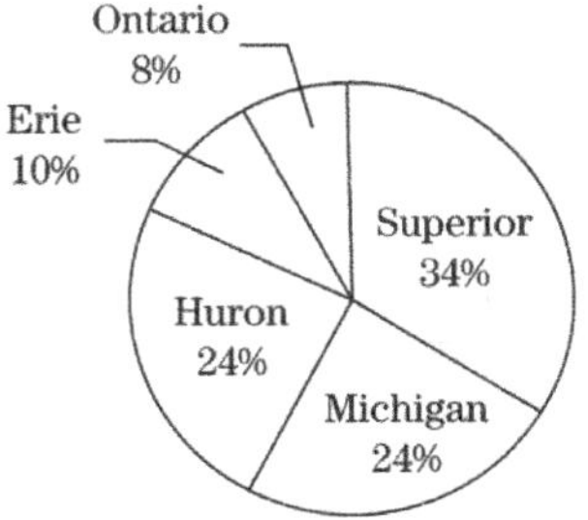

11. Pr(Master or Doctorate) = 0.421 + 0.191 = 0.612

13.

0	卌 ||	7
1	||	2
2	||	2
3	||	2
4	||||	4
5		0
6		0
7	||	2
8	|	1
9		0
10	|	1

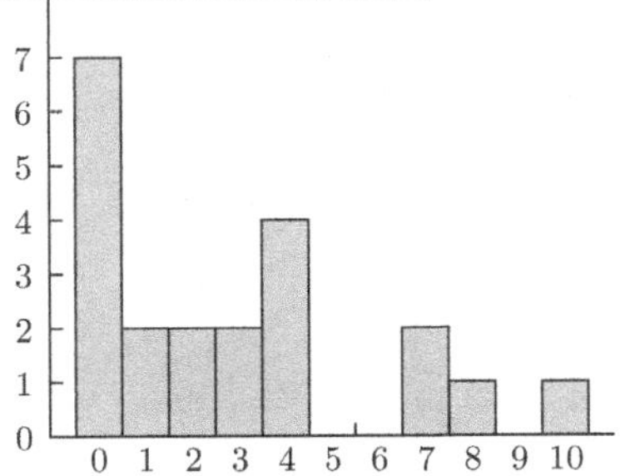

Number of Tie-Breaking Votes

Copyright © 2018 Pearson Education, Inc.

15. The median is $\frac{2+3}{2} = 2.5$.

17.

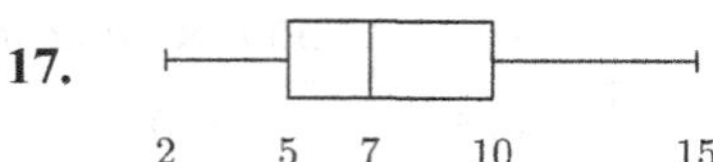

19. $\min = 10,\ Q_1 = 13,\ Q_2 = 17,\ Q_3 = 21,$
$\max = 24;$
$\text{IQR} = Q_3 - Q_1 = 21 - 13 = 8;$

10 13 17 21 24

21. $\min = 20,\ Q_1 = 28,\ Q_2 = 42,\ Q_3 = 52.5,$
$\max = 56;$
$\text{IQR} = Q_3 - Q_1 = 52.5 - 28 = 24.5;$

20 28 42 52.5 56

23. (A) – (c), (B) – (d), (C) – (a), (D) – (b)

25. **a.** $\min = 200,\ Q_1 = 400,\ Q_2 = 600,$
$Q_3 = 700,\ \max = 800$

b. 25%

c. 25%

d. 50%

e. 75%

27. Blue Jays: 0.188, 0.313, 0.297, 0.304, 0.119, 0.045, 0.091, 0.214, 0.200
Red Sox: 0.239, 0.302, 0.262, 0.333, 0.293, 0.333, 0.267, 0.429, 0.290

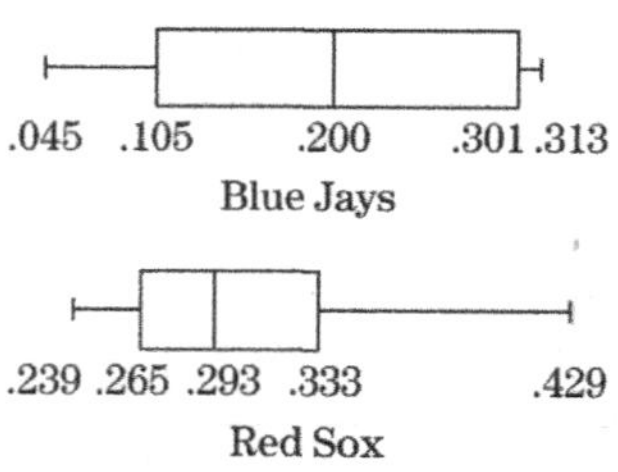

Possible prediction: The Red Sox seem more likely to win because their hitting is more consistent with a higher median.

29.

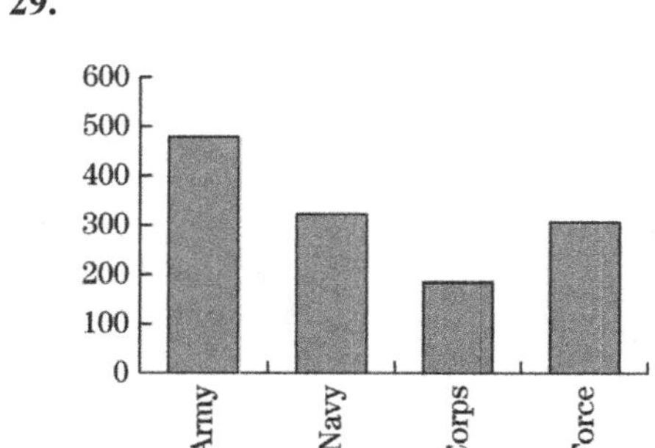

31.

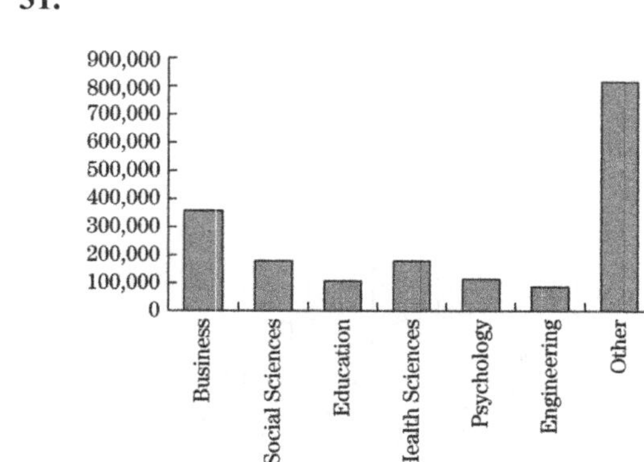

Exercises 7.2

1.

Grade	Relative Frequency
0	$\frac{2}{25} = 0.08$
1	$\frac{4}{25} = 0.16$
2	$\frac{8}{25} = 0.32$
3	$\frac{6}{25} = 0.24$
4	$\frac{5}{25} = 0.20$

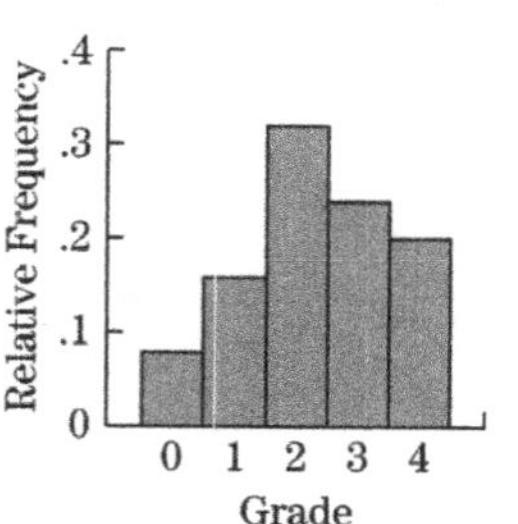

Copyright © 2018 Pearson Education, Inc.

3.

Number of calls during minute	Relative Frequency
20	$\frac{3}{60} = 0.05$
21	$\frac{3}{60} = 0.05$
22	$\frac{0}{60} = 0$
23	$\frac{6}{60} = 0.10$
24	$\frac{18}{60} = 0.30$
25	$\frac{12}{60} = 0.20$
26	$\frac{0}{60} = 0$
27	$\frac{9}{60} = 0.15$
28	$\frac{6}{60} = 0.10$
29	$\frac{3}{60} = 0.05$

5. 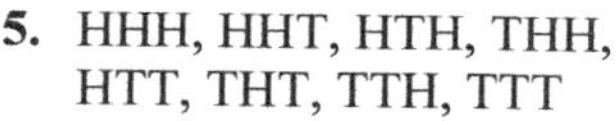
HHH, HHT, HTH, THH,
HTT, THT, TTH, TTT

Number of Heads	Probability
0	$\frac{\binom{3}{0}}{2^3} = \frac{1}{8}$
1	$\frac{\binom{3}{1}}{2^3} = \frac{3}{8}$
2	$\frac{\binom{3}{2}}{2^3} = \frac{3}{8}$
3	$\frac{\binom{3}{3}}{2^3} = \frac{1}{8}$

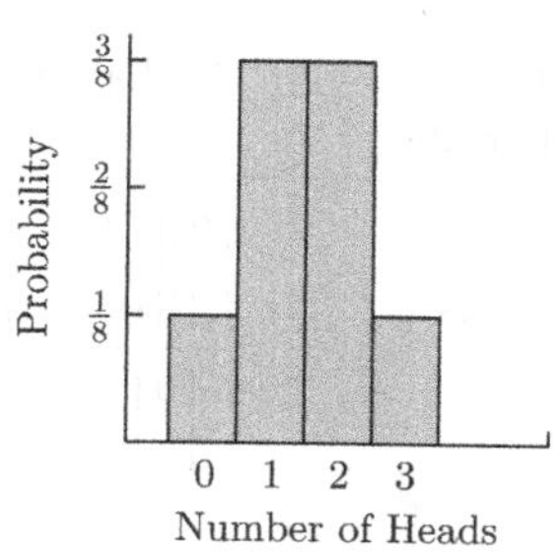

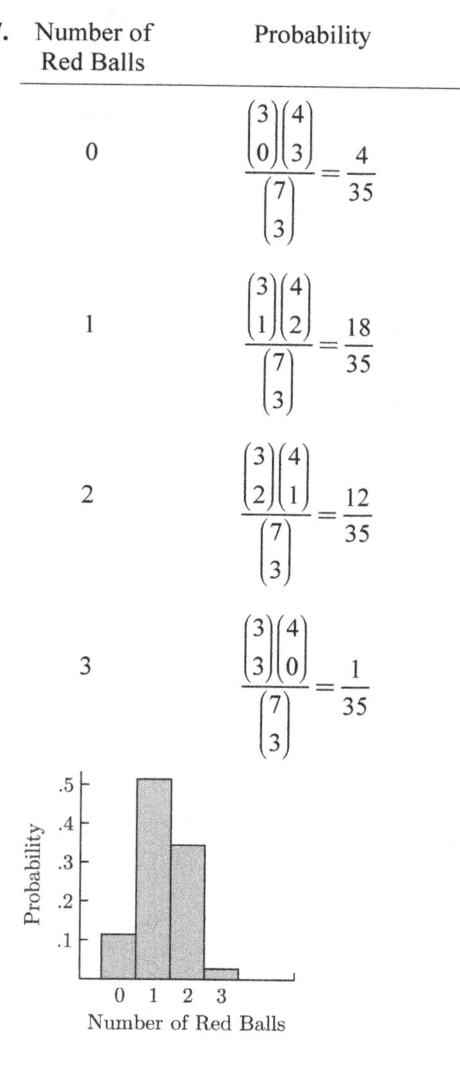

7.

Number of Red Balls	Probability
0	$\frac{\binom{3}{0}\binom{4}{3}}{\binom{7}{3}} = \frac{4}{35}$
1	$\frac{\binom{3}{1}\binom{4}{2}}{\binom{7}{3}} = \frac{18}{35}$
2	$\frac{\binom{3}{2}\binom{4}{1}}{\binom{7}{3}} = \frac{12}{35}$
3	$\frac{\binom{3}{3}\binom{4}{0}}{\binom{7}{3}} = \frac{1}{35}$

Copyright © 2018 Pearson Education, Inc.

9.

No. Red Balls	Player's Earnings	Probability
2	\$5	$\dfrac{\binom{2}{2}\binom{4}{0}}{\binom{6}{2}} = \dfrac{1}{15}$
1	\$1	$\dfrac{\binom{2}{1}\binom{4}{1}}{\binom{6}{2}} = \dfrac{8}{15}$
0	–1\$	$\dfrac{\binom{2}{0}\binom{4}{2}}{\binom{6}{2}} = \dfrac{6}{15}$

11. $\Pr(5 \le X \le 7)$
$= \Pr(X = 5) + \Pr(X = 6) + \Pr(X = 7)$
$= .2 + .1 + .3$
$= .6$

13.

k	$\Pr(X^2 = k)$
0	0.1
1	0.2
4	0.4
9	0.1
16	0.2

15.

k	$\Pr(X - 1 = k)$
–1	0.1
0	0.2
1	0.4
2	0.1
3	0.2

17.

k	$\Pr\left(\frac{1}{5}Y = k\right)$
1	0.3
2	0.4
3	0.1
4	0.1
5	0.1

19.

$(X+1)^2 = k$	$\Pr((X+1)^2 = k)$
$(0+1)^2 = 1$	0.1
$(1+1)^2 = 4$	0.2
$(2+1)^2 = 9$	0.4
$(3+1)^2 = 16$	0.1
$(4+1)^2 = 25$	0.2

21.

	Relative Frequency	
Grade	9 A.M. class	10 A.M. class
F	$\frac{10}{60} \approx 0.17$	$\frac{17}{100} = 0.17$
D	$\frac{15}{60} = 0.25$	$\frac{20}{100} = 0.20$
C	$\frac{20}{60} \approx 0.33$	$\frac{17}{100} = 0.17$
B	$\frac{10}{60} \approx 0.17$	$\frac{20}{100} = 0.20$
A	$\frac{5}{60} \approx 0.08$	$\frac{26}{100} = 0.26$

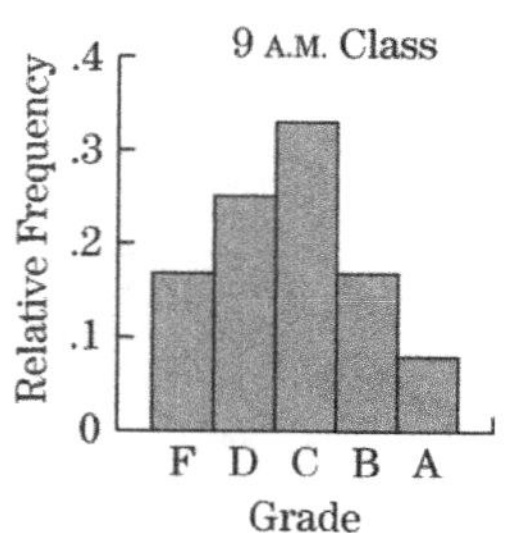

Copyright © 2018 Pearson Education, Inc.

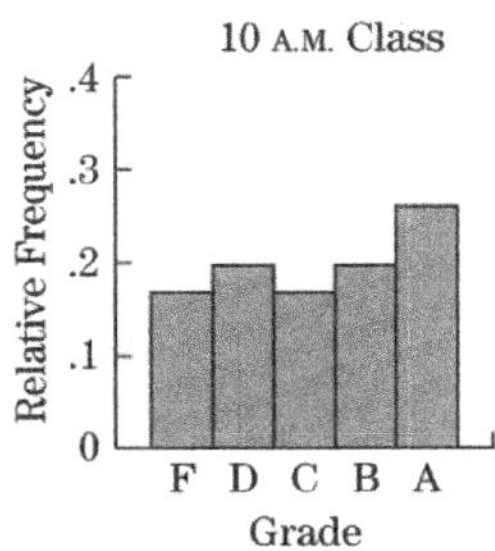

The 9 A.M. class has the distribution centered on the C grade with relatively few A's. The 10A.M. class has a large percentage of A's and D's with fewer C's.

23. percentage with C or higher:

$$\left(\frac{8+6+5}{25}\right)\times 100\% = 76\%$$

25. **a.** less than 22: $3 + 3 = 6$
more than 27: $6 + 3 = 9$

combined: $\left(\frac{6+9}{60}\right)\times 100\% = 25\%$

b. between 23 and 25:

$$\left(\frac{6+18+12}{60}\right)\times 100\% = 60\%$$

27. **a.**

$$\begin{aligned}\Pr(U = 4) &= 1-\left(\frac{4}{15}+\frac{2}{15}+\frac{4}{15}+\frac{3}{15}\right)\\ &= 1-\frac{13}{15}\\ &= \frac{2}{15}\end{aligned}$$

b.

$$\begin{aligned}\Pr(U \geq 2) &= \Pr(U=2) + \Pr(U=3) + \Pr(U=4)\\ &= \frac{4}{15}+\frac{3}{15}+\frac{2}{15}\\ &= \frac{9}{15}\\ &= \frac{3}{5}\end{aligned}$$

c.

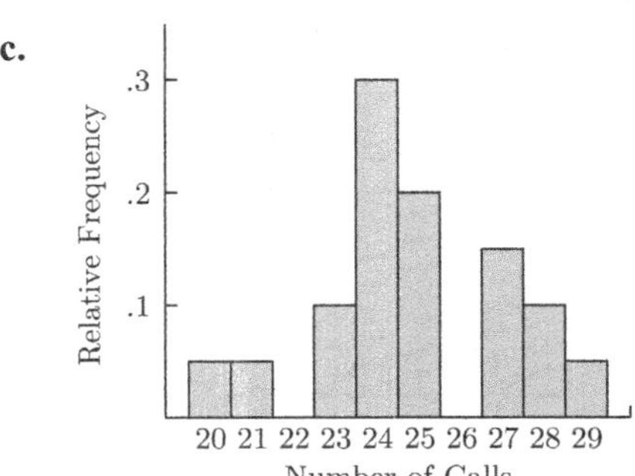

d. Estimated average number of calls would be 24 since that number has the highest frequency of occurrence.
It is actually ≈ 25.

c. $\Pr(U \le 3) = 1 - \Pr(U = 4)$

$$= 1 - \frac{2}{15}$$

$$= \frac{13}{15}$$

d.

$U + 2 = K$	$\Pr(U + 2 = K)$
$0 + 2 = 2$	$\frac{4}{15}$
$1 + 2 = 3$	$\frac{2}{15}$
$2 + 2 = 4$	$\frac{4}{15}$
$3 + 2 = 5$	$\frac{3}{15}$
$4 + 2 = 6$	$\frac{2}{15}$

$\Pr(U + 2 < 4)$
$= \Pr(U + 2 = 2) + \Pr(U + 2 = 3)$
$= \frac{4}{15} + \frac{2}{15}$
$= \frac{6}{15}$
$= \frac{2}{5}$

e.

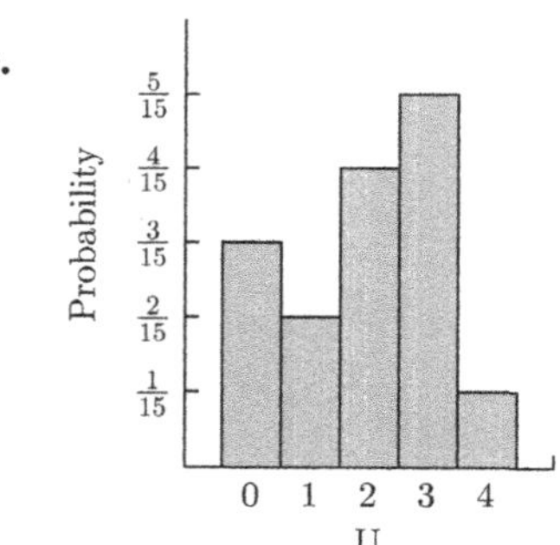

29. a.

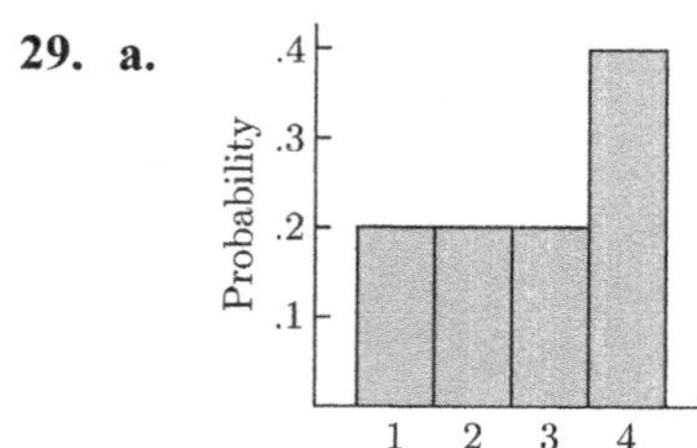

b. $\Pr(Y = 2) + \Pr(Y = 3) = 0.20 + 0.20 = 0.40$

Copyright © 2018 Pearson Education, Inc.

c.

$Y^2 = k$	$\Pr(Y^2 = k)$
$1^2 = 1$	0.20
$2^2 = 4$	0.20
$3^2 = 9$	0.20
$4^2 = 16$	0.40

$\Pr(Y^2 \le 9) = \Pr(Y^2 = 1) + \Pr(Y^2 = 4) + \Pr(Y^2 = 9) = 0.20 + 0.20 + 0.20 = 0.60$

d. $\Pr(Y \le 10) = \Pr(Y = 1) + P(Y = 2) + P(Y = 3) + P(Y = 4) = 0.20 + 0.20 + 0.20 + 0.40 = 1$

e.

$(Y+2)^2 = k$	$\Pr\left((Y+2)^2 = k\right)$
$(1+2)^2 = 9$	0.20
$(2+2)^2 = 16$	0.20
$(3+2)^2 = 25$	0.20
$(4+2)^2 = 36$	0.40

Copyright © 2018 Pearson Education, Inc.

Exercises 7.3

1. $\binom{5}{3}(0.3)^3(0.7)^2 = (10)(0.027)(0.49) = 0.1323$

3. $\binom{4}{3}\left(\frac{1}{3}\right)^3\left(\frac{2}{3}\right)^1 = (4)\left(\frac{1}{27}\right)\left(\frac{2}{3}\right) = \frac{8}{81} \approx 0.0988$

5. $\Pr(X=3) = \binom{10}{3}\left(\frac{1}{2}\right)^3\left(\frac{1}{2}\right)^7 = \frac{15}{128} \approx 0.1172$

7. $\Pr(X=7) = \binom{10}{7}\left(\frac{1}{2}\right)^7\left(\frac{1}{2}\right)^3 = \frac{15}{128} \approx 0.1171875$

$\Pr(X=8) = \binom{10}{8}\left(\frac{1}{2}\right)^8\left(\frac{1}{2}\right)^2 = \frac{45}{1024} \approx 0.043945$

$\Pr(X = 7 \text{ or } 8) = 0.117188 + 0.04395 = 0.1611$

9. $\Pr(\text{At least } 1) = 1 - \Pr(X=0)$
$= 1 - 0.0009766$
$= 0.9990$

11. $\Pr(X=4) = \binom{4}{4}\left(\frac{1}{6}\right)^4\left(\frac{5}{6}\right)^0 = \frac{1}{1296} \approx 0.000772$

13. $\Pr(X=3) = \binom{4}{3}\left(\frac{1}{6}\right)^3\left(\frac{5}{6}\right)^1 = \frac{5}{324} \approx 0.015432$

$\Pr(X = 2 \text{ or } 3) = 0.11574 + 0.01543 = 0.1312$

15. $\Pr(X = 2 \text{ or } 1 \text{ or } 0)$
$= 0.1157 + 0.3858 + 0.4823 = 0.9838$

17. $\Pr(X=2) = \binom{8}{2}(0.14)^2(0.86)^6 \approx 0.2220$

19. $\Pr(X=4) = \binom{8}{4}(0.14)^4(0.86)^4 \approx 0.01471$

$\Pr(X=5) = \binom{8}{5}(0.14)^5(0.86)^3 \approx 0.00192$

$\Pr(X = 4 \text{ or } 5) = 0.01471 + 0.00192 = 0.01663$

21. $\Pr(\text{At least } 3) = 1 - \Pr(X = 0 \text{ or } 1 \text{ or } 2)$
$= 1 - 0.2992 - 0.3897 - 0.2220$
$= 0.08908$

23. $\Pr(X=7) = \binom{7}{7}(0.761)^7(0.239)^0 \approx 0.1478$

25. $\Pr(X=6) = \binom{7}{6}(0.761)^6(0.239)^1 \approx 0.3249$

$\Pr(X = 6 \text{ or } 7) = 0.3249 + 0.1478 = 0.4727$

27. **a.** $\Pr(X=2) = \binom{4}{2}\left(\frac{1}{2}\right)^2\left(\frac{1}{2}\right)^2 = \frac{3}{8} = 0.375$

b. $\Pr(X=3) = 2\binom{4}{3}\left(\frac{1}{2}\right)^3\left(\frac{1}{2}\right)^1 = \frac{8}{16} = 0.5$

c. $\Pr(X=4) = 2\binom{4}{4}\left(\frac{1}{2}\right)^4\left(\frac{1}{2}\right)^0 = \frac{2}{16} = 0.125$

29. $\Pr(X=3) = \binom{5}{3}(0.76)^3(0.24)^2 \approx 0.2529$

$\Pr(X=5) = \binom{5}{5}(0.76)^5(0.24)^0 \approx 0.2536$

The probability of getting 5 successes is higher.

31. Based on the histogram, the probability of getting 10 successes is higher. Out of a group of 40 cattle, it is more likely that exactly 10 recover than exactly 9 recover.

33. The probability that the salesman sells cars to three or four of the customers.

35. **a.** $1 - 0.4602 - 0.0796 = 0.4602$

b. $0.4602 + 0.0796 = 0.5398$

37. The probability of success is .5 therefore the histogram will be symmetrical about the middle value, or 5.

39. 1; because the sum of all individual probabilities is 1.

 Copyright © 2018 Pearson Education, Inc.

41. Let "success" = "lives to 100." Then $p = 0.075$, $q = 0.925$, $n = 77$.

$$\Pr(X \ge 2) = 1 - \Pr(X = 0) - \Pr(X = 1)$$
$$= 1 - (0.925)^{77} - 77(0.075)^1(0.925)^{76}$$
$$\approx 0.9821$$

43. Let "success" = "adverse reaction." Then $p = 0.02$, $q = 0.98$, $n = 56$.

$$\Pr(X \ge 3) = 1 - \Pr(X = 0) - \Pr(X = 1) - \Pr(X = 2)$$
$$= 1 - (0.98)^{56} - \binom{56}{1}(0.02)^1(0.98)^{55} - \binom{56}{2}(0.02)^2(0.98)^{54}$$
$$\approx 0.1018$$

45. Let "success" be "a vote for" the candidate. Then $p = 0.6$, $q = 0.4$, $n = 5$.

$\Pr(X \le 2)$

$= \Pr(X = 0) + \Pr(X = 1) + \Pr(X = 2)$

$= \binom{5}{0}(0.6)^0(0.4)^5 + \binom{5}{1}(0.6)^1(0.4)^4 + \binom{5}{2}(0.6)^2(0.4)^3$

$= 0.01024 + 0.0768 + 0.2304$

≈ 0.3174

47. Let "success" = "defective." Then $p = 0.03$, $q = 0.97$, $n = 20$.

$$\Pr(X \ge 2) = 1 - \Pr(X = 0) - \Pr(X = 1)$$
$$= 1 - (0.97)^{20} - 20(0.03)^1(0.97)^{19}$$
$$\approx 0.1198$$

49.

		Mother A	Mother a
Father	A	AA	Aa
	a	Aa	aa

Child's genes	Probability
AA	$\frac{1}{4}$
Aa	$\frac{2}{4}$
aa	$\frac{1}{4}$

Let "success" be "aa." Then $p = \frac{1}{4}$, $q = \frac{3}{4}$, $n = 3$.

$$\Pr(X \ge 1) = 1 - \Pr(X = 0)$$
$$= 1 - \binom{3}{0}\left(\frac{1}{4}\right)^0\left(\frac{3}{4}\right)^3$$
$$\approx 1 - 0.4219$$
$$= 0.5781$$

51. Let "success" be "gets a hit." Then $p = 0.3$, $q = 0.7$, $n = 4$.

$$\Pr(X = 0) = \binom{4}{0}(0.3)^0(0.7)^4 = 0.2401$$
$$\Pr(X = 3) = \binom{4}{3}(0.3)^3(0.7)^1 = 0.0756$$

53. Here $p = 0.82$. The expected value is $\mu = np = 10(0.82) = 8.2$

$\Pr(X = 8) = C(10,8)(0.82)^8(0.18)^2 = 0.2980$

$\Pr(X = 9) = C(10,9)(0.82)^9(0.18)^1 = 0.3017$

So 9 is the most likely number.

55. **a.** For the underdog to win the series in five sets, the underdog must win two of the first four sets and win the fifth set. The probability that this occurs is

$$\binom{4}{2}(1-p)^2 p^2 (1-p) = \binom{4}{2}(1-p)^3 p^2.$$

Copyright © 2018 Pearson Education, Inc.

b. Using the same reasoning,

$$\Pr(\text{underdog wins in 3 sets}) = (1-p)^3$$

$$\Pr(\text{underdog wins in 4 sets}) = \binom{3}{2}(1-p)^2 p(1-p) = \binom{3}{2}(1-p)^3 p$$

c.

$$\Pr(\text{underdog wins set}) = (1-p)^3\left(1+\binom{3}{1}p+\binom{4}{2}p^2\right)$$
$$= (1-p)^3(1+3p+6p^2)$$
$$= (1-.7)^3(1+3(.7)+6(.7)^2)$$
$$= .16308$$

d. Suppose all five sets of the match are played even if one of the players wins three sets before the fifth set. Then the probability that the underdog wins at least three sets is

$$\binom{5}{0}(1-p)^5+\binom{5}{1}p(1-p)^4+\binom{5}{2}p^2(1-p)^3.$$

Substituting $p = .7$ gives a probability ≈ 0.16308.

57. 17; because $1-\binom{17}{0}\left(\frac{1}{6}\right)^0\left(\frac{5}{6}\right)^{17} = 0.9549$

59. 114; because

$$1-\binom{114}{0}(0.00045)^0(0.99955)^{114} = 0.0500$$

61. Using a TI83 graphing calculator input the following steps:

binomcdf(100, .5, 60) – binomcdf(100, .5, 39) the answer will be approximately 0.9648.

63. Using a TI83 graphing calculator input the following steps:
1 – binomcdf(100, 0.6, 59) $\approx$ 0.5433

65. a. Substituting $p = 0.6$ in this formula gives a probability ≈ 0.2898.

b. Suppose as in Problem 56(d) above that all $2n-1$ games are played. Then the probability that the favorite wins at least n games is

$$\binom{2n-1}{n}p^n(1-p)^{n-1}+\binom{2n-1}{n+1}p^{n+1}(1-p)^{n-2}+\text{L}+\binom{2n-1}{2n-1}p^{2n-1}.$$

Some experimenting shows that for $p = .6$, $n = 21$ is the smallest value of n for which this probability exceeds .9. Since the number of games is $2n-1 = 2(21)-1 = 41$.

67. Using a TI84 graphing calculator input the following steps:
1-binomcdf(80, 0.55, 39) $\approx$ 0.8441

69. Using a TI84 graphing calculator input the following steps:
binomcdf(50, 0.25, 20) – binomcdf(50, 0.25, 9)
$\approx$ 0.8301

Exercises 7.4

1. Population mean

3. Expected value

5. E(*X*) = 0(0.25) + 1(0.2) + 2(0.1) + 3(0.25) + 4(0.2) = 1.95

7. a.

$$\text{GPA} = \frac{4+4+4+4+3+3+2+2+2+1}{10} = \frac{29}{10} = 2.9$$

Copyright © 2018 Pearson Education, Inc.

b.

Grade	Relative Frequency
4	$\frac{4}{10} = 0.4$
3	$\frac{2}{10} = 0.2$
2	$\frac{3}{10} = 0.3$
1	$\frac{1}{10} = 0.1$

c. $E(X) = 4(0.4) + 3(0.2) + 2(0.3) + 1(0.1)$
$= 2.9$

9. $\bar{x}_A = 0(0.3) + 1(0.3) + 2(0.2) + 3(0.1) + 4(0) + 5(0.1)$
$= 1.5$
$\bar{x}_B = 0(0.2) + 1(0.3) + 2(0.3) + 3(0.1) + 4(0.1) + 5(0)$
$= 1.6$
Group A had fewer cavities.

11. $E(X) = 1(0.2) + 2(0.4) + 3(0.3) + 4(0.1)$
$= 2.3$

13. $E(X) = 7(0.25) + 8(0.25) + 9(0.25) + 10(0.25)$
$= 8.5$

15.

Earnings	Probability
–$1	$\frac{37}{38}$
$35	$\frac{1}{38}$

$$E(X) = -1\left(\frac{37}{38}\right) + 35\left(\frac{1}{38}\right) \approx -\$0.0526$$

17.

Earnings	Probability
–50¢	$\frac{2}{6} = \frac{1}{3}$
0¢	$\left(\frac{4}{6}\right)\left(\frac{2}{5}\right) = \frac{4}{15}$
50¢	$\left(\frac{4}{6}\right)\left(\frac{3}{5}\right)\left(\frac{2}{4}\right) = \frac{1}{5}$
$1	$\left(\frac{4}{6}\right)\left(\frac{3}{5}\right)\left(\frac{2}{4}\right)\left(\frac{2}{3}\right) = \frac{2}{15}$
$1.50	$\left(\frac{4}{6}\right)\left(\frac{3}{5}\right)\left(\frac{2}{4}\right)\left(\frac{1}{3}\right)\left(\frac{2}{2}\right) = \frac{1}{15}$

$$E(X) = -.5\left(\frac{1}{3}\right) + 0\left(\frac{4}{15}\right) + .5\left(\frac{1}{5}\right) + 1\left(\frac{2}{15}\right) + 1.5\left(\frac{1}{15}\right)$$
$$\approx \$.1667$$

19. Let x be the cost of the policy.
$E(X) = (-x)(0.9) + (10{,}000 - x)(0.1) = -x + 1000$
The expected value is zero if $x = 1000$.
He should be willing to pay up to $1000.

21. $\dfrac{5 \times 0.300 + 4 \times 0.350}{9} \approx 0.322$

Copyright © 2018 Pearson Education, Inc.

23.

Recorded Value	Probability
1	$\frac{1}{36}$
2	$\frac{3}{36}$
3	$\frac{5}{36}$
4	$\frac{7}{36}$
5	$\frac{9}{36}$
6	$\frac{11}{36}$

$$\mathrm{E}(X)=1\left(\frac{1}{36}\right)+2\left(\frac{3}{36}\right)+3\left(\frac{5}{36}\right)+4\left(\frac{7}{36}\right)+5\left(\frac{9}{36}\right)+6\left(\frac{11}{36}\right)\approx 4.47$$

25. $\mathrm{E}(X)=-1\left(\frac{125}{216}\right)+1\left(\frac{75}{216}\right)+3\left(\frac{15}{216}\right)+5\left(\frac{1}{216}\right)=\frac{0}{216}=0$

27. $\mathrm{E}(X)=15(0.205)=3.075$

29. The expected number of times that a 5 or a 6 will appear is $\mu=np=30\left(\frac{1}{3}\right)=10$

31. The expected value of points when taking one three point shot is $\mu=3(0.40)=1.2$ points. The expected value of taking three free-throws when the success probability is 60% is $\mu=3(0.60)=1.8$, which is the greater expected value.

33.

Number of Republicans	Probability
0	$\frac{10}{84}$
1	$\frac{40}{84}$
2	$\frac{30}{84}$
3	$\frac{4}{84}$

$$\mathrm{E}(\text{Rep.})=0\left(\frac{10}{84}\right)+1\left(\frac{40}{84}\right)+2\left(\frac{30}{84}\right)+3\left(\frac{4}{84}\right)=\frac{4}{3}\approx 1.333$$

$$\mathrm{E}(\text{Dem.})=3-E(\text{Rep.})=3-\frac{4}{3}=\frac{5}{3}\approx 1.667$$

Copyright © 2018 Pearson Education, Inc.

35. No Replacing

Number of Red	Probability
0	$\frac{1}{35}$
1	$\frac{12}{35}$
2	$\frac{18}{35}$
3	$\frac{4}{35}$

$$\begin{aligned} E(\text{Red}) &= 0\left(\frac{1}{35}\right)+1\left(\frac{12}{35}\right)+2\left(\frac{18}{35}\right)+3\left(\frac{4}{35}\right) \\ &= \frac{12}{7} \approx 1.714 \end{aligned}$$

Replacing

Number of Red	Probability
0	$\frac{27}{343}$
1	$\frac{108}{343}$
2	$\frac{144}{343}$
3	$\frac{64}{343}$

$$\begin{aligned} E(\text{Red}) &= 0\left(\frac{27}{343}\right)+1\left(\frac{108}{343}\right)+2\left(\frac{144}{343}\right)+3\left(\frac{64}{343}\right) \\ &= \frac{12}{7} \approx 1.714 \end{aligned}$$

37. Let x = chance of rain.

$$\begin{aligned} -8000+40{,}000x &= 0 \\ x &= 0.20 \to 20\% \end{aligned}$$

39. $\frac{7x+4y}{x+y}$

Answer (b) is correct.

41. Solve $\frac{16\cdot 54+14x}{30} = 56.1$

$$\begin{aligned} 14x+864 &= 1683 \\ 14x &= 819 \\ x &= 58.5^\circ \end{aligned}$$

43. $\frac{5+6+x}{3} = \frac{2+7+9}{3}$

$$\begin{aligned} 11+x &= 18 \\ x &= 7 \end{aligned}$$

45. Solve $\frac{12\cdot 5+24x}{5+x} = 19$

$$\begin{aligned} 19x+95 &= 60+24x \\ -5x &= -35 \\ x &= 7 \end{aligned}$$

47. Let x = number of cases.

$$\begin{aligned} 20\left(\frac{1}{2}x\right)+30\left(\frac{1}{2}x\right) &= 75{,}000 \\ 25x &= 75{,}000 \\ x &= 3000 \text{ cases} \end{aligned}$$

Copyright © 2018 Pearson Education, Inc.

Exercises 7.5

1. $m = 70(0.5) + 71(0.2) + 72(0.1) + 73(0.2) = 71$

$$\begin{aligned}\sigma^2 &= (70-71)^2(0.5)+(71-71)^2(0.2)+(72-71)^2(0.1)+(73-71)^2(0.2)\\ &= 0.5+0+0.1+0.8\\ &= 1.4\end{aligned}$$

3. *B*

5. a. $\mu_A = -10(0.2)+20(0.2)+25(0.6)=17$

$\mu_B = 0(0.3)+10(0.4)+30(0.3)=13$

$\sigma_A^2 = (-10-17)^2(0.2)+(20-17)^2(0.2)+(25-17)^2(0.6)=145.8+1.8+38.4=186$

$\sigma_B^2 = (0-13)^2(0.3)+(10-13)^2(0.4)+(30-13)^2(0.3)=50.7+3.6+86.7=141$

b. Investment A

c. Investment B

7. a. $\mu_A = 100(0.1)+101(0.2)+102(0.3)+103(0)+104(0)+105(0.2)+106(0.2)=103$

$\sigma_A^2 = (100-103)^2(0.1)+(101-103)^2(0.2)+\cdots+(106-103)^2(0.2)=4.6$

$\mu_B = 100(0)+101(0.2)+102(0)+103(0.2)+104(0.1)+105(0.2)+106(0.3)=104$

$\sigma_B^2 = (100-104)^2(0)+(101-104)^2(0.2)+\cdots+(106-104)^2(0.3)=3.4$

b. Business B

c. Business B

9. The number of heads is a binomial random variable *X* with $n = 10, p = 0.5$ so $\mu_X = 10\times 0.5 = 5$, $\sigma_X = \sqrt{10\times 0.5\times 0.5} \approx 1.581$.

11. The number of smart thermostats is a binomial random variable *X* with $n = 200, p = 0.015$ so $\mu_X = 200\times 0.015 = 3$, $\sigma_X = \sqrt{200\times 0.015\times 0.985} \approx 1.719$.

13. a. $35 - c = 25$ and $35 + c = 45$ $c = 10$.

Probability $\geq 1-\dfrac{5^2}{10^2} = 1-\dfrac{25}{100} = .75$

b. $35 - c = 20$ and $35 + c = 50$ $c = 15$

Probability $\geq 1-\dfrac{5^2}{15^2} \approx 0.89$

c. $35 - c = 29$ and $35 + c = 41$ $c = 6$

Probability $\geq 1-\dfrac{5^2}{6^2} \approx 0.31$

Copyright © 2018 Pearson Education, Inc.

15. $\mu = 3000, \sigma = 250$

$3000 - c = 2000$ and $3000 + c = 4000 \qquad c = 1000$

$\text{Probability} \geq 1 - \frac{250^2}{1000^2} = 0.9375$

Number of bulbs to replace: $\geq 5000(0.9375) \approx 4688$

17. $\text{Probability} = 1 - \frac{6^2}{c^2} = \frac{7}{16}$

$16c^2 - 576 = 7c^2$

$9c^2 = 576$

$c = \sqrt{64} = 8$

19. a. $\mu = 2\left(\frac{1}{36}\right) + 3\left(\frac{2}{36}\right) + \text{L} \; + 12\left(\frac{1}{36}\right) = 7$

$\sigma^2 = (2-7)^2\left(\frac{1}{36}\right) + \text{L} \; + (12-7)^2\left(\frac{1}{36}\right)$

$= \frac{210}{36}$

$= \frac{35}{6}$

b. $\Pr(4 \leq X \leq 10)$

$= \frac{3}{36} + \frac{4}{36} + \frac{5}{36} + \frac{6}{36} + \frac{5}{36} + \frac{4}{36} + \frac{3}{36}$

$= \frac{30}{36}$

$= \frac{5}{6}$

c. $7 - c = 4$ and $7 + c = 10 \qquad c = 3$

$\text{Probability } \geq 1 - \frac{\frac{35}{6}}{3^2} = \frac{19}{54}$

21. $E(X) = (-2 - 1 + 0 + 1 + 2)(.2) = 0$

$E(X^2) = (4 + 1 + 0 + 1 + 4)(.2) = 2$

$\text{Var}(X) = E(X^2) - E(X)^2 = 2 - 0 = 2$

Copyright © 2018 Pearson Education, Inc.

23.

$2X$	Probability
–2	$\frac{1}{8}$
–1	$\frac{3}{8}$
0	$\frac{1}{8}$
1	$\frac{1}{8}$
2	$\frac{2}{8}$

$$\mu = -2\left(\frac{1}{8}\right) - 1\left(\frac{3}{8}\right) + \cdots + 2\left(\frac{2}{8}\right) = 0$$

$$\sigma_{2X}^2 = (-2-0)^2\left(\frac{1}{8}\right) + \cdots + (2-0)^2\left(\frac{2}{8}\right) = 2$$

$$\sigma_{2X}^2 = 4\sigma_X^2 = 4\left(\frac{1}{2}\right) = 2$$

25. $\mu = \dfrac{60{,}168 + 59{,}770 + \cdots + 46{,}817}{10} \approx 53{,}227$

$$\sigma^2 \approx \frac{(60{,}168 - 53{,}227)^2 + \cdots + (46{,}817 - 53{,}227)^2}{10}$$

$\sigma \approx 4453.54$

The schools within one standard deviation of the mean are University of Florida, Michigan State University, University of Texas at Austin, University of Minnesota, Texas A & M University, and Florida International University

27. a. $\mu_A = \dfrac{(3)(5) + (4)(7) + (5)(8) + (6)(2) + (7)(1) + (8)(2)}{25} = 4.72$

$$\sigma_A^2 = \frac{(3 - 4.72)^2(5) + \cdots + (8 - 4.72)^2(2)}{25} = 1.9616$$

$\sigma_A \approx 1.40$

$$\mu_B = \frac{(3)(5) + (4)(10) + (5)(3) + (6)(3) + (7)(0) + (8)(4)}{25} = 4.8$$

$$\sigma_B^2 = \frac{(3 - 4.8)^2(5) + \cdots + (8 - 4.8)^2(4)}{25} = 2.72$$

$\sigma_B \approx 1.65$

b. University B

c. University A

 Copyright © 2018 Pearson Education, Inc.

Exercises 7.6

1. $A(1.25) = 0.8944$

3. $1 - A(.25) = 1 - 0.5987 = 0.4013$

5. $A(1.5) - A(.5) = 0.9332 - 0.6915 = 0.2417$

7. $A(-0.5) + (1 - A(0.5)) = 0.3085 + (1 - 0.6915)$
$= 0.6170$

9. $\Pr(Z \geq z) = 0.0401$
$A(z) = 1 - 0.0401 = 0.9599$
$z = 1.75$

11. $\Pr(-z \leq Z \leq z) = 0.5468$
$A(-z) = \dfrac{1-0.5468}{2} = 0.2266$
$-z = -0.75$
$z = 0.75$

13. The 80th percentile of the standard normal distribution is approximately 0.84 (use table or InvNorm(0.80) on TI 83)

15. $\mu = 6, \sigma \approx 2$

17. $\mu = 9, \sigma \approx 1$

19. $\dfrac{4-8}{\frac{3}{4}} = -\dfrac{4}{1}\cdot\dfrac{4}{3} = -\dfrac{16}{3}$

21. $\dfrac{x-8}{\frac{3}{4}} = 10$

$x = \dfrac{30}{4} + 8 = \dfrac{62}{4} = \dfrac{31}{2} = 15\dfrac{1}{2}$

23. $\Pr(X \geq 9) = \Pr\left(Z \geq \dfrac{9-10}{\frac{1}{2}}\right)$
$= \Pr(Z \geq -2)$
$= 1 - \Pr(Z \leq -2)$
$= 1 - 0.0228$
$= 0.9772$

25. $\Pr(6 \leq X \leq 10) = \Pr\left(\dfrac{6-7}{2} \leq Z \leq \dfrac{10-7}{2}\right)$
$= \Pr(-0.50 \leq Z \leq 1.50)$
$= 0.9332 - 0.3085$
$= 0.6247$

27. $\Pr(-2 \leq Z \leq 2) = A(2) - A(-2)$
$= 0.9772 - 0.0228$
$= 0.9544$

29. From Table 2 we see that $\Pr(Z \leq 2) = .9772$ for standard normal Z. Solve $\dfrac{6-5}{\sigma} = 2 : 2\sigma = 1$, $\sigma = .5$.

31. $\mu = 3.3, \sigma \approx .2$
$\Pr(X \geq 4) = \Pr\left(Z \geq \dfrac{4-3.3}{.2}\right)$
$= \Pr(Z \geq 3.5)$
$= 1 - \Pr(Z \leq 3.5)$
$= 1 - 0.9998$
$= 0.0002$

33. $\mu = 6, \sigma = 0.04$
$\Pr(5.95 \leq X \leq 6.05)$
$= \Pr\left(\dfrac{5.95-6}{0.04}x \leq Z \leq \dfrac{6.05-6}{0.04}\right)$
$= \Pr(-1.25 \leq Z \leq 1.25)$
$= 0.8944 - 0.1056$
$= 0.7888$

Copyright © 2018 Pearson Education, Inc.

35. $\mu = 5.4, \sigma = .6$

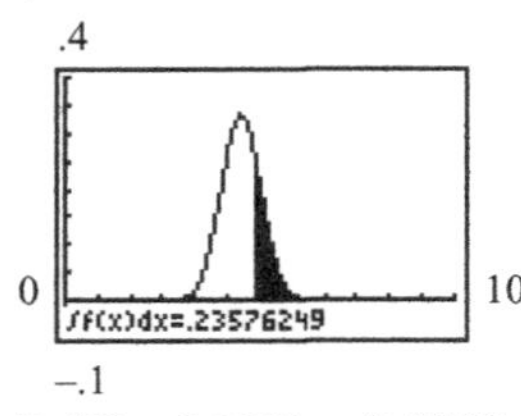

$\Pr(X > 5.832) \approx 0.2358$

37. $\mu = 7500, \sigma = 1000$
$\Pr(X > 9750)$
$= \Pr\left(Z > \frac{9750-7500}{1000}\right)$
$= \Pr(Z > 2.25)$
$= 1 - \Pr(Z \le 2.25)$
$= 1 - 0.9878$
$= 0.0122$

39. $\mu = 520, \sigma = 75$

a. $x_{90} = 520 + 75z_{90}$
$= 520 + 75 \times 1.28$
$= 616$

b. $\Pr(-z \le Z \le z) = 0.90$
$\Pr(Z \le -z) = 0.05 \Rightarrow z_{05} \approx -1.65$
$\frac{x-\mu}{\sigma} = \frac{x-520}{75} = -1.65$
$\Rightarrow x_{05} = 396.25 \approx 396$
$\frac{x-\mu}{\sigma} = \frac{x-520}{75} = 1.65$
$\Rightarrow x_{95} = 643.75 \approx 644$
Between 396 and 644

c. $x_{98} = 520 + 75z_{98} = 520 + 75 \times 2.05 = 674$

41. $\mu = 30,000, \sigma = 5000$
$\Pr(Z \le z) = .02 \Rightarrow z_{02} \approx -2.05$
$\frac{x-\mu}{\sigma} = \frac{x-30,000}{5000} = -2.05 \Rightarrow x_{02} = 19,750$
19,750 miles

43. True; As σ, increases, the normal curve flattens out.

45. normalcdf(1,5,3,2)

NORMDIST (5,3,2,TRUE) – NORMDIST (1,3,2,TRUE)

$P(1 < x < 5)$ for x normal (3,2)

47. normalcdf(0,1000,1200,100)

NORMDIST (1000,1200,100,TRUE)

$P(x < 1000)$ for x normal (1200,100)

Exercises 7.7

1. $n = 25,\ p = \frac{1}{5}$

$\mu = np = 25\left(\frac{1}{5}\right) = 5,$

$\sigma = \sqrt{npq} = \sqrt{25\left(\frac{1}{5}\right)\left(\frac{4}{5}\right)} = 2$

a. $\Pr(X = 5) \approx \Pr\left(\frac{4.5-5}{2} \le Z \le \frac{5.5-5}{2}\right)$
$= \Pr(-.25 \le Z \le .25)$
$= 0.5987 - 0.4013$
$= 0.1974$

b. $\Pr(3 \le X \le 7)$
$\approx \Pr\left(\frac{2.5-5}{2} \le Z \le \frac{7.5-5}{2}\right)$
$= \Pr(-1.25 \le Z \le 1.25)$
$= 0.8944 - 0.1056$
$= 0.7888$

c. $\Pr(X < 10) \approx \Pr\left(Z \le \frac{9.5-5}{2}\right)$
$= \Pr(Z \le 2.25)$
$= 0.9878$

 Copyright © 2018 Pearson Education, Inc.

3. $n = 20,\ p = \dfrac{1}{6}$

$$\mu = 20\left(\frac{1}{6}\right) = \frac{10}{3},\ \sigma = \sqrt{20\left(\frac{1}{6}\right)\left(\frac{5}{6}\right)} = \frac{5}{3}$$

$$\begin{aligned}\Pr(X \ge 8) &\approx \Pr\left(Z \ge \frac{7.5 - \frac{10}{3}}{\frac{5}{3}}\right)\\ &= \Pr(Z \ge 2.5)\\ &= 1 - 0.9938\\ &= 0.0062\end{aligned}$$

5. $n = 100,\ p = \dfrac{1}{2}$

$$\mu = 100\left(\frac{1}{2}\right) = 50,\ \sigma = \sqrt{100\left(\frac{1}{2}\right)\left(\frac{1}{2}\right)} = 5$$

$$\begin{aligned}\Pr(X \ge 63) &\approx \Pr\left(Z \ge \frac{62.5 - 50}{5}\right)\\ &= \Pr(Z \ge 2.5)\\ &= 1 - 0.9938\\ &= 0.0062\end{aligned}$$

7. $n = 75,\ p = \dfrac{3}{4}$

$$\mu = 75\left(\frac{3}{4}\right) = 56.25, \quad \sigma = \sqrt{75\left(\frac{3}{4}\right)\left(\frac{1}{4}\right)} = 3.75$$

$$\begin{aligned}\Pr(X \ge 68) &\approx \Pr\left(Z \ge \frac{67.5 - 56.25}{3.75}\right)\\ &= \Pr(Z \ge 3)\\ &= 1 - 0.9987\\ &= 0.0013\end{aligned}$$

9. $n = 20, p = 0.310$

$\mu = 20(0.310) = 6.2,$

$\sigma = \sqrt{20(0.31)(0.69)} \approx 2.068$

$$\begin{aligned}\Pr(X \ge 6) &\approx \Pr\left(Z \ge \frac{5.5 - 6.2}{2.068}\right)\\ &\approx \Pr(Z \ge -0.34)\\ &\approx \Pr(Z \ge -0.35)\\ &= 1 - 0.3632\\ &= 0.6368\end{aligned}$$

11. $n = 1000, p = 0.02$

$\mu = 1000(0.02) = 20,$

$\sigma = \sqrt{1000(0.02)(0.98)} \approx 4.427$

$$\begin{aligned}\Pr(X < 15) &\approx \Pr\left(Z \le \frac{14.5 - 20}{4.427}\right)\\ &\approx \Pr(Z \le -1.24)\\ &\approx \Pr(Z \le -1.25)\\ &= 0.1056\end{aligned}$$

13. probability of failure $= (0.01)(0.02)(0.01) =$ 0.000002

$n = 1{,}000{,}000,$

$E(X) = \mu = 1{,}000{,}000(0.000002) = 2$

$\sigma = \sqrt{1{,}000{,}000(0.000002)(0.999998)} \approx 1.414$

$$\begin{aligned}\Pr(X > 3) &\approx \Pr\left(Z \ge \frac{3.5 - 2}{1.414}\right)\\ &\approx \Pr(Z \ge 1.06)\\ &\approx \Pr(Z \ge 1.05)\\ &= 1 - 0.8531\\ &= 0.1469\end{aligned}$$

15. $n = 100, p = 0.35$

$\mu = 100(0.35) = 35,$

$\sigma = \sqrt{100(0.35)(0.65)} \approx 4.770$

$\Pr(30 \le X \le 40)$

$$\begin{aligned}&= \Pr\left(\frac{29.5 - 35}{4.77} \le Z \le \frac{40.5 - 35}{4.77}\right)\\ &\approx \Pr(-1.15 \le Z \le 1.15)\\ &= 0.8749 - 0.1251\\ &= 0.7498\end{aligned}$$

17. $n = 1000, p = 0.03$

$\mu = 1000(0.03) = 30,$

$\sigma = \sqrt{1000(0.03)(0.97)} \approx 5.394$

$$\begin{aligned}\Pr(X \ge 29) &\approx \Pr\left(Z \ge \frac{28.5 - 30}{5.394}\right)\\ &\approx \Pr(Z \ge -0.278)\\ &\approx \Pr(Z \ge -0.28)\\ &= 1 - 0.3897\\ &= 0.61\end{aligned}$$

Copyright © 2018 Pearson Education, Inc.

19. $n = 100,\ p = \frac{1}{2}$

Exact: $\Pr(49 \le X \le 51) \approx .2356$

Normal Approximation:

$\mu = 100\left(\frac{1}{2}\right) = 50,\ \sigma = \sqrt{100\left(\frac{1}{2}\right)\left(\frac{1}{2}\right)} = 5$

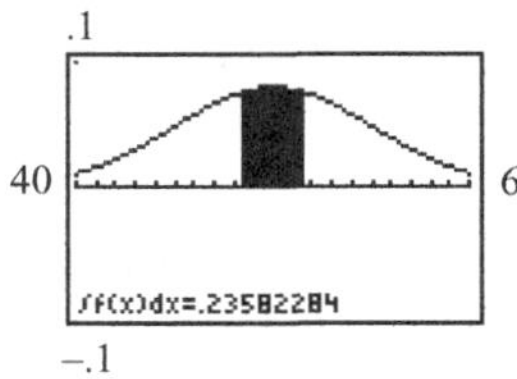

$\Pr(48.5 \le X \le 51.5) \approx 0.2358$

21. $n = 150, p = 0.2$

Exact: $\Pr(X = 30) \approx 0.0812$

Normal Approximation: $\mu = 150(0.2) = 30,$

$\sigma = \sqrt{150(0.2)(0.8)} \approx 4.899$

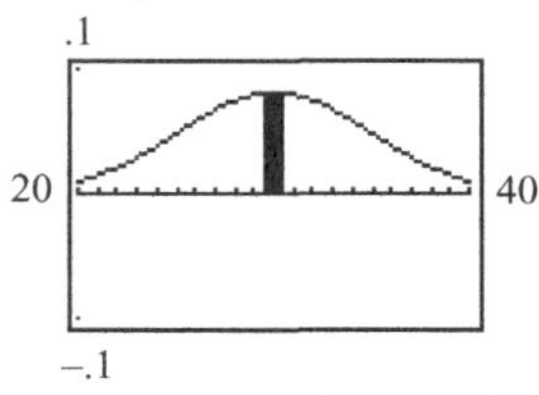

$\Pr(29.5 \le X \le 30.5) \approx 0.0813$

23. 0.0410

1 – binomcdf (300, 0.02, 10)

1 – BINOMDIST (10, 300, 0.02, 1)

binomial probabilities $n = 300, p = 0.02,$

endpoint = 10

25. 0.0796

binompdf (100,0.5,50)

BINOMDIST (50, 100, 0.5, 0)

binomial probabilities $n = 100, p = 0.5,$

endpoint = 50

Chapter 7 Review Exercises

1.

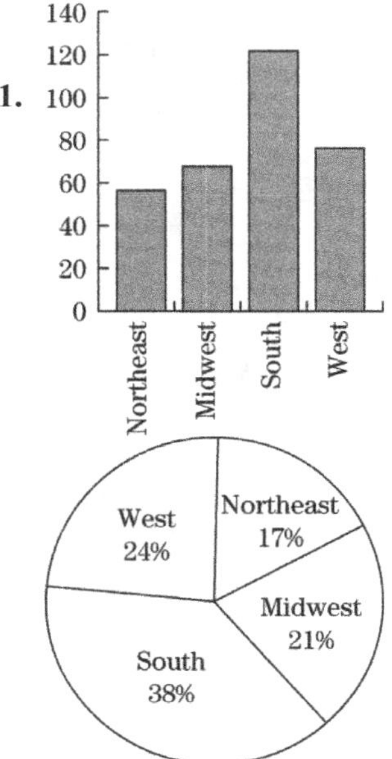

2. min = 1,

$Q_1 = 2.5, Q_2 = 4.5, Q_3 = 11.5, \max = 23;$

$\text{IQR} = Q_3 - Q_1 = 11.5 - 2.5 = 9;$

2.5 11.5 23
1 4.5

3.

Number Waiting in Line	Relative Frequency
0	0.04
1	0.10
2	0.18
3	0.26
4	0.22
5	0.14
6	0.06

Pr(at most 3 customers in line)

$= 0.04 + 0.10 + 0.18 + 0.26$

$= 0.58$

Copyright © 2018 Pearson Education, Inc.

4. a. Possible outcomes are HH, HT, TH, TT

Number of Heads, k	$\Pr(X = k)$
0	0.25
1	0.50
2	0.25

b.

k	$\Pr(2X + 5 = k)$
5	0.25
7	0.50
9	0.25

5. $n = 3,\ p = \frac{1}{3}$

a.

k	$\Pr(X = k)$
0	$\binom{3}{0}\left(\frac{1}{3}\right)^0\left(\frac{2}{3}\right)^3 = \frac{8}{27}$
1	$\binom{3}{1}\left(\frac{1}{3}\right)^1\left(\frac{2}{3}\right)^2 = \frac{12}{27}$
2	$\binom{3}{2}\left(\frac{1}{3}\right)^2\left(\frac{2}{3}\right)^1 = \frac{6}{27}$
3	$\binom{3}{3}\left(\frac{1}{3}\right)^3\left(\frac{2}{3}\right)^0 = \frac{1}{27}$

b. $\mu = 0\left(\frac{8}{27}\right) + 1\left(\frac{12}{27}\right) + 2\left(\frac{6}{27}\right) + 3\left(\frac{1}{27}\right) = 1$

$$\sigma^2 = (0-1)^2\left(\frac{8}{27}\right) + (1-1)^2\left(\frac{12}{27}\right) + (2-1)^2\left(\frac{6}{27}\right) + (3-1)^2\left(\frac{1}{27}\right)$$

$$= \frac{2}{3}$$

6. $n = 4, p = .3$

$$\Pr(X = 2) = \binom{4}{2}(0.3)^2(0.7)^2 = 0.2646$$

7. The student has a .6 probability of guessing correctly on the six questions with answer *true* and a 0.4 probability of guessing correctly on the four questions with answer *false*. Therefore the student's expected score is $6(0.6) + 4(0.4) = 5.2$ correct answers which gives 52 points or 52%.
A better strategy is to choose true for all the questions which guarantees a score of 60%.

Copyright © 2018 Pearson Education, Inc.

8. **a.** $\Pr(\text{get 7 twice}) = \binom{12}{2}\left(\frac{1}{6}\right)^2\left(\frac{5}{6}\right)^{10} \approx 0.2961$

b. $\Pr(\text{get 7 at least twice})$
$= 1 - \Pr(\text{get 7 zero or one time})$
$= 1 - \binom{12}{0}\left(\frac{1}{6}\right)^0\left(\frac{5}{6}\right)^{12} - \binom{12}{1}\left(\frac{1}{6}\right)^1\left(\frac{5}{6}\right)^{11}$
≈ 0.6187

c. The expected number of 7's is $12 \cdot \frac{1}{6} = 2.$

9. $\mu = 0(0.2) + 1(0.3) + 5(0.1) + 10(0.4) = 4.8$
$\sigma^2 = (0-4.8)^2(0.2) + (1-4.8)^2(0.3) + (5-4.8)^2(0.1) + (10-4.8)^2(0.4)$
$= 19.76$

10. Let X be the number of red balls.

k	$\Pr(X = k)$
0	$\frac{\binom{4}{0}\binom{4}{4}}{\binom{8}{4}} = \frac{1}{70}$
1	$\frac{\binom{4}{1}\binom{4}{3}}{\binom{8}{4}} = \frac{16}{70}$
2	$\frac{\binom{4}{2}\binom{4}{2}}{\binom{8}{4}} = \frac{36}{70}$
3	$\frac{\binom{4}{3}\binom{4}{1}}{\binom{8}{4}} = \frac{16}{70}$
4	$\frac{\binom{4}{4}\binom{4}{0}}{\binom{8}{4}} = \frac{1}{70}$

Copyright © 2018 Pearson Education, Inc.

$$\mu = 0\left(\frac{1}{70}\right) + 1\left(\frac{16}{70}\right) + 2\left(\frac{36}{70}\right) + 3\left(\frac{16}{70}\right) + 4\left(\frac{1}{70}\right) = 2$$

$$\sigma^2 = (0-2)^2\left(\frac{1}{70}\right) + (1-2)^2\left(\frac{16}{70}\right) + (2-2)^2\left(\frac{36}{70}\right) + (3-2)^2\left(\frac{16}{70}\right) + (4-2)^2\left(\frac{1}{70}\right)$$

$$= \frac{4}{7}$$

11. X has mean

$\mu = (-2)(0.3) + 0(0.1) + 1(0.4) + 3(0.2) = 0.4,$

variance

$\sigma^2 = (-2-0.4)^2(0.3) + (0-0.4)^2(0.1) + (1-0.4)^2(0.4) + (3-0.4)^2(0.2)$

$= 3.24,$

and standard deviation

$\sigma = \sqrt{3.24} = 1.8.$

12. When a pair of fair dice is rolled, the probabilities that the result is 7 or 11 are $\frac{1}{6}$ and $\frac{1}{18}$ respectively. Hence Lucy's expected winnings are $(-10)\frac{2}{9} + 3\cdot\frac{7}{9} = \frac{1}{9} \approx .11$, or 11 cents per roll.

13. $\mu = 10,\ \sigma = \frac{1}{3}$

$10 - c = 9$ and $10 + c = 11 \qquad c = 1$

Probability: $\geq 1 - \frac{\left(\frac{1}{3}\right)^2}{1^2} = \frac{8}{9}$

14. $\mu = 50, \sigma = 8$

$50 - c = 38$ and $50 + c = 62 \qquad c = 12$

Probability $\geq 1 - \frac{8^2}{12^2} = \frac{5}{9}$

15. $\Pr(6.5 \leq X \leq 11) = \Pr\left(\frac{6.5-5}{3} \leq Z \leq \frac{11-5}{3}\right)$

$= A(2) - A(0.5)$

$= 0.9772 - 0.6915$

$= 0.2857$

16. $\Pr(Z \geq 0.75) = 1 - 0.7734 = 0.2266$

Copyright © 2018 Pearson Education, Inc.

17. $\mu = 5.75, \sigma = 0.2$

$$\begin{aligned}\Pr(X \geq 6) &= \Pr\left(Z \geq \frac{6-5.75}{0.2}\right)\\ &= \Pr(Z \geq 1.25)\\ &= 1 - 0.8944\\ &= 0.1056\end{aligned}$$

10.56%

18. $\Pr(Z \geq z) = 0.7734$
$\Pr(Z < z) = 1 - 0.7734 = 0.2266$
$z = -0.75$

19. $\mu = 80, \sigma = 15$
$\Pr(80 - h \leq X \leq 80 + h) = 0.8664$
$\frac{1 - 0.8664}{2} = 0.0668 \Rightarrow$ (area left of $80 - h$)
$\Pr(Z \leq z) = 0.0668$ when $z = -1.5$
$\Pr(-1.5 \leq Z \leq 1.5) = 0.8664$
Therefore, $\frac{x-\mu}{\sigma} = -1.5$ and $\frac{x+\mu}{\sigma} = 1.5.$
$\frac{(80-h)-80}{15} = -1.5$ and $\frac{(80+h)-80}{15} = 1.5$
$h = 22.5$

20. a. $$\begin{aligned}\Pr(133 \leq X) &\approx \Pr\left(\frac{132.5-100}{15} \leq Z\right)\\ &\approx \Pr(2.167 \leq Z)\\ &\approx \Pr(2.20 \leq Z)\\ &= 1 - 0.9861\\ &= 0.0139\\ &= 1.39\%\end{aligned}$$

b. $$\begin{aligned}x_{95} &= 100 + 15z_{95}\\ &= 100 + 15 \cdot 1.65\\ &= 124.75\end{aligned}$$

21. $n = 54,\ p = \frac{2}{5}$

$\mu = 54\left(\frac{2}{5}\right) = 21.6$

$\sigma = \sqrt{54\left(\frac{2}{5}\right)\left(\frac{3}{5}\right)} = 3.6$

$\Pr(X \leq 13) \approx \Pr\left(Z \leq \frac{13.5-21.6}{3.6}\right) = \Pr(Z \leq -2.25) = 0.0122$

Copyright © 2018 Pearson Education, Inc.

22. $n = 75,\ p = \frac{1}{4}$

$$\mu = 75\left(\frac{1}{4}\right) = 18.75$$

$$\sigma = \sqrt{75\left(\frac{1}{4}\right)\left(\frac{3}{4}\right)} = 3.75$$

$$\Pr(8 \le X \le 22) \approx \Pr\left(\frac{7.5-18.75}{3.75} \le Z \le \frac{22.5-18.75}{3.75}\right) = \Pr(-3 \le Z \le 1) = 0.8413 - 0.0013 = 0.84$$

Conceptual Exercises

23. a. scoring in the third quartile is not very good: 100, 40, 40, 40,

b. scoring in the third quartile corresponds to a perfect grade: 100, 100, 90, 80, 70

24. a. The mean and median are equal: 1, 2, 3, 4, 5, 6, 7, 8, 9, 10 : The mean is 5.5; the median is 5.5

b. the mean is less than the median: 1, 1, 1, 1, 4, 5, 6, 7, 8, 9 : The mean is 4.3; the median is 4.5

c. the median is less than the mean 1, 2, 3, 4, 5, 6, 10, 12, 14, 100. The median is 5.5; the mean is 15.7

25. A population mean is the average of all the data in the entire population. When a sample is taken from a population, the sample mean is the average of all the data in that particular sample. Sample means vary whereas the population mean is fixed.

26. A sample mean is taken from the population and is the average of all the data values in a particular sample, which is a subset of the population.

27. Yes; in general, if we add a constant to each number in a set, then the mean will increase by that constant.

28. Yes; in general, if we multiply each number in a set by some constant, then the standard deviation will be multiplied by that constant.

29. The binomial probability distribution applies when there is a fixed number of independent trials when the probability of success is constant. The outcome of each trial is classified as either a "success" or a "failure".

30. Repeated trials that do not produce a binomial distribution: 1) tossing a coin until a head appears. 2) Having children until a girl is born.

Copyright © 2018 Pearson Education, Inc.

Chapter 8

Exercises 8.1

1. Yes; the matrix is square, all entries are ≥ 0, and the sum of the entries in each column is 1.

3. No; the matrix is not square.

5. Yes; the matrix is square, all entries are ≥ 0, and the sum of the entries in each column is 1.

7. $\begin{array}{c} \\ A \\ B \end{array}\begin{array}{c} \begin{matrix} A & B \end{matrix} \\ \begin{bmatrix} 0.3 & 0.5 \\ 0.7 & 0.5 \end{bmatrix} \end{array}$

9. $\begin{array}{c} \\ A \\ B \\ C \end{array}\begin{array}{c} \begin{matrix} A & B & C \end{matrix} \\ \begin{bmatrix} \frac{1}{3} & \frac{2}{9} & \frac{1}{3} \\ \frac{1}{3} & \frac{4}{9} & \frac{1}{6} \\ \frac{1}{3} & \frac{1}{3} & \frac{1}{2} \end{bmatrix} \end{array}$

11. $\begin{array}{c} \\ A \\ B \\ C \end{array}\begin{array}{c} \begin{matrix} A & B & C \end{matrix} \\ \begin{bmatrix} 0.4 & 0.2 & 0 \\ 0.5 & 0 & 0 \\ 0.1 & 0.8 & 1 \end{bmatrix} \end{array}$

13.

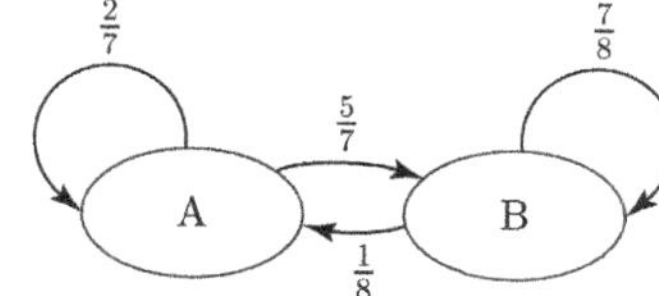

15.

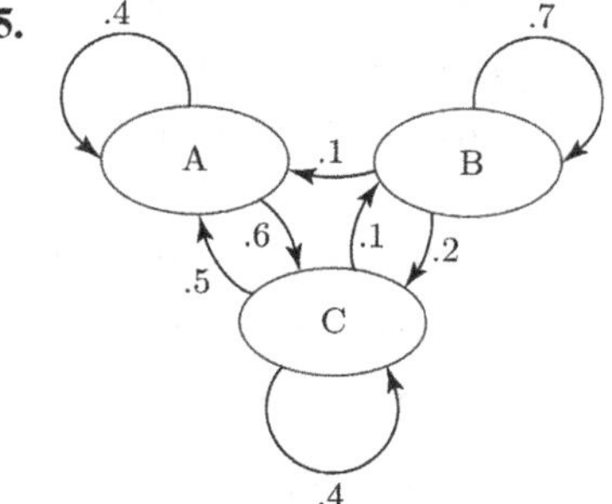

17.

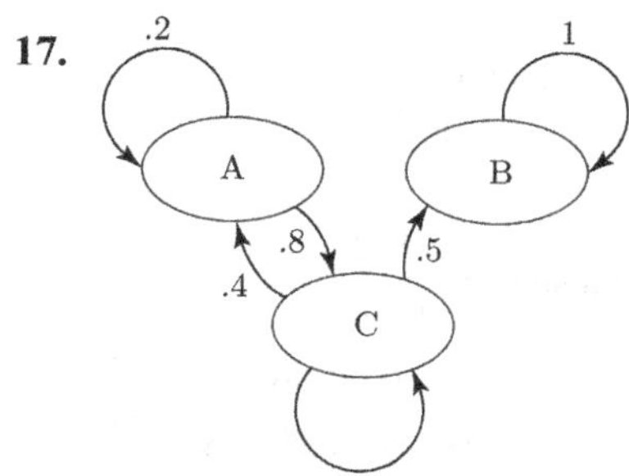

19. initial state matrix $= \begin{bmatrix} 0.47 \\ 0.53 \end{bmatrix}_0$

next state $= A\begin{bmatrix} 0.47 \\ 0.53 \end{bmatrix}_0 = \begin{bmatrix} 0.8 & 0.3 \\ 0.2 & 0.7 \end{bmatrix}\begin{bmatrix} 0.47 \\ 0.53 \end{bmatrix}_0$

$= \begin{bmatrix} 0.535 \\ 0.465 \end{bmatrix}_1$

next state $= A^2\begin{bmatrix} 0.47 \\ 0.53 \end{bmatrix}_0 = \begin{bmatrix} 0.70 & 0.45 \\ 0.30 & 0.55 \end{bmatrix}\begin{bmatrix} 0.47 \\ 0.53 \end{bmatrix}_0$

$= \begin{bmatrix} 0.5675 \\ 0.4325 \end{bmatrix}_2$

After one generation, about 54% of French women will work. After two generations, about 57% of French women will work outside the home..

21. a. $A = \begin{array}{c} \\ L \\ H \end{array}\begin{array}{c} \begin{matrix} L & H \end{matrix} \\ \begin{bmatrix} 0.8 & 0.3 \\ 0.2 & 0.7 \end{bmatrix} \end{array}$

b. initial state matrix $= \begin{bmatrix} 0.5 \\ 0.5 \end{bmatrix}_0$

next state $= A\begin{bmatrix} 0.5 \\ 0.5 \end{bmatrix}_0 = \begin{bmatrix} 0.8 & 0.3 \\ 0.2 & 0.7 \end{bmatrix}\begin{bmatrix} 0.5 \\ 0.5 \end{bmatrix}_0 = \begin{bmatrix} 0.55 \\ 0.45 \end{bmatrix}$

Therefore, 55% of the customers would be Low users in February

next $= A\begin{bmatrix} 0.55 \\ 0.45 \end{bmatrix}_1 = \begin{bmatrix} 0.8 & 0.3 \\ 0.2 & 0.7 \end{bmatrix}\begin{bmatrix} 0.55 \\ 0.45 \end{bmatrix}_1 = \begin{bmatrix} 0.575 \\ 0.425 \end{bmatrix}$

Therefore, 57.5% of the customers would be Low users in March.

Copyright © 2018 Pearson Education, Inc.

23. a.

$$\begin{array}{c} \\ S \\ O \end{array}\begin{array}{c} \begin{array}{cc} S & O \end{array} \\ \begin{bmatrix} 0.992 & 0.007 \\ 0.008 & 0.993 \end{bmatrix} \end{array}$$

b. initial state $= \begin{bmatrix} 0.12 \\ 0.88 \end{bmatrix}_0$

$$\text{next} = \begin{bmatrix} 0.992 & 0.007 \\ 0.008 & 0.993 \end{bmatrix}\begin{bmatrix} 0.12 \\ 0.88 \end{bmatrix}_0 = \begin{bmatrix} 0.1252 \\ 0.8748 \end{bmatrix}_1$$

$$\text{next state} = \begin{bmatrix} 0.992 & 0.007 \\ 0.008 & 0.993 \end{bmatrix}\begin{bmatrix} 0.1252 \\ 0.8748 \end{bmatrix}_1 = \begin{bmatrix} 0.1303 \\ 0.8697 \end{bmatrix}_2$$

Therefore, 12.52% and 13.03% will be the estimate of people living in the Southwest at the beginning of 2016 and 2017 respectively.

25. a.

$$\begin{array}{c} \\ L \\ R \end{array}\begin{array}{c} \begin{array}{cc} L & R \end{array} \\ \begin{bmatrix} 0.9 & 0.7 \\ 0.1 & 0.3 \end{bmatrix} \end{array}$$

b. $A^2 = \begin{bmatrix} 0.9 & 0.7 \\ 0.1 & 0.3 \end{bmatrix}\begin{bmatrix} 0.9 & 0.7 \\ 0.1 & 0.3 \end{bmatrix} = \begin{bmatrix} 0.88 & 0.84 \\ 0.12 & 0.16 \end{bmatrix}$

c. initial state $= \begin{bmatrix} 0.5 \\ 0.5 \end{bmatrix}_0$

$$\begin{bmatrix} \; \\ \; \end{bmatrix}_1 = \begin{bmatrix} 0.9 & 0.7 \\ 0.1 & 0.3 \end{bmatrix}\begin{bmatrix} 0.5 \\ 0.5 \end{bmatrix}_0 = \begin{bmatrix} 0.8 \\ 0.2 \end{bmatrix}_1$$

$$\begin{bmatrix} \; \\ \; \end{bmatrix}_2 = A^2\begin{bmatrix} 0.5 \\ 0.5 \end{bmatrix}_0 = \begin{bmatrix} 0.88 & 0.84 \\ 0.12 & 0.16 \end{bmatrix}\begin{bmatrix} 0.5 \\ 0.5 \end{bmatrix}_0 = \begin{bmatrix} 0.86 \\ 0.14 \end{bmatrix}_2$$

d. Answers will vary. Correct answer is 87.5%.

27. initial state matrix $= \begin{bmatrix} 0.40 \\ 0.40 \\ 0.20 \end{bmatrix}_0$

$$\text{next state} = \begin{bmatrix} 0.5 & 0.4 & 0.2 \\ 0.4 & 0.3 & 0.6 \\ 0.1 & 0.3 & 0.2 \end{bmatrix}\begin{bmatrix} 0.4 \\ 0.4 \\ 0.2 \end{bmatrix} = \begin{bmatrix} 0.4 \\ 0.4 \\ 0.2 \end{bmatrix}_1$$

40% will be Zone I, 40% will be in Zone II, 20% will be in Zone III.

29. a. 4%

b. 96% of the freshmen who held middle-of-the-road political views continued to hold these views as sophomores.

c.

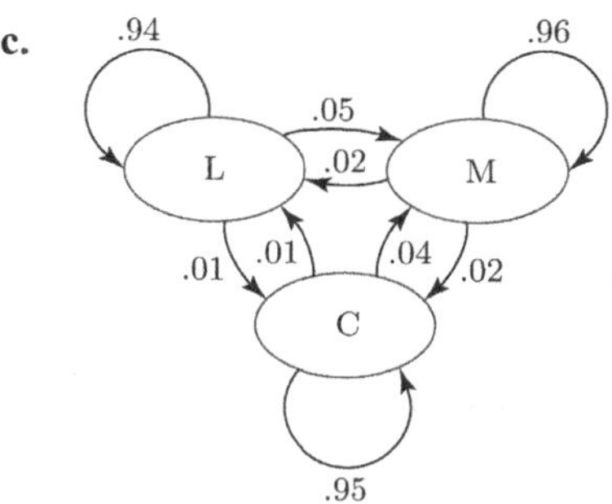

d. initial state $= \begin{bmatrix} 0.33 \\ 0.45 \\ 0.22 \end{bmatrix}_0$

$$\text{next} = \begin{bmatrix} 0.94 & 0.02 & 0.01 \\ 0.05 & 0.96 & 0.04 \\ 0.01 & 0.02 & 0.95 \end{bmatrix}\begin{bmatrix} 0.33 \\ 0.45 \\ 0.22 \end{bmatrix}_0 = \begin{bmatrix} 0.3214 \\ 0.4573 \\ 0.2213 \end{bmatrix}_1$$

$$\text{next state} = \begin{bmatrix} 0.94 & 0.02 & 0.01 \\ 0.05 & 0.96 & 0.04 \\ 0.01 & 0.02 & 0.95 \end{bmatrix}\begin{bmatrix} 0.3214 \\ 0.4573 \\ 0.2213 \end{bmatrix}_1 \approx \begin{bmatrix} 0.3135 \\ 0.4639 \\ 0.2226 \end{bmatrix}_2$$

45.73% of the students held middle-of-the-road political views as sophomores, 46.39% as juniors.

Copyright © 2018 Pearson Education, Inc.

31. a.
$$\begin{array}{c} \\ U \\ S \\ R \end{array}\begin{array}{c} \begin{array}{ccc} U & S & R \end{array} \\ \begin{bmatrix} 0.86 & 0.05 & 0.03 \\ 0.08 & 0.86 & 0.05 \\ 0.06 & 0.09 & 0.92 \end{bmatrix} \end{array}$$

b. Since we are concerned with only the people who live in urban areas of 2017, we use

$$\begin{bmatrix} \ \\ \ \\ \ \end{bmatrix}_0 = \begin{bmatrix} 1 \\ 0 \\ 0 \end{bmatrix}_0.$$

$$\begin{bmatrix} \ \\ \ \\ \ \end{bmatrix}_2 = A^2\begin{bmatrix} \ \\ \ \\ \ \end{bmatrix}_0$$

$$= \begin{bmatrix} 0.86 & 0.05 & 0.03 \\ 0.08 & 0.86 & 0.05 \\ 0.06 & 0.09 & 0.92 \end{bmatrix}\begin{bmatrix} 0.86 & 0.05 & 0.03 \\ 0.08 & 0.86 & 0.05 \\ 0.06 & 0.09 & 0.92 \end{bmatrix}\begin{bmatrix} 1 \\ 0 \\ 0 \end{bmatrix}_0$$

$$= \begin{bmatrix} 0.7454 \\ 0.1406 \\ 0.114 \end{bmatrix}_2$$

11.4% of people who live in urban areas in 2017 will live in rural areas in 2019.

33. a. initial state $= \begin{bmatrix} 0.4 \\ 0.4 \\ 0.2 \end{bmatrix}_0$

$$\begin{bmatrix} \ \\ \ \\ \ \end{bmatrix}_1 = \begin{bmatrix} 0.5 & 0.25 & 0.2 \\ 0.4 & 0.5 & 0.2 \\ 0.1 & 0.25 & 0.6 \end{bmatrix}\begin{bmatrix} 0.4 \\ 0.4 \\ 0.2 \end{bmatrix}_0$$

$$= \begin{bmatrix} 0.34 \\ 0.40 \\ 0.26 \end{bmatrix}_1$$

b. $\begin{bmatrix} \ \\ \ \end{bmatrix}_2 = A\begin{bmatrix} \ \\ \ \end{bmatrix}_1$

$$= \begin{bmatrix} 0.5 & 0.25 & 0.2 \\ 0.4 & 0.5 & 0.2 \\ 0.1 & 0.25 & 0.6 \end{bmatrix}\begin{bmatrix} 0.34 \\ 0.4 \\ 0.26 \end{bmatrix}_1$$

$$= \begin{bmatrix} 0.322 \\ 0.388 \\ 0.29 \end{bmatrix}_2$$

0.322 is the probability that a woman that had high birth weight will have a granddaughter of high birth weight and 0.388 is the probability that her granddaughter will have average birth weights.

35. a.
$$\begin{array}{c} \\ 0 \\ 1 \\ 2 \\ 3 \\ 4 \end{array}\begin{array}{c} \begin{array}{ccccc} 0 & 1 & 2 & 3 & 4 \end{array} \\ \begin{bmatrix} 0 & 0.25 & 0 & 0 & 0 \\ 1 & 0 & 0.5 & 0 & 0 \\ 0 & 0.75 & 0 & 0.75 & 0 \\ 0 & 0 & 0.5 & 0 & 1 \\ 0 & 0 & 0 & 0.25 & 0 \end{bmatrix} \end{array}$$

b. Using the initial matrix as follows:

$$\begin{bmatrix} 0 \\ 0 \\ 0 \\ 1 \\ 0 \end{bmatrix}_0$$

Calculating $A^2 \cdot (\text{initial})$ will yield:

$$\begin{bmatrix} 0 \\ 0.375 \\ 0 \\ 0.625 \\ 0 \end{bmatrix}_2$$

Therefore, there is a .625 probability that there will be 3 balls in Urn A after two time periods.

37. $A^2 = \begin{bmatrix} 0.45 & 0.44 \\ 0.55 & 0.56 \end{bmatrix}$, $A^3 = \begin{bmatrix} 0.445 & 0.444 \\ 0.555 & 0.556 \end{bmatrix}$,

$A^4 = \begin{bmatrix} 0.4445 & 0.4444 \\ 0.5555 & 0.5556 \end{bmatrix}$

Copyright © 2018 Pearson Education, Inc.

$$\begin{bmatrix} \\ \end{bmatrix}_3 = \begin{bmatrix} 0.445 & 0.444 \\ 0.555 & 0.556 \end{bmatrix} \begin{bmatrix} 0.3 \\ 0.7 \end{bmatrix}_0 \approx \begin{bmatrix} 0.44 \\ 0.56 \end{bmatrix}_3$$

$$\begin{bmatrix} \\ \end{bmatrix}_4 = \begin{bmatrix} 0.4445 & 0.4444 \\ 0.5555 & 0.5556 \end{bmatrix} \begin{bmatrix} 0.3 \\ 0.7 \end{bmatrix}_0$$

$$= \begin{bmatrix} 0.44443 \\ 0.55557 \end{bmatrix}_4$$

$$\approx \begin{bmatrix} 0.44 \\ 0.56 \end{bmatrix}_4$$

39. $A^1 = A = \begin{bmatrix} \frac{1}{3} & \frac{1}{3} \\ \frac{2}{3} & \frac{2}{3} \end{bmatrix} \approx \begin{bmatrix} 0.33 & 0.33 \\ 0.67 & 0.67 \end{bmatrix}$

$A^2 = A \cdot A = \begin{bmatrix} \frac{1}{3} & \frac{1}{3} \\ \frac{2}{3} & \frac{2}{3} \end{bmatrix} \begin{bmatrix} \frac{1}{3} & \frac{1}{3} \\ \frac{2}{3} & \frac{2}{3} \end{bmatrix}$

$= \begin{bmatrix} \frac{1}{3} & \frac{1}{3} \\ \frac{2}{3} & \frac{2}{3} \end{bmatrix} \approx \begin{bmatrix} 0.33 & 0.33 \\ 0.67 & 0.67 \end{bmatrix}$

The pattern continues. All powers are $\begin{bmatrix} \frac{1}{3} & \frac{1}{3} \\ \frac{2}{3} & \frac{2}{3} \end{bmatrix}$.

41. $A^1 = A = \begin{bmatrix} 0.1 & 0.3 \\ 0.9 & 0.7 \end{bmatrix}$

$A^2 = A \cdot A = \begin{bmatrix} 0.1 & 0.3 \\ 0.9 & 0.7 \end{bmatrix} \begin{bmatrix} 0.1 & 0.3 \\ 0.9 & 0.7 \end{bmatrix}$

$= \begin{bmatrix} 0.28 & 0.24 \\ 0.72 & 0.76 \end{bmatrix}$

$A^3 = A^2 \cdot A = \begin{bmatrix} 0.28 & 0.24 \\ 0.72 & 0.76 \end{bmatrix} \begin{bmatrix} 0.1 & 0.3 \\ 0.9 & 0.7 \end{bmatrix}$

$= \begin{bmatrix} 0.244 & 0.252 \\ 0.756 & 0.748 \end{bmatrix} \approx \begin{bmatrix} 0.24 & 0.25 \\ 0.76 & 0.75 \end{bmatrix}$

$A^4 = A^3 \cdot A = \begin{bmatrix} 0.244 & 0.252 \\ 0.756 & 0.748 \end{bmatrix} \begin{bmatrix} 0.1 & 0.3 \\ 0.9 & 0.7 \end{bmatrix}$

$= \begin{bmatrix} 0.2512 & 0.2496 \\ 0.7488 & 0.7504 \end{bmatrix} \approx \begin{bmatrix} 0.25 & 0.25 \\ 0.75 & 0.75 \end{bmatrix}$

$A^5 = A^4 \cdot A = \begin{bmatrix} 0.2512 & 0.2496 \\ 0.7488 & 0.7504 \end{bmatrix} \begin{bmatrix} 0.1 & 0.3 \\ 0.9 & 0.7 \end{bmatrix}$

$= \begin{bmatrix} .24976 & .25008 \\ .75024 & .74992 \end{bmatrix}$

$\approx \begin{bmatrix} .25 & .25 \\ .75 & .75 \end{bmatrix}$

43. $A^1 = A = \begin{bmatrix} 0.3 & 0.3 & 0.3 \\ 0.1 & 0.1 & 0.1 \\ 0.6 & 0.6 & 0.6 \end{bmatrix}$

$A^2 = A \cdot A = \begin{bmatrix} 0.3 & 0.3 & 0.3 \\ 0.1 & 0.1 & 0.1 \\ 0.6 & 0.6 & 0.6 \end{bmatrix} \begin{bmatrix} 0.3 & 0.3 & 0.3 \\ 0.1 & 0.1 & 0.1 \\ 0.6 & 0.6 & 0.6 \end{bmatrix}$

$= \begin{bmatrix} 0.3 & 0.3 & 0.3 \\ 0.1 & 0.1 & 0.1 \\ 0.6 & 0.6 & 0.6 \end{bmatrix}$

The pattern continues. All powers are $\begin{bmatrix} 0.3 & 0.3 & 0.3 \\ 0.1 & 0.1 & 0.1 \\ 0.6 & 0.6 & 0.6 \end{bmatrix}$.

45. No; all powers have a zero entry in the upper right corner, so the matrix is not regular.

47. a. Use the method described in the text to generate the next four distribution matrices. $\begin{bmatrix} 0.35 \\ 0.65 \end{bmatrix}, \begin{bmatrix} 0.425 \\ 0.575 \end{bmatrix}, \begin{bmatrix} 0.3875 \\ 0.6125 \end{bmatrix}, \begin{bmatrix} 0.40625 \\ 0.59375 \end{bmatrix}$

b. $A^4 B = \begin{bmatrix} 0.4375 & 0.375 \\ 0.5625 & 0.625 \end{bmatrix} \begin{bmatrix} 0.5 \\ 0.5 \end{bmatrix} = \begin{bmatrix} 0.40625 \\ 0.59375 \end{bmatrix}$

49. The distribution matrix gets closer and closer to $\begin{bmatrix} 0.4 \\ 0.6 \end{bmatrix}$.

51. The matrices get closer and closer to $\begin{bmatrix} 0.4 & 0.4 \\ 0.6 & 0.6 \end{bmatrix}$.

Each column of this matrix is the same as the 2 by 1 matrix found in Exercise 49.

Exercises 8.2

1. Yes; the matrix is regular, since all entries are positive.

 Copyright © 2018 Pearson Education, Inc.

3. Yes; the matrix is regular, since the second power is $\begin{bmatrix}0.79 & 0.3\\ 0.21 & 0.7\end{bmatrix}$, which has all positive entries.

5. No; all powers have zero entries in the second columns.

7. $\begin{cases} x+y=1 \\ \begin{bmatrix}0.5 & 0.1\\ 0.5 & 0.9\end{bmatrix}\begin{bmatrix}x\\ y\end{bmatrix}=\begin{bmatrix}x\\ y\end{bmatrix}\end{cases}$

$$\begin{cases} x + y = 1 \\ 0.5x + 0.1y = x \\ 0.5x + 0.9y = y \end{cases}$$

$$\begin{cases} x + y = 1 \\ -0.5x + 0.1y = 0 \\ 0.5x - 0.1y = 0 \end{cases}$$

The second and third equations in this system are equivalent.

$$\left[\begin{array}{cc|c} 1 & 1 & 1 \\ -0.5 & 0.1 & 0 \end{array}\right] \xrightarrow{[2]+0.5[1]} \left[\begin{array}{cc|c} 1 & 1 & 1 \\ 0 & 0.6 & 0.5 \end{array}\right]$$

$$\xrightarrow{\frac{5}{3}[2]} \left[\begin{array}{cc|c} 1 & 1 & 1 \\ 0 & 1 & \frac{5}{6} \end{array}\right]$$

$$\xrightarrow{[1]+(-1)[2]} \left[\begin{array}{cc|c} 1 & 0 & \frac{1}{6} \\ 0 & 1 & \frac{5}{6} \end{array}\right]$$

$x=\frac{1}{6},\ y=\frac{5}{6}$

The stable distribution is $\begin{bmatrix}x\\ y\end{bmatrix}=\begin{bmatrix}\frac{1}{6}\\ \frac{5}{6}\end{bmatrix}$.

9. $\begin{cases} x+y=1 \\ \begin{bmatrix}0.8 & 0.3\\ 0.2 & 0.7\end{bmatrix}\begin{bmatrix}x\\ y\end{bmatrix}=\begin{bmatrix}x\\ y\end{bmatrix}\end{cases}$

$$\begin{cases} x + y = 1 \\ 0.8x + 0.3y = x \\ 0.2x + 0.7y = y \end{cases}$$

$$\begin{cases} x + y = 1 \\ -0.2x + 0.3y = 0 \\ 0.2x - 0.3y = 0 \end{cases}$$

The second and third equations in this system are equivalent.

$$\left[\begin{array}{cc|c} 1 & 1 & 1 \\ -0.2 & 0.3 & 0 \end{array}\right] \xrightarrow{[2]+0.2[1]} \left[\begin{array}{cc|c} 1 & 1 & 1 \\ 0 & 0.5 & 0.2 \end{array}\right]$$

$$\xrightarrow{2[2]} \left[\begin{array}{cc|c} 1 & 1 & 1 \\ 0 & 1 & 0.4 \end{array}\right]$$

$$\xrightarrow{[1]+(-1)[2]} \left[\begin{array}{cc|c} 1 & 0 & 0.6 \\ 0 & 1 & 0.4 \end{array}\right]$$

$x=.6,\ y=.4$

The stable distribution is $\begin{bmatrix}x\\ y\end{bmatrix}=\begin{bmatrix}0.6\\ 0.4\end{bmatrix}$.

11. $\begin{cases} x+y+z=1 \\ \begin{bmatrix}0.1 & 0.4 & 0.7\\ 0.6 & 0.4 & 0.2\\ 0.3 & 0.2 & 0.1\end{bmatrix}\begin{bmatrix}x\\ y\\ z\end{bmatrix}=\begin{bmatrix}x\\ y\\ z\end{bmatrix}\end{cases}$

$$\begin{cases} x + y + z = 1 \\ 0.1x + 0.4y + 0.7z = x \\ 0.6x + 0.4y + 0.2z = y \\ 0.3x + 0.2y + 0.1z = z \end{cases}$$

$$\begin{cases} x + y + z = 1 \\ -0.9x + 0.4y + 0.7z = 0 \\ 0.6x - 0.6y + 0.2z = 0 \\ 0.3x + 0.2y - 0.9z = 0 \end{cases}$$

$$\left[\begin{array}{ccc|c} 1 & 1 & 1 & 1 \\ -0.9 & 0.4 & 0.7 & 0 \\ 0.6 & -0.6 & 0.2 & 0 \\ 0.3 & 0.2 & -0.9 & 0 \end{array}\right]$$

$$\xrightarrow{\substack{10[2]\\ 10[3]\\ 10[4]}} \left[\begin{array}{ccc|c} 1 & 1 & 1 & 1 \\ -9 & 4 & 7 & 0 \\ 6 & -6 & 2 & 0 \\ 3 & 2 & -9 & 0 \end{array}\right]$$

$$\xrightarrow{\substack{[2]+9[1]\\ [3]+(-6)[1]\\ [4]+(-3)[1]}} \left[\begin{array}{ccc|c} 1 & 1 & 1 & 1 \\ 0 & 13 & 16 & 9 \\ 0 & -12 & -4 & -6 \\ 0 & -1 & -12 & -3 \end{array}\right]$$

$$\xrightarrow{\substack{\frac{1}{13}[2]\\ [1]+(-1)[2]\\ [3]+12[2]\\ [4]+1[2]}} \left[\begin{array}{ccc|c} 1 & 0 & -\frac{3}{13} & \frac{4}{13} \\ 0 & 1 & \frac{16}{13} & \frac{9}{13} \\ 0 & 0 & \frac{140}{13} & \frac{30}{13} \\ 0 & 0 & -\frac{140}{13} & -\frac{30}{13} \end{array}\right]$$

Copyright © 2018 Pearson Education, Inc.

$$\xrightarrow[\substack{[1]+\frac{3}{13}[3] \\ [2]-\frac{16}{13}[3] \\ [4]+\frac{140}{13}[3]}]{\frac{13}{140}[3]} \left[\begin{array}{ccc|c} 1 & 0 & 0 & \frac{5}{14} \\ 0 & 1 & 0 & \frac{3}{7} \\ 0 & 0 & 1 & \frac{3}{14} \\ 0 & 0 & 0 & 0 \end{array}\right]$$

$x = \dfrac{5}{14},\ y = \dfrac{3}{7},\ z = \dfrac{3}{14}$

The stable distribution is $\begin{bmatrix} x \\ y \\ z \end{bmatrix} = \begin{bmatrix} \frac{5}{14} \\ \frac{3}{7} \\ \frac{3}{14} \end{bmatrix}$.

13. The stochastic matrix is $\begin{array}{c} \\ L \\ H \end{array}\begin{array}{c} \begin{array}{cc} L & H \end{array} \\ \begin{bmatrix} 0.8 & 0.3 \\ 0.2 & 0.7 \end{bmatrix} \end{array}$.

Find the stable distribution.

$$\begin{cases} x+y=1 \\ \begin{bmatrix} 0.8 & 0.3 \\ 0.2 & 0.7 \end{bmatrix}\begin{bmatrix} x \\ y \end{bmatrix} = \begin{bmatrix} x \\ y \end{bmatrix} \end{cases}$$

$$\begin{cases} x + y = 1 \\ 0.8x + 0.3y = x \\ 0.2x + 0.7y = y \end{cases}$$

$$\begin{cases} x + y = 1 \\ -0.2x + 0.3y = 0 \\ 0.2x - 0.3y = 0 \end{cases}$$

The second and third equations in this system are equivalent.

$$\left[\begin{array}{cc|c} 1 & 1 & 1 \\ -0.2 & 0.3 & 0 \end{array}\right] \xrightarrow{[2]+0.2[1]} \left[\begin{array}{cc|c} 1 & 1 & 1 \\ 0 & 0.5 & 0.2 \end{array}\right]$$

$$\xrightarrow{2[2]} \left[\begin{array}{cc|c} 1 & 1 & 1 \\ 0 & 1 & 0.4 \end{array}\right]$$

$$\xrightarrow{[1]+(-1)[2]} \left[\begin{array}{cc|c} 1 & 0 & 0.6 \\ 0 & 1 & 0.4 \end{array}\right]$$

$x = 0.6,\ y = 0.4$

The stable distribution is $\begin{bmatrix} x \\ y \end{bmatrix} = \begin{bmatrix} 0.6 \\ 0.4 \end{bmatrix}$.

In the long run, 40% of the customers will be High users.

15. The stochastic matrix is $\begin{array}{c} \\ L \\ R \end{array}\begin{array}{c} \begin{array}{cc} L & R \end{array} \\ \begin{bmatrix} 0.9 & 0.7 \\ 0.1 & 0.3 \end{bmatrix} \end{array}$.

Find the stable distribution.

$$\begin{cases} x+y=1 \\ \begin{bmatrix} 0.9 & 0.7 \\ 0.1 & 0.3 \end{bmatrix}\begin{bmatrix} x \\ y \end{bmatrix} = \begin{bmatrix} x \\ y \end{bmatrix} \end{cases}$$

$$\begin{cases} x + y = 1 \\ 0.9x + 0.7y = x \\ 0.1x + 0.3y = y \end{cases}$$

$$\begin{cases} x + y = 1 \\ -0.1x + 0.7y = 0 \\ 0.1x - 0.7y = 0 \end{cases}$$

The second and third equations in this system are equivalent.

$$\left[\begin{array}{cc|c} 1 & 1 & 1 \\ -0.1 & 0.7 & 0 \end{array}\right] \xrightarrow{[2]+0.1[1]} \left[\begin{array}{cc|c} 1 & 1 & 1 \\ 0 & 0.8 & 0.1 \end{array}\right]$$

$$\xrightarrow{1.25[2]} \left[\begin{array}{cc|c} 1 & 1 & 1 \\ 0 & 1 & 0.125 \end{array}\right]$$

$$\xrightarrow{[1]+(-1)[2]} \left[\begin{array}{cc|c} 1 & 0 & 0.875 \\ 0 & 1 & 0.125 \end{array}\right]$$

$x = 0.785,\ y = 0.125$

The stable distribution is $\begin{bmatrix} x \\ y \end{bmatrix} = \begin{bmatrix} 0.875 \\ 0.125 \end{bmatrix}$.

After many days, 87.5% of the mice will be going to the left.

17. The stochastic matrix is $\begin{array}{c} \\ \text{GM} \\ \text{non-GM} \end{array}\begin{array}{c} \begin{array}{cc} \text{GM} & \text{non-GM} \end{array} \\ \begin{bmatrix} 0.6 & 0.1 \\ 0.4 & 0.9 \end{bmatrix} \end{array}$.

Find the stable distribution.

$$\begin{cases} x+y=1 \\ \begin{bmatrix} 0.6 & 0.1 \\ 0.4 & 0.9 \end{bmatrix}\begin{bmatrix} x \\ y \end{bmatrix} = \begin{bmatrix} x \\ y \end{bmatrix} \end{cases}$$

$$\begin{cases} x+y=1 \\ 0.6x+0.1y=x \\ 0.4x+0.9y=y \end{cases}$$

$$\begin{cases} x+y=1 \\ -0.4x+0.1y=0 \\ 0.4x-0.1y=0 \end{cases}$$

Copyright © 2018 Pearson Education, Inc.

The second and third equations in the system are equivalent.

$$\left[\begin{array}{cc|c} 1 & 1 & 1 \\ -0.4 & 0.1 & 0 \end{array}\right] \xrightarrow{[2]+0.4[1]} \left[\begin{array}{cc|c} 1 & 1 & 1 \\ 0 & 0.5 & 0.4 \end{array}\right]$$

$$\xrightarrow{2[2]} \left[\begin{array}{cc|c} 1 & 1 & 1 \\ 0 & 1 & 0.8 \end{array}\right] \xrightarrow{[1]+(-1)[2]} \left[\begin{array}{cc|c} 1 & 0 & 0.2 \\ 0 & 1 & 0.8 \end{array}\right]$$

The stable distribution is $\begin{bmatrix} 0.2 \\ 0.8 \end{bmatrix}$.

In the long run General Motors market share is 20%.

19. The stochastic matrix is $\begin{array}{c} \\ R \\ S \end{array}\!\!\begin{array}{c} \begin{array}{cc} R & S \end{array} \\ \begin{bmatrix} 0.1 & 0.6 \\ 0.9 & 0.4 \end{bmatrix} \end{array}$.

Find the stable distribution.

$$\begin{cases} x+y=1 \\ \begin{bmatrix} 0.1 & 0.6 \\ 0.9 & 0.4 \end{bmatrix}\begin{bmatrix} x \\ y \end{bmatrix} = \begin{bmatrix} x \\ y \end{bmatrix} \end{cases}$$

$$\begin{cases} x + y = 1 \\ 0.1x + 0.6y = x \\ 0.9x + 0.4y = y \end{cases}$$

$$\begin{cases} x + y = 1 \\ -0.9x + 0.6y = 0 \\ 0.9x - 0.6y = 0 \end{cases}$$

The second and third equations in this system are equivalent.

$$\left[\begin{array}{cc|c} 1 & 1 & 1 \\ -0.9 & 0.6 & 0 \end{array}\right] \xrightarrow{[2]+0.9[1]} \left[\begin{array}{cc|c} 1 & 1 & 1 \\ 0 & 1.5 & 0.9 \end{array}\right]$$

$$\xrightarrow{\frac{2}{3}[2]} \left[\begin{array}{cc|c} 1 & 1 & 1 \\ 0 & 1 & 0.6 \end{array}\right]$$

$$\xrightarrow{[1]+(-1)[2]} \left[\begin{array}{cc|c} 1 & 0 & 0.4 \\ 0 & 1 & 0.6 \end{array}\right]$$

$x = .4, y = .6$

The stable distribution is $\begin{bmatrix} x \\ y \end{bmatrix} = \begin{bmatrix} 0.4 \\ 0.6 \end{bmatrix}$.

In the log run, the daily likelihood of rain is 40% or $\frac{2}{5}$.

21. a.

$$\begin{array}{c} \\ A \\ B \\ C \end{array}\!\!\begin{array}{c} \begin{array}{ccc} A & B & C \end{array} \\ \begin{bmatrix} 0.7 & 0.1 & 0.1 \\ 0.2 & 0.8 & 0.3 \\ 0.1 & 0.1 & 0.6 \end{bmatrix} \end{array}$$

b.

$$\begin{bmatrix} \\ \\ \\ \end{bmatrix}_1 = \begin{bmatrix} 0.7 & 0.1 & 0.1 \\ 0.2 & 0.8 & 0.3 \\ 0.1 & 0.1 & 0.6 \end{bmatrix}\begin{bmatrix} 0.4 \\ 0.3 \\ 0.3 \end{bmatrix}_0 = \begin{bmatrix} 0.34 \\ 0.41 \\ 0.25 \end{bmatrix}_1$$

$$\begin{bmatrix} \\ \\ \\ \end{bmatrix}_2 = \begin{bmatrix} 0.7 & 0.1 & 0.1 \\ 0.2 & 0.8 & 0.3 \\ 0.1 & 0.1 & 0.6 \end{bmatrix}\begin{bmatrix} 0.34 \\ 0.41 \\ 0.25 \end{bmatrix}_1 = \begin{bmatrix} 0.304 \\ 0.471 \\ 0.225 \end{bmatrix}_2$$

34% of the cars are at location A after one day, and 30.4% after two days.

c. Find the stable distribution.

$$\begin{cases} x+y+z=1 \\ 0.7x+0.1y+0.1z=x \\ 0.2x+0.8y+0.3z=y \\ 0.1x+0.1y+0.6z=z \end{cases}$$

$$\begin{cases} x+y+z=1 \\ -0.3x+0.1y+0.1z=0 \\ 0.2x-0.2y+0.3z=0 \\ 0.1x+0.1y-0.4z=0 \end{cases}$$

The fourth equation is equivalent to the second plus the third and so is redundant.

$$\left[\begin{array}{ccc|c} 1 & 1 & 1 & 1 \\ -.3 & .1 & .1 & 0 \\ .2 & -.2 & .3 & 0 \end{array}\right] \xrightarrow[{[3]-.2[1]}]{[2]+.3[1]} \left[\begin{array}{ccc|c} 1 & 1 & 1 & 1 \\ 0 & .4 & .4 & .3 \\ 0 & -.4 & .1 & -.2 \end{array}\right]$$

$$\xrightarrow[{-\frac{10}{4}[3]}]{\frac{10}{4}[2]} \left[\begin{array}{ccc|c} 1 & 1 & 1 & 1 \\ 0 & 1 & 1 & 0.75 \\ 0 & 1 & -0.25 & 0.5 \end{array}\right]$$

$$\xrightarrow[{[3]-[2]}]{[1]-[2]} \left[\begin{array}{ccc|c} 1 & 0 & 0 & 0.25 \\ 0 & 1 & 1 & 0.75 \\ 0 & 0 & -1.25 & -0.25 \end{array}\right]$$

$$\xrightarrow[{-\frac{8}{10}[3]}]{[2]+\frac{8}{10}[3]} \left[\begin{array}{ccc|c} 1 & 0 & 0 & 0.25 \\ 0 & 1 & 0 & 0.55 \\ 0 & 0 & 1 & 0.2 \end{array}\right]$$

Copyright © 2018 Pearson Education, Inc.

In the long run there are $\frac{1}{4}$ at location A, $\frac{11}{20}$ at location B and $\frac{1}{5}$ at location C.

23. The stochastic matrix is $\begin{matrix} & D & R & H \\ D & 0.5 & 0 & 0.25 \\ R & 0 & 0.5 & 0.25 \\ H & 0.5 & 0.5 & 0.5 \end{matrix}$.

Find the stable distribution.

$$\begin{cases} x+y+z=1 \\ \begin{bmatrix} 0.5 & 0 & 0.25 \\ 0 & 0.5 & 0.25 \\ 0.5 & 0.5 & 0.5 \end{bmatrix}\begin{bmatrix} x \\ y \\ z \end{bmatrix} = \begin{bmatrix} x \\ y \\ z \end{bmatrix} \end{cases}$$

$$\begin{cases} x + y + z = 1 \\ 0.5x + 0.25z = x \\ 0.5y + 0.25z = y \\ 0.5x + 0.5y + 0.5z = z \end{cases}$$

$$\begin{cases} x + y + z = 1 \\ -0.5x + 0.25z = 0 \\ -0.5y + 0.25z = 0 \\ 0.5x + 0.5y - 0.5z = 0 \end{cases}$$

$$\left[\begin{array}{ccc|c} 1 & 1 & 1 & 1 \\ -0.5 & 0 & 0.25 & 0 \\ 0 & -0.5 & 0.25 & 0 \\ 0.5 & 0.5 & -0.5 & 0 \end{array}\right]$$

$$\xrightarrow[2[4]]{4[2] \\ 4[3]} \left[\begin{array}{ccc|c} 1 & 1 & 1 & 1 \\ -2 & 0 & 1 & 0 \\ 0 & -2 & 1 & 0 \\ 1 & 1 & -1 & 0 \end{array}\right]$$

$$\xrightarrow[[4]+(-1)[1]]{[2]+2[1]} \left[\begin{array}{ccc|c} 1 & 1 & 1 & 1 \\ 0 & 2 & 3 & 2 \\ 0 & -2 & 1 & 0 \\ 0 & 0 & -2 & -1 \end{array}\right]$$

$$\xrightarrow[[3]+2[2]]{.5[2] \\ [1]+(-1)[2]} \left[\begin{array}{ccc|c} 1 & 0 & -.5 & 0 \\ 0 & 1 & 1.5 & 1 \\ 0 & 0 & 4 & 2 \\ 0 & 0 & -2 & -1 \end{array}\right]$$

$$\xrightarrow[[2]+(-1.5)[3] \\ [4]+2[3]]{0.25[3] \\ [1]+(0.5)[3]} \left[\begin{array}{ccc|c} 1 & 0 & 0 & 0.25 \\ 0 & 1 & 0 & 0.25 \\ 0 & 0 & 1 & 0.5 \\ 0 & 0 & 0 & 0 \end{array}\right]$$

$x = .25, y = .25, z = .5$

The stable distribution is $\begin{bmatrix} x \\ y \\ z \end{bmatrix} = \begin{bmatrix} 0.25 \\ 0.25 \\ 0.5 \end{bmatrix}$.

In the long run, 25% will be dominant.

25. $\begin{bmatrix} 0.6 \\ 0 \\ 0.4 \end{bmatrix}$ and $\begin{bmatrix} 0.3 \\ 0.5 \\ 0.2 \end{bmatrix}$ are stable distributions for the matrix $A = \begin{bmatrix} 0.4 & 0 & 0.9 \\ 0 & 1 & 0 \\ 0.6 & 0 & 0.1 \end{bmatrix}$ because their values add to 1and

$\begin{bmatrix} .4 & 0 & .9 \\ 0 & 1 & 0 \\ .6 & 0 & .1 \end{bmatrix}\begin{bmatrix} .6 \\ 0 \\ .4 \end{bmatrix} = \begin{bmatrix} .6 \\ 0 \\ .4 \end{bmatrix}$ and $\begin{bmatrix} .4 & 0 & .9 \\ 0 & 1 & 0 \\ .6 & 0 & .1 \end{bmatrix}\begin{bmatrix} .3 \\ .5 \\ .2 \end{bmatrix} = \begin{bmatrix} .3 \\ .5 \\ .2 \end{bmatrix}$.

However, given an arbitrary initial distribution $\begin{bmatrix} \\ \\ \\ \end{bmatrix}_0 \neq \begin{bmatrix} 0.6 \\ 0 \\ 0.4 \end{bmatrix}$ or $\begin{bmatrix} 0.3 \\ 0.5 \\ 0.2 \end{bmatrix}$, $A^n \begin{bmatrix} \\ \\ \\ \end{bmatrix}_0$ will not approach $\begin{bmatrix} 0.6 \\ 0 \\ 0.4 \end{bmatrix}$ or $\begin{bmatrix} 0.3 \\ 0.5 \\ 0.2 \end{bmatrix}$ as n gets large, so the existence of a stable distribution for A does not contradict the main premise of this section and the matrix is not regular.

27. The stochastic matrix is

$$\begin{matrix} & H & A & L \\ L & 0.5 & 0.25 & 0.2 \\ A & 0.4 & 0.5 & 0.2 \\ H & 0.1 & 0.25 & 0.6 \end{matrix}.$$

Find the stable distribution.

$$\begin{cases} x+y+z=1 \\ 0.5x+0.25y+0.2z=x \\ 0.4x+0.5y+0.2z=y \\ 0.1x+0.25y+0.6z=z \end{cases}$$

Copyright © 2018 Pearson Education, Inc.

$$\begin{cases} x+y+z=1 \\ -0.5x+0.25y+0.2z=0 \\ 0.4x-0.5y+0.2z=0 \\ 0.1x+0.25y-0.4z=0 \end{cases}$$

The fourth equation is equivalent to the second plus the third and so is redundant.

$$\left[\begin{array}{ccc|c} 1 & 1 & 1 & 1 \\ -.5 & .25 & .2 & 0 \\ .4 & -.5 & .2 & 0 \end{array}\right] \xrightarrow[\text{[3]}-.4\text{[1]}]{\text{[2]}+.5\text{[1]}} \left[\begin{array}{ccc|c} 1 & 1 & 1 & 1 \\ 0 & .75 & .7 & .5 \\ 0 & -.9 & -.2 & -.4 \end{array}\right]$$

$$\xrightarrow{\frac{1}{.75}\text{[2]}} \left[\begin{array}{ccc|c} 1 & 1 & 1 & 1 \\ 0 & 1 & \frac{14}{15} & \frac{2}{3} \\ 0 & -.9 & -.2 & -.4 \end{array}\right]$$

$$\xrightarrow[\text{[3]}+0.9\text{[2]}]{\text{[1]}-\text{[2]}} \left[\begin{array}{ccc|c} 1 & 0 & \frac{1}{15} & \frac{1}{3} \\ 0 & 1 & \frac{14}{15} & \frac{2}{3} \\ 0 & 0 & \frac{16}{25} & \frac{1}{5} \end{array}\right]$$

$$\xrightarrow{\frac{25}{16}\text{[3]}} \left[\begin{array}{ccc|c} 1 & 0 & \frac{1}{15} & \frac{1}{3} \\ 0 & 1 & \frac{14}{15} & \frac{2}{3} \\ 0 & 0 & 1 & \frac{5}{16} \end{array}\right]$$

$$\xrightarrow[\text{[2]}-\frac{14}{15}\text{[3]}]{\text{[1]}-\frac{1}{15}\text{[3]}} \left[\begin{array}{ccc|c} 1 & 0 & 0 & \frac{5}{16} \\ 0 & 1 & 0 & \frac{3}{8} \\ 0 & 0 & 1 & \frac{5}{16} \end{array}\right]$$

In the long run, the $\frac{5}{16}$ of female babies will have High birth weight and $\frac{3}{8}$ will have Average.

29. Calculating [A] ^ 255 gives $\begin{bmatrix} 0.7 & 0.7 \\ 0.3 & 0.3 \end{bmatrix}$.

This suggests that the stable distribution is $\begin{bmatrix} 0.7 \\ 0.3 \end{bmatrix}$.

Check: $x = .7, y = .3$ is indeed a solution to the system $\begin{cases} x+y=1 \\ \begin{bmatrix} 0.85 & 0.35 \\ 0.15 & 0.65 \end{bmatrix}\begin{bmatrix} x \\ y \end{bmatrix} = \begin{bmatrix} x \\ y \end{bmatrix} \end{cases}$.

Also, $\begin{bmatrix} 0.85 & 0.35 \\ 0.15 & 0.65 \end{bmatrix}\begin{bmatrix} 0.7 \\ 0.3 \end{bmatrix} = \begin{bmatrix} 0.7 \\ 0.3 \end{bmatrix}$.

31. Calculating [A] ^ 255 → Frac gives $\begin{bmatrix} \frac{8}{35} & \frac{8}{35} & \frac{8}{35} \\ \frac{3}{7} & \frac{3}{7} & \frac{3}{7} \\ \frac{12}{35} & \frac{12}{35} & \frac{12}{35} \end{bmatrix}$. This suggests that the stable distribution is $\begin{bmatrix} \frac{8}{35} \\ \frac{3}{7} \\ \frac{12}{35} \end{bmatrix}$.

Check: $x = \frac{8}{35}$, $y = \frac{3}{7}$, $z = \frac{12}{35}$ is indeed a solution to the system $\begin{cases} x+y+z=1 \\ \begin{bmatrix} 0.1 & 0.4 & 0.1 \\ 0.3 & 0.2 & 0.8 \\ 0.6 & 0.4 & 0.1 \end{bmatrix}\begin{bmatrix} x \\ y \\ z \end{bmatrix} = \begin{bmatrix} x \\ y \\ z \end{bmatrix} \end{cases}$.

Also, $\begin{bmatrix} 0.1 & 0.4 & 0.1 \\ 0.3 & 0.2 & 0.8 \\ 0.6 & 0.4 & 0.1 \end{bmatrix}\begin{bmatrix} \frac{8}{35} \\ \frac{3}{7} \\ \frac{12}{35} \end{bmatrix} = \begin{bmatrix} \frac{8}{35} \\ \frac{3}{7} \\ \frac{12}{35} \end{bmatrix}$.

Exercises 8.3

1. Yes; the states *A* and *B* are absorbing, and it's possible to get *A* and *B* from states *C* and *D*.

3. No; the state *A* is absorbing, but it's not possible to get to *A* from *C* or *D*.

5. No; states 1 and 2 are absorbing states, but states 3 and 4 do not lead to absorbing states.

7. Yes; state 1 is an absorbing state, and state 3 leads to state 1. Furthermore, it is possible to go from state 2 to state 1 through an intermediate step (state 2 to state 3 to state 1).

9.
$$\begin{array}{c} \\ B \\ A \\ C \end{array}\begin{array}{c} \begin{array}{ccc} B & A & C \end{array} \\ \begin{bmatrix} 1 & 0.3 & 0.4 \\ 0 & 0.2 & 0.5 \\ 0 & 0.5 & 0.1 \end{bmatrix} \end{array}$$

Copyright © 2018 Pearson Education, Inc.

11.
$$\begin{array}{c|cccc} & D & A & B & C \\ \hline D & 1 & 0.4 & 0 & 0.1 \\ A & 0 & 0.1 & 1 & 0.6 \\ B & 0 & 0.2 & 0 & 0.1 \\ C & 0 & 0.3 & 0 & 0.2 \end{array}$$

13.
$$\left[\begin{array}{cc|c} 1 & 0 & 0.3 \\ 0 & 1 & 0.2 \\ \hline 0 & 0 & 0.5 \end{array}\right] = \left[\begin{array}{c|c} I & S \\ \hline 0 & R \end{array}\right]$$

$R = [0.5]$; $S = \begin{bmatrix} 0.3 \\ 0.2 \end{bmatrix}$;

$F = (I - R)^{-1} = [1 - 0.5]^{-1} = [2]$

Find the stable matrix:

$$S(I-R)^{-1} = \begin{bmatrix} 0.3 \\ 0.2 \end{bmatrix}[2] = \begin{bmatrix} 0.6 \\ 0.4 \end{bmatrix}$$

$$\left[\begin{array}{c|c} I & S(I-R)^{-1} \\ \hline 0 & 0 \end{array}\right] = \left[\begin{array}{cc|c} 1 & 0 & 0.6 \\ 0 & 1 & 0.4 \\ \hline 0 & 0 & 0 \end{array}\right]$$

15.
$$\left[\begin{array}{cc|cc} 1 & 0 & \frac{1}{4} & \frac{1}{6} \\ 0 & 1 & \frac{1}{6} & 0 \\ \hline 0 & 0 & \frac{1}{4} & \frac{1}{2} \\ 0 & 0 & \frac{1}{3} & \frac{1}{3} \end{array}\right] = \left[\begin{array}{c|c} I & S \\ \hline 0 & R \end{array}\right]$$

$$R = \begin{bmatrix} \frac{1}{4} & \frac{1}{2} \\ \frac{1}{3} & \frac{1}{3} \end{bmatrix}; S = \begin{bmatrix} \frac{1}{4} & \frac{1}{6} \\ \frac{1}{6} & 0 \end{bmatrix}$$

Find the fundamental matrix:

$$I - R = \begin{bmatrix} 1 & 0 \\ 0 & 1 \end{bmatrix} - \begin{bmatrix} \frac{1}{4} & \frac{1}{2} \\ \frac{1}{3} & \frac{1}{3} \end{bmatrix} = \begin{bmatrix} \frac{3}{4} & -\frac{1}{2} \\ -\frac{1}{3} & \frac{2}{3} \end{bmatrix} = \begin{bmatrix} a & b \\ c & d \end{bmatrix}$$

$$\Delta = ad - bc = \left(\frac{3}{4}\right)\left(\frac{2}{3}\right) - \left(-\frac{1}{2}\right)\left(-\frac{1}{3}\right) = \frac{1}{3}$$

$$F = (I-R)^{-1} = \begin{bmatrix} \frac{d}{\Delta} & -\frac{b}{\Delta} \\ -\frac{c}{\Delta} & \frac{a}{\Delta} \end{bmatrix} = \begin{bmatrix} \frac{2/3}{1/3} & -\frac{-1/2}{1/3} \\ -\frac{-1/3}{1/3} & \frac{3/4}{1/3} \end{bmatrix} = \begin{bmatrix} 2 & \frac{3}{2} \\ 1 & \frac{9}{4} \end{bmatrix}$$

Find the stable matrix:

$$S(I-R)^{-1} = \begin{bmatrix} \frac{1}{4} & \frac{1}{6} \\ \frac{1}{6} & 0 \end{bmatrix}\begin{bmatrix} 2 & \frac{3}{2} \\ 1 & \frac{9}{4} \end{bmatrix} = \begin{bmatrix} \frac{2}{3} & \frac{3}{4} \\ \frac{1}{3} & \frac{1}{4} \end{bmatrix}$$

$$\left[\begin{array}{c|c} I & S(I-R)^{-1} \\ \hline 0 & 0 \end{array}\right] = \left[\begin{array}{cc|cc} 1 & 0 & \frac{2}{3} & \frac{3}{4} \\ 0 & 1 & \frac{1}{3} & \frac{1}{4} \\ \hline 0 & 0 & 0 & 0 \\ 0 & 0 & 0 & 0 \end{array}\right]$$

17.
$$\left[\begin{array}{ccc|cc} 1 & 0 & 0 & 0.1 & 0.2 \\ 0 & 1 & 0 & 0.3 & 0 \\ 0 & 0 & 1 & 0 & 0.2 \\ \hline 0 & 0 & 0 & 0.5 & 0 \\ 0 & 0 & 0 & 0.1 & 0.6 \end{array}\right] = \left[\begin{array}{c|c} I & S \\ \hline 0 & R \end{array}\right]$$

$$R = \begin{bmatrix} 0.5 & 0 \\ 0.1 & 0.6 \end{bmatrix}; S = \begin{bmatrix} 0.1 & 0.2 \\ 0.3 & 0 \\ 0 & 0.2 \end{bmatrix}$$

Find the fundamental matrix:

$$I - R = \begin{bmatrix} 1 & 0 \\ 0 & 1 \end{bmatrix} - \begin{bmatrix} 0.5 & 0 \\ 0.1 & 0.6 \end{bmatrix} = \begin{bmatrix} 0.5 & 0 \\ -0.1 & 0.4 \end{bmatrix} = \begin{bmatrix} a & b \\ c & d \end{bmatrix}$$

$$\Delta = ad - bc = (0.5)(0.4) - (0)(-0.1) = 0.2$$

$$F = (I-R)^{-1} = \begin{bmatrix} \frac{d}{\Delta} & -\frac{b}{\Delta} \\ -\frac{c}{\Delta} & \frac{a}{\Delta} \end{bmatrix} = \begin{bmatrix} \frac{.4}{.2} & -\frac{0}{.2} \\ -\frac{-.1}{.2} & \frac{.5}{.2} \end{bmatrix} = \begin{bmatrix} 2 & 0 \\ 0.5 & 2.5 \end{bmatrix}$$

Find the stable matrix:

$$S(I-R)^{-1} = \begin{bmatrix} 0.1 & 0.2 \\ 0.3 & 0 \\ 0 & 0.2 \end{bmatrix}\begin{bmatrix} 2 & 0 \\ 0.5 & 2.5 \end{bmatrix} = \begin{bmatrix} 0.3 & 0.5 \\ 0.6 & 0 \\ 0.1 & 0.5 \end{bmatrix}$$

$$\left[\begin{array}{c|c} I & S(I-R)^{-1} \\ \hline 0 & 0 \end{array}\right] = \left[\begin{array}{ccc|cc} 1 & 0 & 0 & 0.3 & 0.5 \\ 0 & 1 & 0 & 0.6 & 0 \\ 0 & 0 & 1 & 0.1 & 0.5 \\ \hline 0 & 0 & 0 & 0 & 0 \\ 0 & 0 & 0 & 0 & 0 \end{array}\right]$$

 Copyright © 2018 Pearson Education, Inc.

19. If the gambler begins with \$2, he should have \$1 for an expected number of 0.79 plays.

21. a.

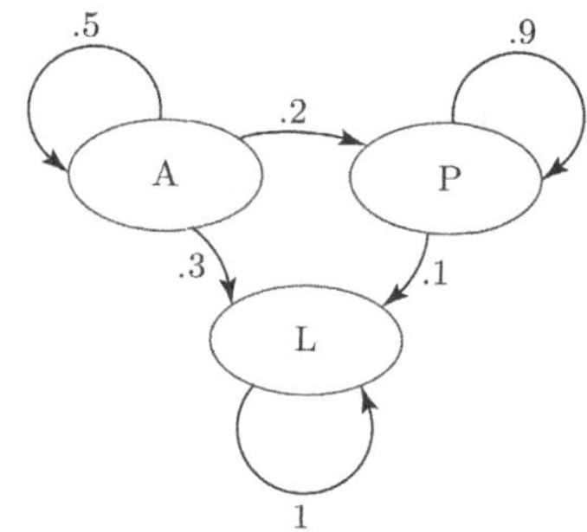

b.
$$\begin{array}{c|c|cc} & L & A & P \\ \hline L & 1 & 0.3 & 0.1 \\ \hline A & 0 & 0.5 & 0 \\ P & 0 & 0.2 & 0.9 \end{array}$$

c. $R = \begin{bmatrix} 0.5 & 0 \\ 0.2 & 0.9 \end{bmatrix}$; $S = [0.3 \quad 0.1]$

$$I - R = \begin{bmatrix} 1 & 0 \\ 0 & 1 \end{bmatrix} - \begin{bmatrix} 0.5 & 0 \\ 0.2 & 0.9 \end{bmatrix} = \begin{bmatrix} 0.5 & 0 \\ -0.2 & 0.1 \end{bmatrix}$$

$$\Delta = (0.5)(0.1) - (0)(-0.2) = 0.05$$

$$(I - R)^{-1} = \begin{bmatrix} \frac{0.1}{0.05} & 0 \\ \frac{0.2}{0.05} & \frac{0.5}{0.05} \end{bmatrix} = \begin{bmatrix} 2 & 0 \\ 4 & 10 \end{bmatrix}$$

$$S(I - R)^{-1} = [0.3 \quad 0.1]\begin{bmatrix} 2 & 0 \\ 4 & 10 \end{bmatrix} = [1 \quad 1]$$

The stable matrix is
$$\begin{array}{c|c|cc} & L & A & P \\ \hline L & 1 & 1 & 1 \\ \hline A & 0 & 0 & 0 \\ P & 0 & 0 & 0 \end{array}.$$

d. Add the numbers in the A column of the fundamental matrix: $2 + 4 = 6$ yrs.

e. In the long term, all the lawyers will be partners, therefore 0% will be associates.

23. a.

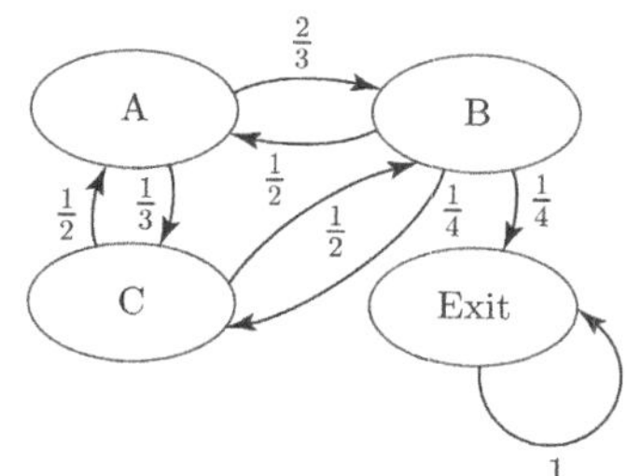

b.
$$\begin{array}{c|c|ccc} & E & A & B & C \\ \hline E & 1 & 0 & \frac{1}{4} & 0 \\ \hline A & 0 & 0 & \frac{1}{2} & \frac{1}{2} \\ B & 0 & \frac{2}{3} & 0 & \frac{1}{2} \\ C & 0 & \frac{1}{3} & \frac{1}{4} & 0 \end{array}$$

c. $R = \begin{bmatrix} 0 & \frac{1}{2} & \frac{1}{2} \\ \frac{2}{3} & 0 & \frac{1}{2} \\ \frac{1}{3} & \frac{1}{4} & 0 \end{bmatrix}$; $S = \begin{bmatrix} 0 & \frac{1}{4} & 0 \end{bmatrix}$

$$I - R = \begin{bmatrix} 1 & 0 & 0 \\ 0 & 1 & 0 \\ 0 & 0 & 1 \end{bmatrix} - \begin{bmatrix} 0 & \frac{1}{2} & \frac{1}{2} \\ \frac{2}{3} & 0 & \frac{1}{2} \\ \frac{1}{3} & \frac{1}{4} & 0 \end{bmatrix} = \begin{bmatrix} 1 & -\frac{1}{2} & -\frac{1}{2} \\ -\frac{2}{3} & 1 & -\frac{1}{2} \\ -\frac{1}{3} & -\frac{1}{4} & 1 \end{bmatrix}$$

$$(I - R)^{-1} = \begin{bmatrix} 4.2 & 3 & 3.6 \\ 4 & 4 & 4 \\ 2.4 & 2 & 3.2 \end{bmatrix}$$

$$S(I - R)^{-1} = \begin{bmatrix} 0 & \frac{1}{4} & 0 \end{bmatrix}\begin{bmatrix} 4.2 & 3 & 3.6 \\ 4 & 4 & 4 \\ 2.4 & 2 & 3.2 \end{bmatrix} = [1 \quad 1 \quad 1]$$

The stable matrix is
$$\begin{array}{c|c|ccc} & E & A & B & C \\ \hline E & 1 & 1 & 1 & 1 \\ \hline A & 0 & 0 & 0 & 0 \\ B & 0 & 0 & 0 & 0 \\ C & 0 & 0 & 0 & 0 \end{array}.$$

d. Add the numbers in the A column of the fundamental matrix: 10.6 minutes.

Copyright © 2018 Pearson Education, Inc.

25. a.

$$\begin{array}{c} \\ D \\ G \\ F \\ S \end{array}\begin{array}{c} \begin{array}{cccc} D & G & F & S \end{array} \\ \left[\begin{array}{cc|cc} 1 & 0 & 0.2 & 0.1 \\ 0 & 1 & 0 & 0.9 \\ \hline 0 & 0 & 0 & 0 \\ 0 & 0 & 0.8 & 0 \end{array}\right] \end{array}$$

b. $R = \begin{bmatrix} 0 & 0 \\ 0.8 & 0 \end{bmatrix}; S = \begin{bmatrix} 0.2 & 0.1 \\ 0 & 0.9 \end{bmatrix}$

$$I - R = \begin{bmatrix} 1 & 0 \\ 0 & 1 \end{bmatrix} - \begin{bmatrix} 0 & 0 \\ 0.8 & 0 \end{bmatrix} = \begin{bmatrix} 1 & 0 \\ -0.8 & 1 \end{bmatrix} = \begin{bmatrix} a & b \\ c & d \end{bmatrix}$$

$\Delta = ad - bc = (1)(1) - (0)(-.8) = 1$

$$(I - R)^{-1} = \begin{bmatrix} \frac{d}{\Delta} & -\frac{b}{\Delta} \\ -\frac{c}{\Delta} & \frac{a}{\Delta} \end{bmatrix} = \begin{bmatrix} \frac{1}{1} & -\frac{0}{1} \\ -\frac{-0.8}{1} & \frac{1}{1} \end{bmatrix} = \begin{bmatrix} 1 & 0 \\ 0.8 & 1 \end{bmatrix}$$

$$S(I - R)^{-1} = \begin{bmatrix} 0.2 & 0.1 \\ 0 & 0.9 \end{bmatrix}\begin{bmatrix} 1 & 0 \\ 0.8 & 1 \end{bmatrix} = \begin{bmatrix} 0.28 & 0.1 \\ 0.72 & 0.9 \end{bmatrix}$$

$$\left[\begin{array}{c|c} I & S(I-R)^{-1} \\ \hline 0 & 0 \end{array}\right] = \begin{array}{c} \\ D \\ G \\ F \\ S \end{array}\begin{array}{c} \begin{array}{cccc} D & G & F & S \end{array} \\ \left[\begin{array}{cc|cc} 1 & 0 & 0.28 & 0.1 \\ 0 & 1 & 0.72 & 0.9 \\ \hline 0 & 0 & 0 & 0 \\ 0 & 0 & 0 & 0 \end{array}\right] \end{array}$$

c. From the F column of the stable matrix, the probability that a freshman will eventually graduate is .72.

d. The fundamental matrix is

$$F = (I - R)^{-1} = \begin{array}{c} \\ F \\ S \end{array}\begin{array}{c} \begin{array}{cc} F & S \end{array} \\ \begin{bmatrix} 1 & 0 \\ 0.8 & 1 \end{bmatrix} \end{array}.$$

Add the numbers in the F column:
$1 + 0.8 = 1.8.$
The student will attend an expected number of 1.8 years.

27. First, arrange the matrix so that the absorbing states come first and calculate the fundamental and stable matrices.

$$\begin{array}{c} \\ \text{Paid} \\ \text{Bad} \\ \le 30 \\ < 60 \end{array}\begin{array}{c} \begin{array}{cccc} \text{Paid} & \text{Bad} & \le 30 & < 60 \end{array} \\ \left[\begin{array}{cc|cc} 1 & 0 & 0.4 & 0.1 \\ 0 & 1 & 0 & 0.1 \\ \hline 0 & 0 & 0.4 & 0.4 \\ 0 & 0 & 0.2 & 0.4 \end{array}\right] \end{array} = \left[\begin{array}{c|c} I & S \\ \hline 0 & R \end{array}\right]$$

$R = \begin{bmatrix} 0.4 & 0.4 \\ 0.2 & 0.4 \end{bmatrix}; S = \begin{bmatrix} 0.4 & 0.1 \\ 0 & 0.1 \end{bmatrix}$

$$I - R = \begin{bmatrix} 1 & 0 \\ 0 & 1 \end{bmatrix} - \begin{bmatrix} 0.4 & 0.4 \\ 0.2 & 0.4 \end{bmatrix} = \begin{bmatrix} 0.6 & -0.4 \\ -0.2 & 0.6 \end{bmatrix} = \begin{bmatrix} a & b \\ c & d \end{bmatrix}$$

$\Delta = ad - bc = (.6)(.6) - (-.4)(-.2) = .28$

$$F = (I - R)^{-1} = \begin{bmatrix} \frac{d}{\Delta} & -\frac{b}{\Delta} \\ -\frac{c}{\Delta} & \frac{a}{\Delta} \end{bmatrix} = \begin{bmatrix} \frac{0.6}{0.28} & -\frac{-0.4}{0.28} \\ -\frac{-0.2}{0.28} & \frac{0.6}{0.28} \end{bmatrix} = \begin{bmatrix} \frac{15}{7} & \frac{10}{7} \\ \frac{5}{7} & \frac{15}{7} \end{bmatrix}$$

$$S(I - R)^{-1} = \begin{bmatrix} 0.4 & 0.1 \\ 0 & 0.1 \end{bmatrix}\begin{bmatrix} \frac{15}{7} & \frac{10}{7} \\ \frac{5}{7} & \frac{15}{7} \end{bmatrix} = \begin{bmatrix} \frac{13}{14} & \frac{11}{14} \\ \frac{1}{14} & \frac{3}{14} \end{bmatrix}$$

The stable matrix is

$$\begin{array}{c} \\ \text{Paid} \\ \text{Bad} \\ \le 30 \\ < 60 \end{array}\begin{array}{c} \begin{array}{cccc} \text{Paid} & \text{Bad} & \le 30 & < 60 \end{array} \\ \left[\begin{array}{cc|cc} 1 & 0 & \frac{13}{14} & \frac{11}{14} \\ 0 & 1 & \frac{1}{14} & \frac{3}{14} \\ \hline 0 & 0 & 0 & 0 \\ 0 & 0 & 0 & 0 \end{array}\right] \end{array}.$$

a. From the stable matrix, the probability of an account eventually being paid off is $\frac{13}{14}$ if it is currently at most 30 days overdue and $\frac{11}{14}$ if it is less than 60 days overdue (but more than 30 days overdue).

Copyright © 2018 Pearson Education, Inc.

b. The fundamental matrix is

$$\begin{array}{c} \\ \le 30 \\ <60 \end{array}\begin{array}{c} \le 30 \quad <60 \\ \begin{bmatrix} \frac{15}{7} & \frac{10}{7} \\ \frac{5}{7} & \frac{15}{7} \end{bmatrix} \end{array}$$

. Add the numbers in the first column: $\frac{15}{7}+\frac{5}{7}=\frac{20}{7}$. An account that is overdue at most 30 days is expected to reach an absorbing state (paid or bad) after $\frac{20}{7}$ months.

c. From the stable matrix, about $\frac{13}{14}$ of the "≤30 day" debt will be paid, and about $\frac{11}{14}$ of the "<60 day" debt will be paid.

$\frac{13}{14}(\$2000)+\frac{11}{14}(\$5000)\approx \$5786.$ About \$5786 will be paid and about \$1214 will become bad debt.

29. First, find the absorbing stochastic matrix and calculate the fundamental and stable matrices.

$$\begin{array}{c} \\ \$0 \\ \$4 \\ \$1 \\ \$2 \\ \$3 \end{array}\begin{array}{c} \$0 \quad \$4 \quad \$1 \quad \$2 \quad \$3 \\ \left[\begin{array}{cc|ccc} 1 & 0 & \frac{1}{2} & 0 & 0 \\ 0 & 1 & 0 & 0 & \frac{1}{2} \\ \hline 0 & 0 & 0 & \frac{1}{2} & 0 \\ 0 & 0 & \frac{1}{2} & 0 & \frac{1}{2} \\ 0 & 0 & 0 & \frac{1}{2} & 0 \end{array}\right] \end{array} = \left[\begin{array}{c|c} I & S \\ \hline 0 & R \end{array}\right]$$

$$R=\begin{bmatrix} 0 & \frac{1}{2} & 0 \\ \frac{1}{2} & 0 & \frac{1}{2} \\ 0 & \frac{1}{2} & 0 \end{bmatrix}; S=\begin{bmatrix} \frac{1}{2} & 0 & 0 \\ 0 & 0 & \frac{1}{2} \end{bmatrix}$$

$$I-R=\begin{bmatrix} 1 & 0 & 0 \\ 0 & 1 & 0 \\ 0 & 0 & 1 \end{bmatrix}-\begin{bmatrix} 0 & \frac{1}{2} & 0 \\ \frac{1}{2} & 0 & \frac{1}{2} \\ 0 & \frac{1}{2} & 0 \end{bmatrix}$$

$$=\begin{bmatrix} 1 & -\frac{1}{2} & 0 \\ -\frac{1}{2} & 1 & -\frac{1}{2} \\ 0 & -\frac{1}{2} & 1 \end{bmatrix}$$

$$F=(I-R)^{-1}=\begin{bmatrix} 1 & -\frac{1}{2} & 0 \\ -\frac{1}{2} & 1 & -\frac{1}{2} \\ 0 & -\frac{1}{2} & 1 \end{bmatrix}^{-1}$$

$$=\begin{bmatrix} \frac{3}{2} & 1 & \frac{1}{2} \\ 1 & 2 & 1 \\ \frac{1}{2} & 1 & \frac{3}{2} \end{bmatrix}$$

$$S(I-R)^{-1}=\begin{bmatrix} \frac{1}{2} & 0 & 0 \\ 0 & 0 & \frac{1}{2} \end{bmatrix}\begin{bmatrix} \frac{3}{2} & 1 & \frac{1}{2} \\ 1 & 2 & 1 \\ \frac{1}{2} & 1 & \frac{3}{2} \end{bmatrix}$$

$$=\begin{bmatrix} \frac{3}{4} & \frac{1}{2} & \frac{1}{4} \\ \frac{1}{4} & \frac{1}{2} & \frac{3}{4} \end{bmatrix}$$

The stable matrix is

$$\begin{array}{c} \\ \$0 \\ \$4 \\ \$1 \\ \$2 \\ \$3 \end{array}\begin{array}{c} \$0 \quad \$4 \quad \$1 \quad \$2 \quad \$3 \\ \left[\begin{array}{cc|ccc} 1 & 0 & \frac{3}{4} & \frac{1}{2} & \frac{1}{4} \\ 0 & 1 & \frac{1}{4} & \frac{1}{2} & \frac{3}{4} \\ \hline 0 & 0 & 0 & 0 & 0 \\ 0 & 0 & 0 & 0 & 0 \\ 0 & 0 & 0 & 0 & 0 \end{array}\right] \end{array}.$$

a. From the top row of the stable matrix, the probability of eventually going broke is $\frac{3}{4}$ if he starts with \$1, $\frac{1}{2}$ if he starts with \$2, and $\frac{1}{4}$ if he starts with \$3.

b. The fundamental matrix is

$$\begin{array}{c} \\ \$1 \\ \$2 \\ \$3 \end{array}\begin{array}{c} \$1 \quad \$2 \quad \$3 \\ \begin{bmatrix} \frac{3}{2} & 1 & \frac{1}{2} \\ 1 & 2 & 1 \\ \frac{1}{2} & 1 & \frac{3}{2} \end{bmatrix} \end{array}.$$

Add the entries in the middle column:
$1+2+1=4$
Starting with \$2, he will play for an expected number of 4 times.

Copyright © 2018 Pearson Education, Inc.

31. Consider the Markov process with states 1, 2 and 3 corresponding to the number of different quotations received so far. This process has absorbing stochastic matrix

$$A = \begin{array}{c} \\ 3 \\ 2 \\ 1 \end{array}\begin{array}{c} \begin{array}{ccc} 3 & 2 & 1 \end{array} \\ \left[\begin{array}{c|cc} 1 & \frac{1}{3} & 0 \\ \hline 0 & \frac{2}{3} & \frac{2}{3} \\ 0 & 0 & \frac{1}{3} \end{array}\right] \end{array}.$$

The distribution matrix after purchasing one bottle is $B = \begin{bmatrix} 0 \\ 0 \\ 1 \end{bmatrix}$, so the distribution matrix after purchasing four more bottles (for a total of five) is $A^4B = \begin{pmatrix} 1 & \frac{65}{81} & \frac{50}{81} \\ 0 & \frac{16}{81} & \frac{10}{27} \\ 0 & 0 & \frac{1}{81} \end{pmatrix}\begin{pmatrix} 0 \\ 0 \\ 1 \end{pmatrix} = \begin{pmatrix} \frac{50}{81} \\ \frac{10}{27} \\ \frac{1}{81} \end{pmatrix}$.

Therefore, the probability of receiving all three quotations after purchasing five bottles is $\frac{50}{81}$.

We have $R = \begin{bmatrix} \frac{2}{3} & \frac{2}{3} \\ 0 & \frac{1}{3} \end{bmatrix}$ so the fundamental matrix

$$(I-R)^{-1} = \begin{bmatrix} \frac{1}{3} & -\frac{2}{3} \\ 0 & \frac{2}{3} \end{bmatrix}^{-1} = \begin{bmatrix} 3 & 3 \\ 0 & \frac{3}{2} \end{bmatrix}.$$

The expected number of soft drinks you have to purchase to get all three quotations is the sum of the numbers in the "1" column of the fundamental matrix plus one for the first bottle, so it's $3+\frac{3}{2}+1=5\frac{1}{2}$.

33. Calculating [A] ^ 225 → Frac gives

$$\begin{bmatrix} 1 & 0 & 0 & \frac{9}{37} & \frac{24}{37} \\ 0 & 1 & 0 & \frac{53}{74} & \frac{9}{37} \\ 0 & 0 & 1 & \frac{3}{74} & \frac{4}{37} \\ 0 & 0 & 0 & 0 & 0 \\ 0 & 0 & 0 & 0 & 0 \end{bmatrix}.$$

Let $[B] = S = \begin{bmatrix} 0 & 0.6 \\ 0.5 & 0.1 \\ 0 & 0.1 \end{bmatrix}$ and $[C] = R = \begin{bmatrix} 0.2 & 0.2 \\ 0.3 & 0 \end{bmatrix}$.

Then calculating

$[B]*(\text{identity}(2)-[C])^{-1} \rightarrow$ Frac

gives $\begin{bmatrix} \frac{9}{37} & \frac{24}{37} \\ \frac{53}{74} & \frac{9}{37} \\ \frac{3}{74} & \frac{4}{37} \end{bmatrix}$. Therefore, the matrix given above is the exact stable matrix.

Chapter 8 Review Exercises

1. Stochastic, neither; the matrix is stochastic because it is square, the entries are all ≥ 0, and the sum of the entries in each column is 1. It is not regular because all powers of the matrix include zero entries. The first state is an absorbing state, but the matrix is not an absorbing matrix because an object that begins in the third or fourth state will always remain within these two (nonabsorbing) states.

2. Stochastic, regular; the matrix is stochastic because it is square, the entries are all ≥ 0, and the sum of the entries in each column is 1. It is regular because it has no zero entries.

3. Stochastic, regular; the matrix is stochastic because it is square, the entries are all ≥ 0, and the sum of the entries in each column is 1. It is regular because the second power, $\begin{bmatrix} .3 & .21 \\ .7 & .79 \end{bmatrix}$, contains no zero entries.

4. Stochastic, absorbing; the matrix is stochastic because it is square, the entries are all ≥ 0, and the sum of the entries in each column is 1. It is absorbing because the first and second states are absorbing states, and it is possible for an object to get from the third state to an absorbing state.

5. Not stochastic because the sum of the entries in the middle column is not 1.

6. Stochastic, absorbing; the matrix is stochastic because it is square, the entries are all ≥ 0, and the sum of the entries in each column is 1. It is absorbing because the first and second states are absorbing states and it is possible to get from the

Copyright © 2018 Pearson Education, Inc.

third state (indirectly) or the fourth state (directly) to an absorbing state.

7. $\begin{cases} x+y=1 \\ \begin{bmatrix} 0.6 & 0.5 \\ 0.4 & 0.5 \end{bmatrix}\begin{bmatrix} x \\ y \end{bmatrix} = \begin{bmatrix} x \\ y \end{bmatrix} \end{cases}$

$$\begin{cases} x + y = 1 \\ 0.6x + 0.5y = x \\ 0.4x + 0.5y = y \end{cases}$$

$$\begin{cases} x + y = 1 \\ -0.4x + 0.5y = 0 \\ 0.4x - 0.5y = 0 \end{cases}$$

The second and third equations in this system are equivalent.

$$\left[\begin{array}{cc|c} 1 & 1 & 1 \\ -0.4 & 0.5 & 0 \end{array}\right] \xrightarrow{[2]+0.4[1]} \left[\begin{array}{cc|c} 1 & 1 & 1 \\ 0 & 0.9 & 0.4 \end{array}\right]$$

$$\xrightarrow{\frac{10}{9}[2]} \left[\begin{array}{cc|c} 1 & 1 & 1 \\ 0 & 1 & \frac{4}{9} \end{array}\right] \xrightarrow{[1]+(-1)[2]} \left[\begin{array}{cc|c} 1 & 0 & \frac{5}{9} \\ 0 & 1 & \frac{4}{9} \end{array}\right]$$

$x = \frac{5}{9},\ y = \frac{4}{9}$

The stable distribution is $\begin{bmatrix} \frac{5}{9} \\ \frac{4}{9} \end{bmatrix}$.

8. $\left[\begin{array}{ccc|cc} 1 & 0 & 0 & \frac{1}{8} & \frac{1}{4} \\ 0 & 1 & 0 & \frac{1}{8} & 0 \\ 0 & 0 & 1 & 0 & \frac{1}{4} \\ \hline 0 & 0 & 0 & \frac{1}{4} & \frac{1}{2} \\ 0 & 0 & 0 & \frac{1}{2} & 0 \end{array}\right] = \left[\begin{array}{c|c} I & S \\ \hline 0 & R \end{array}\right]$

$$R = \begin{bmatrix} \frac{1}{4} & \frac{1}{2} \\ \frac{1}{2} & 0 \end{bmatrix};\ S = \begin{bmatrix} \frac{1}{8} & \frac{1}{4} \\ \frac{1}{8} & 0 \\ 0 & \frac{1}{4} \end{bmatrix}$$

$$I - R = \begin{bmatrix} 1 & 0 \\ 0 & 1 \end{bmatrix} - \begin{bmatrix} \frac{1}{4} & \frac{1}{2} \\ \frac{1}{2} & 0 \end{bmatrix} = \begin{bmatrix} \frac{3}{4} & -\frac{1}{2} \\ -\frac{1}{2} & 1 \end{bmatrix} = \begin{bmatrix} a & b \\ c & d \end{bmatrix}$$

$$\Delta = ad - bc = \left(\frac{3}{4}\right)(1) - \left(-\frac{1}{2}\right)\left(-\frac{1}{2}\right) = \frac{1}{2}$$

$$(I-R)^{-1} = \begin{bmatrix} \frac{d}{\Delta} & -\frac{b}{\Delta} \\ -\frac{c}{\Delta} & \frac{a}{\Delta} \end{bmatrix} = \begin{bmatrix} \frac{1}{1/2} & -\frac{-1/2}{1/2} \\ -\frac{-1/2}{1/2} & \frac{3/4}{1/2} \end{bmatrix} = \begin{bmatrix} 2 & 1 \\ 1 & \frac{3}{2} \end{bmatrix}$$

$$S(I-R)^{-1} = \begin{bmatrix} \frac{1}{8} & \frac{1}{4} \\ \frac{1}{8} & 0 \\ 0 & \frac{1}{4} \end{bmatrix}\begin{bmatrix} 2 & 1 \\ 1 & \frac{3}{2} \end{bmatrix} = \begin{bmatrix} \frac{1}{2} & \frac{1}{2} \\ \frac{1}{4} & \frac{1}{8} \\ \frac{1}{4} & \frac{3}{8} \end{bmatrix}$$

$$\left[\begin{array}{c|c} I & S(I-R)^{-1} \\ \hline 0 & 0 \end{array}\right] = \left[\begin{array}{ccc|cc} 1 & 0 & 0 & \frac{1}{2} & \frac{1}{2} \\ 0 & 1 & 0 & \frac{1}{4} & \frac{1}{8} \\ 0 & 0 & 1 & \frac{1}{4} & \frac{3}{8} \\ \hline 0 & 0 & 0 & 0 & 0 \\ 0 & 0 & 0 & 0 & 0 \end{array}\right]$$

9. a.

	H	M	L
H	0.5	0.4	0.3
M	0.4	0.3	0.5
L	0.1	0.3	0.2

b. $\begin{bmatrix} 0.5 & 0.4 & 0.3 \\ 0.4 & 0.3 & 0.5 \\ 0.1 & 0.3 & 0.2 \end{bmatrix}\begin{bmatrix} 0.1 \\ 0.6 \\ 0.3 \end{bmatrix} = \begin{bmatrix} 0.38 \\ 0.37 \\ 0.25 \end{bmatrix}$

38% of the children of the current generation will have high incomes.

c. Find the stable distribution.

$$\begin{cases} x+y+z=1 \\ \begin{bmatrix} 0.5 & 0.4 & 0.3 \\ 0.4 & 0.3 & 0.5 \\ 0.1 & 0.3 & 0.2 \end{bmatrix}\begin{bmatrix} x \\ y \\ z \end{bmatrix} = \begin{bmatrix} x \\ y \\ z \end{bmatrix} \end{cases}$$

Copyright © 2018 Pearson Education, Inc.

$$\begin{cases} x + y + z = 1 \\ 0.5x + 0.4y + 0.3z = x \\ 0.4x + 0.3y + 0.5z = y \\ 0.1x + 0.3y + 0.2z = z \end{cases}$$

$$\begin{cases} x + y + z = 1 \\ -0.5x + 0.4y + 0.3z = 0 \\ 0.4x - 0.7y + 0.5z = 0 \\ 0.1x + 0.3y - 0.8z = 0 \end{cases}$$

The fourth equation is equivalent to the second plus third and so is redundant.

$$\left[\begin{array}{ccc|c} 1 & 1 & 1 & 1 \\ -0.5 & 0.4 & 0.3 & 0 \\ 0.4 & -0.7 & 0.5 & 0 \end{array}\right] \xrightarrow[10[3]]{10[2]} \left[\begin{array}{ccc|c} 1 & 1 & 1 & 1 \\ -5 & 4 & 3 & 0 \\ 4 & -7 & 5 & 0 \end{array}\right]$$

$$\xrightarrow[[3]+(-4)[1]]{[2]+5[1]} \left[\begin{array}{ccc|c} 1 & 1 & 1 & 1 \\ 0 & 9 & 8 & 5 \\ 0 & -11 & 1 & -4 \end{array}\right]$$

$$\xrightarrow[\substack{[1]+(-1)[2] \\ [3+(11)[2]}]{\frac{1}{9}[2]} \left[\begin{array}{ccc|c} 1 & 0 & \frac{1}{9} & \frac{4}{9} \\ 0 & 1 & \frac{8}{9} & \frac{5}{9} \\ 0 & 0 & \frac{97}{9} & \frac{19}{9} \end{array}\right]$$

$$\xrightarrow[\substack{[1]+\left(-\frac{1}{9}\right)[3] \\ [2]+\left(-\frac{8}{9}\right)[3]}]{\frac{9}{97}[3]} \left[\begin{array}{ccc|c} 1 & 0 & 0 & \frac{41}{97} \\ 0 & 1 & 0 & \frac{37}{97} \\ 0 & 0 & 1 & \frac{19}{97} \end{array}\right]$$

$x = \frac{41}{97}, y = \frac{37}{97}, z = \frac{19}{97}.$

The stable distribution is $\begin{bmatrix} \frac{41}{97} \\ \frac{37}{97} \\ \frac{19}{97} \end{bmatrix}$. In the long run, $\frac{19}{97}$ of the population will have low incomes.

10. a.

$$\begin{array}{c} \\ P \\ N \end{array} \begin{array}{c} \begin{array}{cc} P & N \end{array} \\ \begin{bmatrix} 0.8 & 0.3 \\ 0.2 & 0.7 \end{bmatrix} \end{array}$$

b. $A^2\begin{bmatrix} \ \\ \ \end{bmatrix}_0 = \begin{bmatrix} 0.8 & 0.3 \\ 0.2 & 0.7 \end{bmatrix}\begin{bmatrix} 0.8 & 0.3 \\ 0.2 & 0.7 \end{bmatrix}\begin{bmatrix} 1 \\ 0 \end{bmatrix}_0 = \begin{bmatrix} .7 \\ .3 \end{bmatrix}_2$

30% will need adjusting after 2 days.

c. Find the stable distribution.

$$\begin{cases} x + y = 1 \\ \begin{bmatrix} 0.8 & 0.3 \\ 0.2 & 0.7 \end{bmatrix}\begin{bmatrix} x \\ y \end{bmatrix} = \begin{bmatrix} x \\ y \end{bmatrix} \end{cases}$$

$$\begin{cases} x + y = 1 \\ 0.8x + 0.3y = x \\ 0.2x + 0.7y = y \end{cases}$$

$$\begin{cases} x + y = 1 \\ -0.2x + 0.3y = 0 \\ 0.2x - 0.3y = 0 \end{cases}$$

The second and third equations in this system are equivalent.

$$\left[\begin{array}{cc|c} 1 & 1 & 1 \\ -0.2 & 0.3 & 0 \end{array}\right] \xrightarrow{[2]+0.2[1]} \left[\begin{array}{cc|c} 1 & 1 & 1 \\ 0 & 0.5 & 0.2 \end{array}\right]$$

$$\xrightarrow{2[2]} \left[\begin{array}{cc|c} 1 & 1 & 1 \\ 0 & 1 & 0.4 \end{array}\right]$$

$$\xrightarrow{[1]+(-1)[2]} \left[\begin{array}{cc|c} 1 & 0 & 0.6 \\ 0 & 1 & 0.4 \end{array}\right]$$

$x = .6, y = .4$

The stable distribution is $\begin{bmatrix} 0.6 \\ 0.4 \end{bmatrix}$. In the long run, 60% will be properly adjusted.

11. $\left[\begin{array}{cc|ccc} 1 & 0 & \frac{1}{6} & \frac{1}{2} & \frac{2}{5} \\ 0 & 1 & 0 & 0 & \frac{2}{5} \\ \hline 0 & 0 & 0 & 0 & 0 \\ 0 & 0 & \frac{2}{3} & \frac{1}{2} & 0 \\ 0 & 0 & \frac{1}{6} & 0 & \frac{1}{5} \end{array}\right] = \left[\begin{array}{c|c} I & S \\ \hline 0 & R \end{array}\right]$

$R = \begin{bmatrix} 0 & 0 & 0 \\ \frac{2}{3} & \frac{1}{2} & 0 \\ \frac{1}{6} & 0 & \frac{1}{5} \end{bmatrix}; S = \begin{bmatrix} \frac{1}{6} & \frac{1}{2} & \frac{2}{5} \\ 0 & 0 & \frac{2}{5} \end{bmatrix}$

Copyright © 2018 Pearson Education, Inc.

$$I-R=\begin{bmatrix}1&0&0\\0&1&0\\0&0&1\end{bmatrix}-\begin{bmatrix}0&0&0\\\frac{2}{3}&\frac{1}{2}&0\\\frac{1}{6}&0&\frac{1}{5}\end{bmatrix}$$

$$=\begin{bmatrix}1&0&0\\-\frac{2}{3}&\frac{1}{2}&0\\-\frac{1}{6}&0&\frac{4}{5}\end{bmatrix}$$

Use the Gauss-Jordan method to find $(I-R)^{-1}$.

$$\left[\begin{array}{ccc|ccc}1&0&0&1&0&0\\-\frac{2}{3}&\frac{1}{2}&0&0&1&0\\-\frac{1}{6}&0&\frac{4}{5}&0&0&1\end{array}\right]$$

$$\xrightarrow[{[3]+\frac{1}{6}[1]}]{[2]+\frac{2}{3}[1]}\left[\begin{array}{ccc|ccc}1&0&0&1&0&0\\0&\frac{1}{2}&0&\frac{2}{3}&1&0\\0&0&\frac{4}{5}&\frac{1}{6}&0&1\end{array}\right]$$

$$\xrightarrow[{\frac{5}{4}[3]}]{2[2]}\left[\begin{array}{ccc|ccc}1&0&0&1&0&0\\0&1&0&\frac{4}{3}&2&0\\0&0&1&\frac{5}{24}&0&\frac{5}{4}\end{array}\right]$$

$$S(I-R)^{-1}=\begin{bmatrix}\frac{1}{6}&\frac{1}{2}&\frac{2}{5}\\0&0&\frac{2}{5}\end{bmatrix}\begin{bmatrix}1&0&0\\\frac{4}{3}&2&0\\\frac{5}{24}&0&\frac{5}{4}\end{bmatrix}$$

$$=\begin{bmatrix}\frac{11}{12}&1&\frac{1}{2}\\\frac{1}{12}&0&\frac{1}{2}\end{bmatrix}$$

$$\left[\begin{array}{c|c}I&S(I-R)^{-1}\\\hline 0&0\end{array}\right]=\left[\begin{array}{cc|ccc}1&0&\frac{11}{12}&1&\frac{1}{2}\\0&1&\frac{1}{12}&0&\frac{1}{2}\\\hline 0&0&0&0&0\\0&0&0&0&0\\0&0&0&0&0\end{array}\right]$$

12. a.

$$\begin{array}{c}\\ \text{I}\\ \text{II}\\ \text{III}\\ \text{IV}\end{array}\begin{array}{c}\begin{array}{cccc}\text{I}&\text{II}&\text{III}&\text{IV}\end{array}\\\left[\begin{array}{cc|cc}1&0&0&\frac{1}{4}\\0&1&\frac{1}{3}&\frac{1}{4}\\\hline 0&0&0&\frac{1}{2}\\0&0&\frac{2}{3}&0\end{array}\right]\end{array}$$

b. The mouse will find the cheese after two minutes if he either goes to room II after one minute, or goes to room III after one minute and then room II after two minutes. Therefore the probability that the mouse finds the cheese after two minutes is $\frac{1}{4}+\frac{1}{2}\cdot\frac{1}{3}=\frac{5}{12}$.

c. Find the stable matrix.

$$R=\begin{bmatrix}0&\frac{1}{2}\\\frac{2}{3}&0\end{bmatrix},\ S=\begin{bmatrix}0&\frac{1}{4}\\\frac{1}{3}&\frac{1}{4}\end{bmatrix}$$

$$I-R=\begin{bmatrix}1&0\\0&1\end{bmatrix}-\begin{bmatrix}0&\frac{1}{2}\\\frac{2}{3}&0\end{bmatrix}$$

$$=\begin{bmatrix}1&-\frac{1}{2}\\-\frac{2}{3}&1\end{bmatrix}$$

$$=\begin{bmatrix}a&b\\c&d\end{bmatrix}$$

$$\Delta=ad-bc=(1)(1)-\left(-\frac{2}{3}\right)\left(-\frac{1}{2}\right)=\frac{2}{3}$$

$$F=(I-R)^{-1}=\begin{bmatrix}\frac{d}{\Delta}&-\frac{b}{\Delta}\\-\frac{c}{\Delta}&\frac{a}{\Delta}\end{bmatrix}$$

$$=\begin{bmatrix}\frac{1}{2/3}&-\frac{-1/2}{2/3}\\-\frac{-2/3}{2/3}&\frac{1}{2/3}\end{bmatrix}$$

$$=\begin{bmatrix}\frac{3}{2}&\frac{3}{4}\\1&\frac{3}{2}\end{bmatrix}$$

$$S(I-R)^{-1}=\begin{bmatrix}0&\frac{1}{4}\\\frac{1}{3}&\frac{1}{4}\end{bmatrix}\begin{bmatrix}\frac{3}{2}&\frac{3}{4}\\1&\frac{3}{2}\end{bmatrix}=\begin{bmatrix}\frac{1}{4}&\frac{3}{8}\\\frac{3}{4}&\frac{5}{8}\end{bmatrix}$$

Copyright © 2018 Pearson Education, Inc.

The stable matrix is

$$\left[\begin{array}{c|c} I & S(I-R)^{-1} \\ \hline 0 & 0 \end{array}\right] = \begin{array}{c} \\ \text{I} \\ \text{II} \\ \text{III} \\ \text{IV} \end{array}\begin{array}{c} \begin{array}{cccc} \text{I} & \text{II} & \text{III} & \text{IV} \end{array} \\ \left[\begin{array}{cc|cc} 1 & 0 & \frac{1}{4} & \frac{3}{8} \\ 0 & 1 & \frac{3}{4} & \frac{5}{8} \\ \hline 0 & 0 & 0 & 0 \\ 0 & 0 & 0 & 0 \end{array}\right] \end{array}$$

From the fourth column, if he starts in room IV the probability of finding cheese in the long run is $\frac{5}{8} = .625$.

d. The fundamental matrix is $\begin{array}{c} \\ \text{III} \\ \text{IV} \end{array}\begin{array}{c} \begin{array}{cc} \text{III} & \text{IV} \end{array} \\ \begin{bmatrix} \frac{3}{2} & \frac{3}{4} \\ 1 & \frac{3}{2} \end{bmatrix} \end{array}$.

Add the entries in the column for room III: $\frac{3}{2} + 1 = \frac{5}{2}$.

A mouse that starts in room III will spend an expected $2\frac{1}{2}$ minutes before finding the cheese or being trapped.

13. (c) is the correct choice because it satisfies the system $\begin{cases} x+y+z=1 \\ \begin{bmatrix} 0.4 & 0.4 & 0.2 \\ 0.1 & 0.1 & 0.3 \\ 0.5 & 0.5 & 0.5 \end{bmatrix}\begin{bmatrix} x \\ y \\ z \end{bmatrix} = \begin{bmatrix} x \\ y \\ z \end{bmatrix} \end{cases}$

14. $A = \begin{array}{c} \\ \text{A} \\ \text{B} \end{array}\begin{array}{c} \begin{array}{cc} \text{A} & \text{B} \end{array} \\ \begin{bmatrix} 0.9 & 0.2 \\ 0.1 & 0.8 \end{bmatrix} \end{array}$

$$A^2\begin{bmatrix} \ \\ \ \end{bmatrix}_0 = \begin{bmatrix} 0.9 & 0.2 \\ 0.1 & 0.8 \end{bmatrix}\begin{bmatrix} 0.9 & 0.2 \\ 0.1 & 0.8 \end{bmatrix}\begin{bmatrix} 0.5 \\ 0.5 \end{bmatrix}_0 = \begin{bmatrix} 0.585 \\ 0.415 \end{bmatrix}_2$$

58.5% of the regular listeners will listen to station A two days from now.

15. a. If the traffic is moderate on a particular day then for the next day the probability of light traffic is 0.2, the probability of moderate traffic is 0.75, and the probability of heavy traffic is 0.05.

b. Find the stable distribution.

$$\begin{cases} x+y+z=1 \\ \begin{bmatrix} 0.70 & 0.20 & 0.10 \\ 0.20 & 0.75 & 0.30 \\ 0.10 & 0.05 & 0.60 \end{bmatrix}\begin{bmatrix} x \\ y \\ z \end{bmatrix} = \begin{bmatrix} x \\ y \\ z \end{bmatrix} \end{cases}$$

$$\begin{cases} x + y + z = 1 \\ 0.7x + 0.2y + 0.1z = x \\ 0.2x + 0.75y + 0.3z = y \\ 0.1x + 0.05y + 0.6z = z \end{cases}$$

$$\begin{cases} x + y + z = 1 \\ -0.3x + 0.2y + 0.1z = 0 \\ 0.2x - 0.25y + 0.3z = 0 \\ 0.1x + 0.05y - 0.4z = 0 \end{cases}$$

The fourth equation is equivalent to the second plus the third and so is redundant.

$$\left[\begin{array}{ccc|c} 1 & 1 & 1 & 1 \\ -0.3 & 0.2 & 0.1 & 0 \\ 0.2 & -.25 & 0.3 & 0 \end{array}\right] \xrightarrow[20[3]]{10[2]} \left[\begin{array}{ccc|c} 1 & 1 & 1 & 1 \\ -3 & 2 & 1 & 0 \\ 4 & -5 & 6 & 0 \end{array}\right]$$

$$\xrightarrow[[3]+(-4)[1]]{[2]+3[1]} \left[\begin{array}{ccc|c} 1 & 1 & 1 & 1 \\ 0 & 5 & 4 & 3 \\ 0 & -9 & 2 & -4 \end{array}\right]$$

$$\xrightarrow[\substack{[1]+(-1)[2] \\ [3]+9[2]}]{\frac{1}{5}[2]} \left[\begin{array}{ccc|c} 1 & 0 & \frac{1}{5} & \frac{2}{5} \\ 0 & 1 & \frac{4}{5} & \frac{3}{5} \\ 0 & 0 & \frac{46}{5} & \frac{7}{5} \end{array}\right]$$

$$\xrightarrow[\substack{[1]+\left(-\frac{1}{5}\right)[3] \\ [2]+\left(-\frac{4}{5}\right)[3]}]{\frac{5}{46}[3]} \left[\begin{array}{ccc|c} 1 & 0 & 0 & \frac{17}{46} \\ 0 & 1 & 0 & \frac{11}{23} \\ 0 & 0 & 1 & \frac{7}{46} \end{array}\right]$$

The stable distribution is $\begin{bmatrix} \frac{17}{46} \\ \frac{11}{23} \\ \frac{7}{46} \end{bmatrix}$ or about

Copyright © 2018 Pearson Education, Inc.

$$\begin{bmatrix} 0.370 \\ 0.478 \\ 0.152 \end{bmatrix}.$$

About 37.0% of workdays will have light traffic, 47.8% will have moderate traffic, and 15.2% will have heavy traffic.

c. $\frac{7}{46} \cdot 20 \approx 3.04$

About 3 workdays will have heavy traffic.

16.
$$\begin{array}{c} \\ C \\ L \\ G \\ S \end{array} \begin{array}{c} \begin{matrix} C & L & G & S \end{matrix} \\ \left[\begin{array}{cc|cc} 1 & 0 & 0.60 & 0.05 \\ 0 & 1 & 0.10 & 0.40 \\ \hline 0 & 0 & 0.20 & 0.50 \\ 0 & 0 & 0.10 & 0.05 \end{array}\right] \end{array}$$

$$R = \begin{bmatrix} 0.20 & 0.50 \\ 0.10 & 0.05 \end{bmatrix}$$

$$I - R = \begin{bmatrix} 1 & 0 \\ 0 & 1 \end{bmatrix} - \begin{bmatrix} 0.20 & 0.50 \\ 0.10 & 0.05 \end{bmatrix}$$

$$= \begin{bmatrix} 0.80 & -0.50 \\ -0.10 & 0.95 \end{bmatrix}$$

$$= \begin{bmatrix} a & b \\ c & d \end{bmatrix}$$

$$\Delta = ad - bc = (0.80)(0.95) - (-0.5)(-0.1) = 0.71$$

$$F = (I - R)^{-1}$$

$$= \begin{bmatrix} \frac{d}{\Delta} & -\frac{b}{\Delta} \\ -\frac{c}{\Delta} & \frac{a}{\Delta} \end{bmatrix}$$

$$= \begin{bmatrix} \frac{0.95}{0.71} & -\frac{-0.50}{0.71} \\ -\frac{-0.10}{0.71} & \frac{0.80}{0.71} \end{bmatrix}$$

$$= \begin{array}{c} \\ G \\ S \end{array} \begin{array}{c} \begin{matrix} G & S \end{matrix} \\ \begin{bmatrix} \frac{95}{71} & \frac{50}{71} \\ \frac{10}{71} & \frac{80}{71} \end{bmatrix} \end{array}$$

Add the entries in each column:

$$\frac{95}{71} + \frac{10}{71} = \frac{105}{71} = 1\frac{34}{71} \approx 1.48$$

$$\frac{50}{71} + \frac{80}{71} = \frac{130}{71} = 1\frac{59}{71} \approx 1.83$$

A patient who begins in state G has an expected number of approximately 1.48 months; a patient who begins in state S has an expected number of approximately 1.83 months.

17. a.
$$\begin{cases} x + y + z + w = 1 \\ \begin{bmatrix} 0.20 & 0.10 & 0.05 & 0.05 \\ 0.30 & 0.20 & 0.20 & 0.30 \\ 0.40 & 0.40 & 0.50 & 0.40 \\ 0.10 & 0.30 & 0.25 & 0.25 \end{bmatrix} \begin{bmatrix} x \\ y \\ z \\ w \end{bmatrix} = \begin{bmatrix} x \\ y \\ z \\ w \end{bmatrix} \end{cases}$$

$$\begin{cases} x + y + z + w = 1 \\ .2x + .1y + .05z + .05w = x \\ .3x + .2y + .2z + .3w = y \\ .4x + .4y + .5z + .4w = z \\ .1x + .3y + .25z + .25w = w \end{cases}$$

$$\begin{cases} x + y + z + w = 1 \\ -0.8x + 0.1y + .05z + .05w = 0 \\ 0.3x - 0.8y + 0.2z + 0.3w = 0 \\ 0.4x + 0.4y - 0.5z + 0.4w = 0 \\ 0.1x + 0.3y + .25z - .75w = 0 \end{cases}$$

The fifth equation is equivalent to the sum of the second, third and fourth and so is redundant.

Copyright © 2018 Pearson Education, Inc.

$$\left[\begin{array}{cccc|c} 1 & 1 & 1 & 1 & 1 \\ -0.8 & 0.1 & 0.05 & 0.05 & 0 \\ 0.3 & -0.8 & 0.2 & 0.3 & 0 \\ 0.4 & 0.4 & -0.5 & 0.4 & 0 \end{array}\right] \xrightarrow[\begin{subarray}{c} 10[3] \\ 10[4] \end{subarray}]{20[2]} \left[\begin{array}{cccc|c} 1 & 1 & 1 & 1 & 1 \\ -16 & 2 & 1 & 1 & 0 \\ 3 & -8 & 2 & 3 & 0 \\ 4 & 4 & -5 & 4 & 0 \end{array}\right] \xrightarrow[\begin{subarray}{c} [3]+(-3)[1] \\ [4]+(-4)[1] \end{subarray}]{[2]+16[1]} \left[\begin{array}{cccc|c} 1 & 1 & 1 & 1 & 1 \\ 0 & 18 & 17 & 17 & 16 \\ 0 & -11 & -1 & 0 & -3 \\ 0 & 0 & -9 & 0 & -4 \end{array}\right]$$

$$\xrightarrow[\begin{subarray}{c} [1]+(-1)[2] \\ [3]+11[2] \end{subarray}]{\frac{1}{18}[2]} \left[\begin{array}{cccc|c} 1 & 0 & \frac{1}{18} & \frac{1}{18} & \frac{1}{9} \\ 0 & 1 & \frac{17}{18} & \frac{17}{18} & \frac{8}{9} \\ 0 & 0 & \frac{169}{18} & \frac{187}{18} & \frac{61}{9} \\ 0 & 0 & -9 & 0 & -4 \end{array}\right] \xrightarrow[\begin{subarray}{c} [1]+\left(-\frac{1}{18}\right)[3] \\ [2]+\left(-\frac{17}{18}\right)[3] \\ [4]+9[3] \end{subarray}]{\frac{18}{169}[3]} \left[\begin{array}{cccc|c} 1 & 0 & 0 & -\frac{1}{169} & \frac{12}{169} \\ 0 & 1 & 0 & -\frac{17}{169} & \frac{35}{169} \\ 0 & 0 & 1 & \frac{187}{169} & \frac{122}{169} \\ 0 & 0 & 0 & \frac{1683}{169} & \frac{422}{169} \end{array}\right]$$

$$\xrightarrow[\begin{subarray}{c} [1]+\frac{1}{169}[4] \\ [2]+\frac{17}{169}[4] \\ [3]+\left(-\frac{187}{169}\right)[4] \end{subarray}]{\frac{169}{1683}[4]} \left[\begin{array}{cccc|c} 1 & 0 & 0 & 0 & \frac{122}{1683} \\ 0 & 1 & 0 & 0 & \frac{23}{99} \\ 0 & 0 & 1 & 0 & \frac{4}{9} \\ 0 & 0 & 0 & 1 & \frac{422}{1683} \end{array}\right]$$

The stable distribution is $\begin{bmatrix} \frac{122}{1683} \\ \frac{23}{99} \\ \frac{4}{9} \\ \frac{422}{1683} \end{bmatrix}$.

In the long run, the probability of having 1, 2, 3, or 4 units of water in the reservoir at any given time will be $\frac{122}{1683}, \frac{23}{99}, \frac{4}{9}$, or $\frac{422}{1683}$, respectively.

b. $\frac{122}{1683}(\$4000)+\frac{23}{99}(\$6000)+\frac{4}{9}(\$10,000)+\frac{422}{1683}(\$3000) \approx \$6881$

The average weekly benefits will be about \$6881.

Conceptual Exercises

18. The entries in each column are the probabilities of the transitions to the various states from the state associated to the column. These must add up to 1, just as the probabilities of all branches emanating from any node in a tree diagram must add up to 1.

19. The probability of going from state *A to* state *B* after four time periods.

20. The stochastic matrix *A* represents a Markov process; A^2 represents this Markov process after two times periods, and so is also a stochastic matrix.

21. True

22. False

 Copyright © 2018 Pearson Education, Inc.

Chapter 9

Exercises 9.1

1. $R:\begin{bmatrix} 6 & \underline{4} \\ 7 & -\underline{5} \end{bmatrix}$, row 1;

$C:\begin{bmatrix} 6 & \underline{4} \\ \underline{7} & -5 \end{bmatrix}$, column 2

Strictly determined with a value = 4.

3. $R:\begin{bmatrix} 4 & -\underline{4} \\ -\underline{1} & 2 \end{bmatrix}$, row 2;

$C:\begin{bmatrix} \underline{4} & -4 \\ -1 & \underline{2} \end{bmatrix}$, column 2

Not strictly determined.

5. $R:\begin{bmatrix} 5 & -\underline{4} & -1 \\ 4 & -3 & -\underline{5} \\ 2 & \underline{1} & 9 \end{bmatrix}$, row 3;

$C:\begin{bmatrix} \underline{5} & -4 & -1 \\ 4 & -3 & -5 \\ 2 & \underline{1} & \underline{9} \end{bmatrix}$, column 2

Strictly determined with a value = 1.

7. $R:\begin{bmatrix} -\underline{4} & -2 \\ 2 & -\underline{3} \\ 1 & \underline{0} \end{bmatrix}$, row 3;

$C:\begin{bmatrix} -4 & -2 \\ \underline{2} & -3 \\ 1 & \underline{0} \end{bmatrix}$, column 2

Strictly determined with a value = 0.

9. $R:\begin{bmatrix} 2 & -\underline{6} & 7 \\ -\underline{5} & 3 & 9 \end{bmatrix}$, row 2;

$C:\begin{bmatrix} \underline{2} & -6 & 7 \\ -5 & \underline{3} & \underline{9} \end{bmatrix}$, column 1

Not strictly determined.

11. $R:\begin{bmatrix} \underline{1} & 2 \\ 5 & -\underline{7} \\ -\underline{2} & 0 \\ 4 & \underline{3} \end{bmatrix}$, row 4;

$C:\begin{bmatrix} 1 & 2 \\ \underline{5} & -7 \\ -2 & 0 \\ 4 & \underline{3} \end{bmatrix}$, column 2

Strictly determined with a value = 3.

13. True

15. $Rosa's:\begin{bmatrix} 50 & \underline{30} \\ -\underline{25} & 15 \end{bmatrix}$, row 1;

$Carlo's:\begin{bmatrix} \underline{50} & \underline{30} \\ -25 & 15 \end{bmatrix}$, column 2

Rosa's Pizzeria should lower their price and Carlo's Pizzeria should improve their décor. It is strictly determined with a value = 30.

17. $Rigel:\begin{bmatrix} -75 & 600 & -\underline{200} \\ 400 & 450 & \underline{300} \\ -\underline{100} & 500 & 150 \end{bmatrix}$, row 2;

$Canis:\begin{bmatrix} -75 & \underline{600} & -200 \\ \underline{400} & 450 & \underline{300} \\ -100 & 500 & 150 \end{bmatrix}$, col 3

Rigelbucks should move to mall B and Canisbucks should move to mall C. It is strictly determined with a value = 300 in Rigelbucks' favor.

19.
$$\begin{array}{c|cc} & H & T \\ \hline H & 2 & -1 \\ T & -1 & -4 \end{array}$$

Row minima: $\begin{bmatrix} 2 & \underline{-1} \\ -1 & \underline{-4} \end{bmatrix}$,

column maxima: $\begin{bmatrix} \underline{2} & \underline{-1} \\ -1 & -4 \end{bmatrix}$

Row 1, column 2 is a saddle point, so the game is strictly determined. *R* should show heads, *C* should show tails.

21. $\begin{array}{c} \\ F \\ A \\ N \end{array}\begin{array}{c} \begin{array}{ccc} F & A & N \end{array} \\ \begin{bmatrix} 8000 & -1000 & 1000 \\ -7000 & 4000 & -2000 \\ 3000 & 3000 & 2000 \end{bmatrix} \end{array}$

Row minima: $\begin{bmatrix} 8000 & \underline{-1000} & 1000 \\ \underline{-7000} & 4000 & -2000 \\ 3000 & 3000 & \underline{2000} \end{bmatrix}$

Column maxima: $\begin{bmatrix} \underline{8000} & -1000 & 1000 \\ -7000 & \underline{4000} & -2000 \\ 3000 & 3000 & \underline{2000} \end{bmatrix}$

Row 3, column 3 is a saddle point, so the game is strictly determined. Both candidates should be neutral.

23. $\begin{array}{c} \\ 5 \\ 10 \end{array}\begin{array}{c} \begin{array}{ccc} 6 & 7 & 8 \end{array} \\ \begin{bmatrix} 1 & -5 & -5 \\ -6 & 3 & 2 \end{bmatrix} \end{array}$

Row minima: $\begin{bmatrix} 1 & \underline{-5} & \underline{-5} \\ \underline{-6} & 3 & 2 \end{bmatrix}$

Column maxima: $\begin{bmatrix} \underline{1} & -5 & -5 \\ -6 & \underline{3} & \underline{2} \end{bmatrix}$

No saddle point, so the game is not strictly determined.

Exercises 9.2

1. a. $[0.5\ \ 0.5]\begin{bmatrix} 3 & -1 \\ -7 & 5 \end{bmatrix}\begin{bmatrix} 0.5 \\ 0.5 \end{bmatrix} = [0]$

b. $[1\ \ 0]\begin{bmatrix} 3 & -1 \\ -7 & 5 \end{bmatrix}\begin{bmatrix} 0.5 \\ 0.5 \end{bmatrix} = [1]$

c. $[0.3\ \ 0.7]\begin{bmatrix} 3 & -1 \\ -7 & 5 \end{bmatrix}\begin{bmatrix} 0.6 \\ 0.4 \end{bmatrix} = [-1.12]$

d. $[0.75\ \ 0.25]\begin{bmatrix} 3 & -1 \\ -7 & 5 \end{bmatrix}\begin{bmatrix} 0.2 \\ 0.8 \end{bmatrix} = [0.5]$

(b) is most advantageous to R.

3. $[0.3\ \ 0.7]\begin{bmatrix} -20,000 & 0 \\ 0 & -50,000 \end{bmatrix}\begin{bmatrix} 0.2 \\ 0.8 \end{bmatrix} = [-29,200]$

$29,200

5. The payoff matrix is $\begin{array}{c} \\ V \\ C \end{array}\begin{array}{c} \begin{array}{cc} V & C \end{array} \\ \begin{bmatrix} 0 & 2 \\ -1 & 0 \end{bmatrix} \end{array}$.

$[0.25\ \ 0.75]\begin{bmatrix} 0 & 2 \\ -1 & 0 \end{bmatrix}\begin{bmatrix} 0.4 \\ 0.6 \end{bmatrix} = [0]$

Zero

 Copyright © 2018 Pearson Education, Inc.

7. a.

$$R\overset{C}{\begin{bmatrix} 3 & -1 \\ -2 & 2 \end{bmatrix}}$$ max min of rows = -1

$$R\overset{C}{\begin{bmatrix} 3 & -1 \\ -2 & 2 \end{bmatrix}}$$ min max of columns = 2

No, there is no saddle point.

b. $[0.3 \quad 0.7]\begin{bmatrix} 3 & -1 \\ -2 & 2 \end{bmatrix}\begin{bmatrix} c_1 \\ c_2 \end{bmatrix} \to [-0.5 \quad 1.1]\begin{bmatrix} c_1 \\ c_2 \end{bmatrix} \to -0.5c_1 + 1.1c_2 = 0 \to c_1 = \frac{1.1c_2}{0.5}$ (to be fair)

Further, $c_1 + c_2 = 1$

$\frac{11}{5}c_2 + c_2 = 1 \to \frac{16}{5}c_2 = 1 \to c_2 = \frac{5}{16} \to c_1 = \frac{11}{16}$

For fair game, Carol's strategy is $\begin{bmatrix} \frac{11}{16} \\ \frac{5}{16} \end{bmatrix}$

9. a.

$$R\overset{C}{\begin{bmatrix} 5 & -1 \\ -2 & 2 \end{bmatrix}}$$ max min of rows = -1

$$R\overset{C}{\begin{bmatrix} 5 & -1 \\ -2 & 2 \end{bmatrix}}$$ min max of columns = 2 Not strictly determined.

b. $[0.7 \quad 0.3]\begin{bmatrix} 5 & -1 \\ -2 & 2 \end{bmatrix} = [2.9 \quad -0.1]$

$[2.9 \quad -0.1]\begin{bmatrix} c_1 \\ c_2 \end{bmatrix} \to 2.9c_1 - 0.1c_2 = 0 \to c_2 = 29c_1$

Carol's strategy: $\begin{bmatrix} \frac{1}{30} \\ \frac{29}{30} \end{bmatrix}$

11. a.

$$R\overset{C}{\begin{bmatrix} 1 & 2 & 4 \\ 1 & 0 & 5 \\ 0 & 1 & -1 \end{bmatrix}}$$ max min of rows = 1

$$R\overset{C}{\begin{bmatrix} 1 & 2 & 4 \\ 1 & 0 & 5 \\ 0 & 1 & -1 \end{bmatrix}}$$ min max of rows = 1

Game strictly determined. Optimal strategy- play row 1, column 1.

Copyright © 2018 Pearson Education, Inc.

$R \rightarrow \begin{bmatrix} 1 & 0 & 0 \end{bmatrix}$

C

$\begin{bmatrix} 1 \\ 0 \\ 0 \end{bmatrix}$

b. Value of game = \$1.00 in favor of Renee'. The game is not fair since the value of the game is different from zero.

$$\begin{bmatrix} 1 & 0 & 0 \end{bmatrix} \begin{bmatrix} -3 & -2 & 6 \\ 2 & 0 & 2 \\ 5 & -2 & -4 \end{bmatrix} = \begin{bmatrix} -3 & -2 & 6 \end{bmatrix}$$

$$\begin{bmatrix} -3 & -2 & 6 \end{bmatrix} \begin{bmatrix} \frac{1}{3} \\ c_2 \\ c_3 \end{bmatrix} = 0 \rightarrow$$

$$-1 - 2c_2 + 6c_3 = 0$$

$$-2c_2 = -6c_3 + 1$$

$$c_2 = 3c_3 - \frac{1}{2}$$

$$c_2 + c_3 = \frac{2}{3}$$

$$3c_3 - \frac{1}{2} + c_3 = \frac{2}{3}$$

$$4c_3 = \frac{7}{6}$$

$$c_3 = \frac{7}{24} \rightarrow c_2 = \frac{9}{24}$$

Strategy: $\begin{bmatrix} \frac{1}{3} & \frac{3}{8} & \frac{7}{24} \end{bmatrix}$

13. a. The payoff matrix is

$$\begin{array}{cc} & \text{Defense} \\ & \begin{array}{cc} R & P \end{array} \\ \text{Offense} \begin{array}{c} R \\ P \end{array} & \begin{bmatrix} 1 & 3 \\ 7 & -4 \end{bmatrix}. \end{array}$$

b. $[0.6 \ 0.4] \begin{bmatrix} 1 & 3 \\ 7 & -4 \end{bmatrix} \begin{bmatrix} 0.7 \\ 0.3 \end{bmatrix} = [2.44]$ The expected number of yards gained is 2.44 yards.

15. a. The payoff matrix is

$$\begin{array}{cc} & \text{Coddy} \\ & \begin{array}{cc} 1 & 2 \end{array} \\ \text{Reven} \begin{array}{c} 1 \\ 2 \end{array} & \begin{bmatrix} 2 & -3 \\ -3 & 4 \end{bmatrix}. \end{array}$$

b. $[0.5 \ 0.5] \begin{bmatrix} 2 & -3 \\ -3 & 4 \end{bmatrix} \begin{bmatrix} 0.25 \\ 0.75 \end{bmatrix} = [0.25]$ The expected value of the game is 0.25 in Reven's favor.

 Copyright © 2018 Pearson Education, Inc.

Exercises 9.3

1. Objective function : Minimize $y_1 + y_2$

Constraints: $\begin{cases} y_1 + 4y_2 \geq 1 \\ 6y_1 + 3y_2 \geq 1 \\ y_1 \geq 0 \\ y_2 \geq 0 \end{cases}$

3. Objective function : Maximize $z_1 + z_2$

Constraints: $\begin{cases} z_1 + 6z_2 \leq 1 \\ 4z_1 + 3z_2 \leq 1 \\ z_1 \geq 0 \\ z_2 \geq 0 \end{cases}$

5. Maximize $M = z_1 + z_2$ subject to

$\begin{cases} 2z_1 + 4z_2 \leq 1 \\ 5z_1 + 3z_2 \leq 1 . \\ z_1 \geq 0,\ z_2 \geq 0 \end{cases}$

$$\begin{array}{c} \\ t \\ u \\ M \end{array} \begin{array}{c} \begin{array}{cccccc} z_1 & z_2 & t & u & M & \end{array} \\ \left[\begin{array}{ccccc|c} 2 & 4 & 1 & 0 & 0 & 1 \\ \underline{5} & 3 & 0 & 1 & 0 & 1 \\ \hline -1 & -1 & 0 & 0 & 1 & 0 \end{array}\right] \end{array}$$

$$\begin{array}{c} \\ t \\ z_1 \\ M \end{array} \begin{array}{c} \begin{array}{cccccc} z_1 & z_2 & t & u & M & \end{array} \\ \left[\begin{array}{ccccc|c} 0 & \underline{\frac{14}{5}} & 1 & -\frac{2}{5} & 0 & \frac{3}{5} \\ 1 & \frac{3}{5} & 0 & \frac{1}{5} & 0 & \frac{1}{5} \\ \hline 0 & -\frac{2}{5} & 0 & \frac{1}{5} & 1 & \frac{1}{5} \end{array}\right] \end{array}$$

$$\begin{array}{c} \\ z_2 \\ z_1 \\ M \end{array} \begin{array}{c} \begin{array}{cccccc} z_1 & z_2 & t & u & M & \end{array} \\ \left[\begin{array}{ccccc|c} 0 & 1 & \frac{5}{14} & -\frac{1}{7} & 0 & \frac{3}{14} \\ 1 & 0 & -\frac{3}{14} & \frac{2}{7} & 0 & \frac{1}{14} \\ \hline 0 & 0 & \frac{1}{7} & \frac{1}{7} & 1 & \frac{2}{7} \end{array}\right] \end{array}$$

$z_1 = \frac{1}{14},\ z_2 = \frac{3}{14},\ M = \frac{2}{7},\ v = \frac{1}{M} = \frac{7}{2},$
and the optimal strategy for C is

$\begin{bmatrix} vz_1 \\ vz_2 \end{bmatrix} = \begin{bmatrix} \frac{1}{4} \\ \frac{3}{4} \end{bmatrix}$. The optimal strategy for R is given by the bottom entries under t and u:

$[vt\ \ vu] = \begin{bmatrix} \frac{7}{2} \cdot \frac{1}{7} & \frac{7}{2} \cdot \frac{1}{7} \end{bmatrix} = \begin{bmatrix} \frac{1}{2} & \frac{1}{2} \end{bmatrix}.$

The value of the game is

$v = \frac{1}{M} = \frac{7}{2}.$

7. Add 7 to each entry to make all the entries positive. We get $\begin{bmatrix} 10 & 1 \\ 2 & 11 \end{bmatrix}$. Then maximize

$M = z_1 + z_2$ subject to $\begin{cases} 10z_1 + z_2 \leq 1 \\ 2z_1 + 11z_2 \leq 1. \\ z_1 \geq 0,\ z_2 \geq 0 \end{cases}$

$$\begin{array}{c} \\ t \\ u \\ M \end{array} \begin{array}{c} \begin{array}{cccccc} z_1 & z_2 & t & u & M & \end{array} \\ \left[\begin{array}{ccccc|c} \underline{10} & 1 & 1 & 0 & 0 & 1 \\ 2 & 11 & 0 & 1 & 0 & 1 \\ \hline -1 & -1 & 0 & 0 & 1 & 0 \end{array}\right] \end{array}$$

$$\begin{array}{c} \\ z_1 \\ u \\ M \end{array} \begin{array}{c} \begin{array}{cccccc} z_1 & z_2 & t & u & M & \end{array} \\ \left[\begin{array}{ccccc|c} 1 & \frac{1}{10} & \frac{1}{10} & 0 & 0 & \frac{1}{10} \\ 0 & \underline{\frac{54}{5}} & -\frac{1}{5} & 1 & 0 & \frac{4}{5} \\ \hline 0 & -\frac{9}{10} & \frac{1}{10} & 0 & 1 & \frac{1}{10} \end{array}\right] \end{array}$$

$$\begin{array}{c} \\ z_1 \\ z_2 \\ M \end{array} \begin{array}{c} \begin{array}{cccccc} z_1 & z_2 & t & u & M & \end{array} \\ \left[\begin{array}{ccccc|c} 1 & 0 & \frac{11}{108} & -\frac{1}{108} & 0 & \frac{5}{54} \\ 0 & 1 & -\frac{1}{54} & \frac{5}{54} & 0 & \frac{2}{27} \\ \hline 0 & 0 & \frac{1}{12} & \frac{1}{12} & 1 & \frac{1}{6} \end{array}\right] \end{array}$$

$z_1 = \frac{5}{54},\ z_2 = \frac{2}{27},\ M = \frac{1}{6},\ v = \frac{1}{M} = 6,$
and the optimal strategy for C is

$\begin{bmatrix} vz_1 \\ vz_2 \end{bmatrix} = \begin{bmatrix} \frac{5}{9} \\ \frac{4}{9} \end{bmatrix}$. The optimal strategy for R is given by the bottom entries under t and u:

$[vt\ \ vu] = \begin{bmatrix} 6 \cdot \frac{1}{12} & 6 \cdot \frac{1}{12} \end{bmatrix} = \begin{bmatrix} \frac{1}{2} & \frac{1}{2} \end{bmatrix}$

The value of the game is

$v - 7 = \frac{1}{M} - 7 = 6 - 7 = -1.$

Copyright © 2018 Pearson Education, Inc.

9. Maximize $M = z_1 + z_2$ subject to

$$\begin{cases} 4z_1 + z_2 \le 1 \\ 2z_1 + 4z_2 \le 1 \\ z_1 \ge 0,\ z_2 \ge 0 \end{cases}.$$

$$\begin{array}{c} \\ t \\ u \\ M \end{array}\begin{array}{c} \begin{array}{cccccc} z_1 & z_2 & t & u & M & \end{array} \\ \left[\begin{array}{ccccc|c} 4 & 1 & 1 & 0 & 0 & 1 \\ 2 & \underline{4} & 0 & 1 & 0 & 1 \\ \hline -1 & -1 & 0 & 0 & 1 & 0 \end{array}\right] \end{array}$$

$$\begin{array}{c} \\ t \\ z_2 \\ M \end{array}\begin{array}{c} \begin{array}{cccccc} z_1 & z_2 & t & u & M & \end{array} \\ \left[\begin{array}{ccccc|c} \frac{7}{2} & 0 & 1 & -\frac{1}{4} & 0 & \frac{3}{4} \\ \frac{1}{2} & 1 & 0 & \frac{1}{4} & 0 & \frac{1}{4} \\ \hline -\frac{1}{2} & 0 & 0 & \frac{1}{4} & 1 & \frac{1}{4} \end{array}\right] \end{array}$$

$$\begin{array}{c} \\ z_1 \\ z_2 \\ M \end{array}\begin{array}{c} \begin{array}{cccccc} z_1 & z_2 & t & u & M & \end{array} \\ \left[\begin{array}{ccccc|c} 1 & 0 & \frac{2}{7} & -\frac{1}{14} & 0 & \frac{3}{14} \\ 0 & 1 & -\frac{1}{7} & \frac{2}{7} & 0 & \frac{1}{7} \\ \hline 0 & 0 & \frac{1}{7} & \frac{3}{14} & 1 & \frac{5}{14} \end{array}\right] \end{array}$$

$z_1 = \frac{3}{14},\ z_2 = \frac{1}{7},\ M = \frac{5}{14},\ v = \frac{1}{M} = \frac{14}{5}$, and the optimal strategy for C is

$$\begin{bmatrix} vz_1 \\ vz_2 \end{bmatrix} = \begin{bmatrix} \frac{3}{5} \\ \frac{2}{5} \end{bmatrix}.$$

The optimal strategy for R is given by the bottom entries under t and u:

$$[vt\ \ vu] = \begin{bmatrix} \frac{14}{5}\cdot\frac{1}{7} & \frac{14}{5}\cdot\frac{3}{14} \end{bmatrix} = \begin{bmatrix} \frac{2}{5} & \frac{3}{5} \end{bmatrix}$$

The value of the game is $v = \frac{1}{M} = \frac{14}{5}$.

11. The original matrix is not strictly determined.

The original matrix is $\begin{bmatrix} 5 & -3 \\ -3 & 1 \end{bmatrix}$. Add 4 to each entry to make all the entries positive. We get $\begin{bmatrix} 9 & 1 \\ 1 & 5 \end{bmatrix}$. Then maximize

$$M = z_1 + z_2 \text{ subject to } \begin{cases} 9z_1 + z_2 \le 1 \\ z_1 + 5z_2 \le 1 \\ z_1 \ge 0,\ z_2 \ge 0 \end{cases}.$$

$$\begin{array}{c} \\ t \\ u \\ M \end{array}\begin{array}{c} \begin{array}{cccccc} z_1 & z_2 & t & u & M & \end{array} \\ \left[\begin{array}{ccccc|c} 9 & \underline{1} & 1 & 0 & 0 & 1 \\ 1 & 5 & 0 & 1 & 0 & 1 \\ \hline -1 & -1 & 0 & 0 & 1 & 0 \end{array}\right] \end{array}$$

$$\begin{array}{c} \\ z_2 \\ u \\ M \end{array}\begin{array}{c} \begin{array}{cccccc} z_1 & z_2 & t & u & M & \end{array} \\ \left[\begin{array}{ccccc|c} 9 & 1 & 1 & 0 & 0 & 1 \\ \underline{-44} & 0 & -5 & 1 & 0 & -4 \\ \hline 8 & 0 & 1 & 0 & 1 & 1 \end{array}\right] \end{array}$$

$$\begin{array}{c} \\ z_2 \\ z_1 \\ M \end{array}\begin{array}{c} \begin{array}{cccccc} z_1 & z_2 & t & u & M & \end{array} \\ \left[\begin{array}{ccccc|c} 0 & 1 & -\frac{1}{44} & \frac{9}{44} & 0 & \frac{2}{11} \\ 1 & 0 & \frac{5}{44} & -\frac{1}{44} & 0 & \frac{1}{11} \\ \hline 0 & 0 & \frac{1}{11} & \frac{2}{11} & 1 & \frac{3}{11} \end{array}\right] \end{array}$$

$$v = \frac{1}{M} = \frac{11}{3}$$

R's optimal strategy is given by the bottom entries under t and u:

$$[vt\ \ vu] = \begin{bmatrix} \frac{1}{3} & \frac{2}{3} \end{bmatrix}$$

C's optimal strategy is $\begin{bmatrix} vz_1 \\ vz_2 \end{bmatrix} = \begin{bmatrix} \frac{1}{3} \\ \frac{2}{3} \end{bmatrix}$.

The value of the game is

$$v - 4 = \frac{1}{M} - 4 = \frac{11}{3} - 4 = -\frac{1}{3}.$$

13. Add 2 to each entry to make all entries of the payoff matrix positive.

$$\begin{bmatrix} 5 & 7 & 1 \\ 6 & 1 & 8 \end{bmatrix}$$

Set the tableaux up to find C's optimal strategy, then read the dual's solution off the bottom row:

$$\begin{array}{c} \\ t \\ u \\ M \end{array}\begin{array}{c} \begin{array}{ccccccc} z_1 & z_2 & z_3 & t & u & M & \end{array} \\ \left[\begin{array}{cccccc|c} 5 & \underline{7} & 1 & 1 & 0 & 0 & 1 \\ 6 & 1 & 8 & 0 & 1 & 0 & 1 \\ \hline -1 & -1 & -1 & 0 & 0 & 1 & 0 \end{array}\right] \end{array}$$

$$\begin{array}{c} \\ z_2 \\ u \\ M \end{array}\begin{array}{c} \begin{array}{ccccccc} z_1 & z_2 & z_3 & t & u & M & \end{array} \\ \left[\begin{array}{cccccc|c} \frac{5}{7} & 1 & \frac{1}{7} & \frac{1}{7} & 0 & 0 & \frac{1}{7} \\ \frac{37}{7} & 0 & \underline{\frac{55}{7}} & -\frac{1}{7} & 1 & 0 & \frac{6}{7} \\ \hline -\frac{2}{7} & 0 & -\frac{6}{7} & \frac{1}{7} & 0 & 1 & \frac{1}{7} \end{array}\right] \end{array}$$

Copyright © 2018 Pearson Education, Inc.

$$\begin{array}{c|cccccc|c} & z_1 & z_2 & z_3 & t & u & M & \\ \hline z_2 & \frac{34}{55} & 1 & 0 & \frac{8}{55} & -\frac{1}{55} & 0 & \frac{7}{55} \\ z_3 & \frac{37}{55} & 0 & 1 & -\frac{1}{55} & \frac{7}{55} & 0 & \frac{6}{55} \\ \hline M & \frac{16}{55} & 0 & 0 & \frac{7}{55} & \frac{6}{55} & 1 & \frac{13}{55} \end{array}$$

$t=\frac{7}{55}$, $u=\frac{6}{55}$, $M=\frac{13}{55}$, $v=\frac{1}{M}=\frac{55}{13}$, and R's optimal strategy is

$[vt\ \ vu]=\begin{bmatrix}\frac{7}{13} & \frac{6}{13}\end{bmatrix}$.

The value of the game is

$v-2=\frac{1}{M}-2=\frac{55}{13}-2=\frac{29}{13}$.

15. Add 4 to each entry to make all entries of the payoff matrix positive.

$\begin{bmatrix} 5 & 3 & 10 \\ 3 & 6 & 1 \end{bmatrix}$

Set the tableaux up to find C's optimal strategy, then read the dual's solution off the bottom row:

$$\begin{array}{c|cccccc|c} & z_1 & z_2 & z_3 & t & u & M & \\ \hline t & 5 & 3 & 10 & 1 & 0 & 0 & 1 \\ u & 3 & 6 & 1 & 0 & 1 & 0 & 1 \\ \hline M & -1 & -1 & -1 & 0 & 0 & 1 & 0 \end{array}$$

$$\begin{array}{c|cccccc|c} & z_1 & z_2 & z_3 & t & u & M & \\ \hline z_1 & 1 & \frac{3}{5} & 2 & \frac{1}{5} & 0 & 0 & \frac{1}{5} \\ u & 0 & \frac{21}{5} & -5 & -\frac{3}{5} & 1 & 0 & \frac{2}{5} \\ \hline M & 0 & -\frac{2}{5} & 1 & \frac{1}{5} & 0 & 1 & \frac{1}{5} \end{array}$$

$$\begin{array}{c|cccccc|c} & z_1 & z_2 & z_3 & t & u & M & \\ \hline z_1 & 1 & 0 & \frac{19}{7} & \frac{2}{7} & -\frac{1}{7} & 0 & \frac{1}{7} \\ z_2 & 0 & 1 & -\frac{25}{21} & -\frac{1}{7} & \frac{5}{21} & 0 & \frac{2}{21} \\ \hline M & 0 & 0 & \frac{11}{21} & \frac{1}{7} & \frac{2}{21} & 1 & \frac{5}{21} \end{array}$$

$t=\frac{1}{7}$, $u=\frac{2}{21}$, $M=\frac{5}{21}$, $v=\frac{1}{M}=\frac{21}{5}$, and R's optimal strategy is

$[vt\ \ vu]=\begin{bmatrix}\frac{3}{5} & \frac{2}{5}\end{bmatrix}$.

The value of the game is

$v-4=\frac{1}{M}-4=\frac{21}{5}-4=\frac{1}{5}$.

17. Add 4 to each entry to make them all positive. Maximize $M=z_1+z_2$ subject

to $\begin{cases} z_1+5z_2\le 1 \\ 8z_1+3z_2\le 1 \\ 5z_1+4z_2\le 1 \\ z_1\ge 0,\ z_2\ge 0 \end{cases}$.

$$\begin{array}{c|cccccc|c} & z_1 & z_2 & s & t & u & M & \\ \hline s & 1 & 5 & 1 & 0 & 0 & 0 & 1 \\ t & 8 & 3 & 0 & 1 & 0 & 0 & 1 \\ u & 5 & 4 & 0 & 0 & 1 & 0 & 1 \\ \hline M & -1 & -1 & 0 & 0 & 0 & 1 & 0 \end{array}$$

$$\begin{array}{c|cccccc|c} & z_1 & z_2 & s & t & u & M & \\ \hline z_2 & \frac{1}{5} & 1 & \frac{1}{5} & 0 & 0 & 0 & \frac{1}{5} \\ t & \frac{37}{5} & 0 & -\frac{3}{5} & 1 & 0 & 0 & \frac{2}{5} \\ u & \frac{21}{5} & 0 & -\frac{4}{5} & 0 & 1 & 0 & \frac{1}{5} \\ \hline M & -\frac{4}{5} & 0 & \frac{1}{5} & 0 & 0 & 1 & \frac{1}{5} \end{array}$$

$$\begin{array}{c|cccccc|c} & z_1 & z_2 & s & t & u & M & \\ \hline z_2 & 0 & 1 & \frac{8}{37} & -\frac{1}{37} & 0 & 0 & \frac{7}{37} \\ z_1 & 1 & 0 & -\frac{3}{37} & \frac{5}{37} & 0 & 0 & \frac{2}{37} \\ u & 0 & 0 & -\frac{17}{37} & -\frac{21}{37} & 1 & 0 & -\frac{1}{37} \\ \hline M & 0 & 0 & \frac{5}{37} & \frac{4}{37} & 0 & 1 & \frac{9}{37} \end{array}$$

$z_1=\frac{2}{37}$, $z_2=\frac{7}{37}$, $M=\frac{9}{37}$, $v=\frac{1}{M}=\frac{37}{9}$,

and C's optimal strategy is $\begin{bmatrix} vz_1 \\ vz_2 \end{bmatrix}=\begin{bmatrix} \frac{2}{9} \\ \frac{7}{9} \end{bmatrix}$.

The value of the game is

$v-4=\frac{1}{M}-4=\frac{37}{9}-4=\frac{1}{9}$..

19. Add 3 to each entry to make them all positive. Maximize $M=z_1+z_2$ subject

to $\begin{cases} 4z_1+3z_2\le 1 \\ z_1+7z_2\le 1 \\ 2z_1+2z_2\le 1 \\ z_1\ge 0,\ z_2\ge 0 \end{cases}$.

Copyright © 2018 Pearson Education, Inc.

$$\begin{array}{c} \\ s \\ t \\ u \\ M \end{array}\begin{array}{c}\begin{matrix} z_1 & z_2 & s & t & u & M \end{matrix}\\ \left[\begin{array}{rrrrrr|r} 4 & 3 & 1 & 0 & 0 & 0 & 1 \\ 1 & 7 & 0 & 1 & 0 & 0 & 1 \\ 2 & 2 & 0 & 0 & 1 & 0 & 1 \\ \hline -1 & -1 & 0 & 0 & 0 & 1 & 0 \end{array}\right]\end{array}$$

$$\begin{array}{c} \\ z_2 \\ t \\ u \\ M \end{array}\begin{array}{c}\begin{matrix} z_1 & z_2 & s & t & u & M \end{matrix}\\ \left[\begin{array}{rrrrrr|r} 1 & \frac{3}{4} & \frac{1}{4} & 0 & 0 & 0 & \frac{1}{4} \\ 0 & \frac{25}{4} & -\frac{1}{4} & 1 & 0 & 0 & \frac{3}{4} \\ 0 & \frac{1}{2} & -\frac{1}{2} & 0 & 1 & 0 & \frac{1}{2} \\ \hline 0 & -\frac{1}{4} & \frac{1}{4} & 0 & 0 & 1 & \frac{1}{4} \end{array}\right]\end{array}$$

$$\begin{array}{c} \\ z_2 \\ z_1 \\ u \\ M \end{array}\begin{array}{c}\begin{matrix} z_1 & z_2 & s & t & u & M \end{matrix}\\ \left[\begin{array}{rrrrrr|r} 1 & 0 & \frac{7}{25} & -\frac{3}{25} & 0 & 0 & \frac{4}{25} \\ 0 & 1 & -\frac{1}{25} & \frac{4}{25} & 0 & 0 & \frac{3}{25} \\ 0 & 0 & -\frac{12}{25} & -\frac{2}{25} & 1 & 0 & \frac{11}{25} \\ \hline 0 & 0 & \frac{6}{25} & \frac{1}{25} & 0 & 1 & \frac{7}{25} \end{array}\right]\end{array}$$

$z_1 = \frac{3}{25},\ z_2 = \frac{4}{25},\ M = \frac{7}{25},\ v = \frac{1}{M} = \frac{25}{7},$

and C's optimal strategy is $\begin{bmatrix} vz_1 \\ vz_2 \end{bmatrix} = \begin{bmatrix} \frac{3}{7} \\ \frac{4}{7} \end{bmatrix}$.

The value of the game is

$$v - 3 = \frac{1}{M} - 3 = \frac{25}{7} - 3 = \frac{4}{7}.$$

21. a. The payoff matrix, with entries in thousands of dollars, is $\begin{array}{c} \\ I \\ II \end{array}\begin{array}{c}\begin{matrix} I & II \end{matrix}\\ \begin{bmatrix} -2 & 7 \\ 7 & -1 \end{bmatrix}\end{array}$.

Add 3 to each entry to make all the entries positive. We get $\begin{bmatrix} 1 & 10 \\ 10 & 2 \end{bmatrix}$.

Then maximize $M = z_1 + z_2$ subject to $\begin{cases} z_1 + 10z_2 \le 1 \\ 10z_1 + 2z_2 \le 1. \\ z_1 \ge 0,\ z_2 \ge 0 \end{cases}$

$$\begin{array}{c} \\ t \\ u \\ M \end{array}\begin{array}{c}\begin{matrix} z_1 & z_2 & t & u & M \end{matrix}\\ \left[\begin{array}{rrrrr|r} 1 & \underline{10} & 1 & 0 & 0 & 1 \\ 10 & 2 & 0 & 1 & 0 & 1 \\ \hline -1 & -1 & 0 & 0 & 1 & 0 \end{array}\right]\end{array}$$

$$\begin{array}{c} \\ z_2 \\ u \\ M \end{array}\begin{array}{c}\begin{matrix} z_1 & z_2 & t & u & M \end{matrix}\\ \left[\begin{array}{rrrrr|r} \frac{1}{10} & 1 & \frac{1}{10} & 0 & 0 & \frac{1}{10} \\ \frac{49}{5} & 0 & -\frac{1}{5} & 1 & 0 & \frac{4}{5} \\ \hline -\frac{9}{10} & 0 & \frac{1}{10} & 0 & 1 & \frac{1}{10} \end{array}\right]\end{array}$$

$$\begin{array}{c} \\ z_2 \\ z_1 \\ M \end{array}\begin{array}{c}\begin{matrix} z_1 & z_2 & t & u & M \end{matrix}\\ \left[\begin{array}{rrrrr|r} 0 & 1 & \frac{5}{49} & -\frac{1}{98} & 0 & \frac{9}{98} \\ 1 & 0 & -\frac{1}{49} & \frac{5}{49} & 0 & \frac{4}{49} \\ \hline 0 & 0 & \frac{4}{49} & \frac{9}{98} & 1 & \frac{17}{98} \end{array}\right]\end{array}$$

$$v = \frac{1}{M} = \frac{98}{17}$$

b. R's optimal strategy is given by the bottom entries under t and u:
$[vt\ \ vu] = \begin{bmatrix} \frac{8}{17} & \frac{9}{17} \end{bmatrix}$

c. C's optimal strategy is $\begin{bmatrix} vz_1 \\ vz_2 \end{bmatrix} = \begin{bmatrix} \frac{8}{17} \\ \frac{9}{17} \end{bmatrix}$.

d. $v - 3 = \frac{1}{M} - 3 = \frac{98}{17} - 3 \approx 2.765$
The value of the game is about \$2765.

23. The payoff matrix is $\begin{array}{c} \\ R \\ P \end{array}\begin{array}{c}\begin{matrix} R & P \end{matrix}\\ \begin{bmatrix} 1 & 3 \\ 7 & -4 \end{bmatrix}\end{array}$. Add 5 to each entry to make all the entries positive. We get $\begin{bmatrix} 6 & 8 \\ 12 & 1 \end{bmatrix}$. Then maximize

$$M = z_1 + z_2 \text{ subject to } \begin{cases} 6z_1 + 8z_2 \le 1 \\ 12z_1 + z_2 \le 1\ . \\ z_1 \ge 0,\ z_2 \ge 0 \end{cases}$$

$$\begin{array}{c} \\ t \\ u \\ M \end{array}\begin{array}{c}\begin{matrix} z_1 & z_2 & t & u & M \end{matrix}\\ \left[\begin{array}{rrrrr|r} 6 & 8 & 1 & 0 & 0 & 1 \\ \underline{12} & 1 & 0 & 1 & 0 & 1 \\ \hline -1 & -1 & 0 & 0 & 1 & 0 \end{array}\right]\end{array}$$

Copyright © 2018 Pearson Education, Inc.

$$\begin{array}{c} \\ t \\ z_1 \\ M \end{array}\begin{array}{c} \begin{array}{cccccc} z_1 & z_2 & t & u & M & \end{array} \\ \left[\begin{array}{ccccc|c} 0 & \frac{12}{5} & 1 & -\frac{1}{2} & 0 & \frac{1}{2} \\ 1 & \frac{1}{12} & 0 & \frac{1}{12} & 0 & \frac{1}{12} \\ \hline 0 & -\frac{11}{12} & 0 & \frac{1}{12} & 1 & \frac{1}{12} \end{array}\right] \end{array}$$

$$\begin{array}{c} \\ z_2 \\ z_1 \\ M \end{array}\begin{array}{c} \begin{array}{cccccc} z_1 & z_2 & t & u & M & \end{array} \\ \left[\begin{array}{ccccc|c} 0 & 1 & \frac{2}{15} & -\frac{1}{15} & 0 & \frac{1}{15} \\ 1 & 0 & -\frac{1}{90} & \frac{91}{180} & 0 & \frac{7}{90} \\ \hline 0 & 0 & \frac{11}{90} & \frac{1}{45} & 1 & \frac{13}{90} \end{array}\right] \end{array}$$

$$v = \frac{1}{M} = \frac{90}{13}$$

R's optimal strategy is given by the bottom entries under t and u:

$$[vt \;\; vu] = \left[\frac{90}{13}\cdot\frac{11}{90} \;\; \frac{90}{13}\cdot\frac{1}{45}\right] = \left[\frac{11}{13} \;\; \frac{2}{13}\right]$$

C's optimal strategy is

$$\begin{bmatrix} vz_1 \\ vz_2 \end{bmatrix} = \begin{bmatrix} \frac{90}{13}\cdot\frac{7}{90} \\ \frac{90}{13}\cdot\frac{1}{15} \end{bmatrix} = \begin{bmatrix} \frac{7}{13} \\ \frac{6}{13} \end{bmatrix}.$$

$$v - 5 = \frac{1}{M} - 5 = \frac{90}{13} - 5 = \frac{25}{13} \approx 1.92$$

The expected yards gained is about 1.92.

25. a. The payoff matrix is $\begin{array}{c} \\ 1 \\ 2 \end{array}\begin{array}{c} \begin{array}{cc} 1 & 2 \end{array} \\ \begin{bmatrix} 2 & -3 \\ -3 & 4 \end{bmatrix} \end{array}.$

Add 4 to each entry to make them all positive, then maximize $M = z_1 + z_2$

subject to $\begin{cases} 6z_1 + z_2 \le 1 \\ z_1 + 8z_2 \le 1 \\ z_1 \ge 0,\ z_2 \ge 0 \end{cases}$.

$$\begin{array}{c} \\ t \\ u \\ M \end{array}\begin{array}{c} \begin{array}{cccccc} z_1 & z_2 & t & u & M & \end{array} \\ \left[\begin{array}{ccccc|c} 6 & 1 & 1 & 0 & 0 & 1 \\ 1 & \underline{8} & 0 & 1 & 0 & 1 \\ \hline -1 & -1 & 0 & 0 & 1 & 0 \end{array}\right] \end{array}$$

$$\begin{array}{c} \\ t \\ z_2 \\ M \end{array}\begin{array}{c} \begin{array}{cccccc} z_1 & z_2 & t & u & M & \end{array} \\ \left[\begin{array}{ccccc|c} \frac{47}{8} & 0 & 1 & -\frac{1}{8} & 0 & \frac{7}{8} \\ \frac{1}{8} & 1 & 0 & \frac{1}{8} & 0 & \frac{1}{8} \\ \hline -\frac{7}{8} & 0 & 0 & \frac{1}{8} & 1 & \frac{1}{8} \end{array}\right] \end{array}$$

$$\begin{array}{c} \\ z_1 \\ z_2 \\ M \end{array}\begin{array}{c} \begin{array}{cccccc} z_1 & z_2 & t & u & M & \end{array} \\ \left[\begin{array}{ccccc|c} 1 & 0 & \frac{8}{47} & -\frac{1}{47} & 0 & \frac{7}{47} \\ 0 & 1 & -\frac{1}{47} & \frac{6}{47} & 0 & \frac{5}{47} \\ \hline 0 & 0 & \frac{7}{47} & \frac{5}{47} & 1 & \frac{12}{47} \end{array}\right] \end{array}$$

$$v = \frac{1}{M} = \frac{47}{12}$$

b. Revin's optimal strategy is given by the bottom entries under t and u:

$$[vt \;\; vu] = \left[\frac{47}{12}\cdot\frac{7}{47} \;\; \frac{47}{12}\cdot\frac{5}{47}\right] = \left[\frac{7}{12} \;\; \frac{5}{12}\right]$$

Coddy's optimal strategy is

$$\begin{bmatrix} vz_1 \\ vz_2 \end{bmatrix} = \begin{bmatrix} \frac{47}{12}\cdot\frac{7}{47} \\ \frac{47}{12}\cdot\frac{5}{47} \end{bmatrix} = \begin{bmatrix} \frac{7}{12} \\ \frac{5}{12} \end{bmatrix}.$$

The value of the game is

$v - 4 = \frac{47}{12} - \frac{48}{12} = -\frac{1}{12}$. Since this is negative, the game favors Coddy.

27. Add 4 to each entry to make them all positive. Maximize $M = z_1 + z_2$

subject to $\begin{cases} 2z_1 + 5z_2 \le 1 \\ 6z_1 + z_2 \le 1 \\ 5z_1 + 2z_2 \le 1 \\ z_1 \ge 0,\ z_2 \ge 0 \end{cases}$.

$$\begin{array}{c} \\ s \\ t \\ u \\ M \end{array}\begin{array}{c} \begin{array}{ccccccc} z_1 & z_2 & s & t & u & M & \end{array} \\ \left[\begin{array}{cccccc|c} 2 & \underline{5} & 1 & 0 & 0 & 0 & 1 \\ 6 & 1 & 0 & 1 & 0 & 0 & 1 \\ 5 & 2 & 0 & 0 & 1 & 0 & 1 \\ \hline -1 & -1 & 0 & 0 & 0 & 1 & 0 \end{array}\right] \end{array}$$

$$\begin{array}{c} \\ z_2 \\ t \\ u \\ M \end{array}\begin{array}{c} \begin{array}{ccccccc} z_1 & z_2 & s & t & u & M & \end{array} \\ \left[\begin{array}{cccccc|c} \frac{2}{5} & 1 & \frac{1}{5} & 0 & 0 & 0 & \frac{1}{5} \\ \frac{28}{5} & 0 & -\frac{1}{5} & 1 & 0 & 0 & \frac{4}{5} \\ \frac{21}{5} & 0 & -\frac{2}{5} & 0 & 1 & 0 & \frac{3}{5} \\ \hline -\frac{3}{5} & 0 & \frac{1}{5} & 0 & 0 & 1 & \frac{1}{5} \end{array}\right] \end{array}$$

Copyright © 2018 Pearson Education, Inc.

$$\begin{array}{c} \\ z_2 \\ z_1 \\ u \\ M \end{array}\begin{array}{c} \begin{array}{cccccc} z_1 & z_2 & s & t & u & M \end{array} \\ \left[\begin{array}{cccccc|c} 0 & 1 & \frac{3}{14} & -\frac{1}{14} & 0 & 0 & \frac{1}{7} \\ 1 & 0 & -\frac{1}{28} & \frac{5}{28} & 0 & 0 & \frac{1}{7} \\ 0 & 0 & -\frac{1}{4} & -\frac{3}{4} & 1 & 0 & 0 \\ \hline 0 & 0 & \frac{5}{28} & \frac{3}{28} & 0 & 1 & \frac{2}{7} \end{array}\right] \end{array}$$

$$v = \frac{1}{M} = \frac{7}{2}$$

Rosedale's optimal strategy is given by the bottom entries under *s*, *t* and *u*:

$$[vs \;\; vt \;\; vu] = \begin{bmatrix} \frac{5}{8} & \frac{3}{8} & 0 \end{bmatrix}$$

Carter's optimal strategy is

$$\begin{bmatrix} vz_1 \\ vz_2 \end{bmatrix} = \begin{bmatrix} \frac{1}{2} \\ \frac{1}{2} \end{bmatrix}.$$

29. Using Wolfram|Alpha or excel, the expected value and the strategies will be: $v = -0.083331$;

Reven: $[0.583333 \quad 0.416667]$

Coddy: $\begin{bmatrix} 0.583333 \\ 0.416667 \end{bmatrix}.$

Chapter 9 Review Exercises

1. Row minima: $\begin{bmatrix} 5 & \underline{-1} & 1 \\ \underline{-3} & 5 & 1 \\ 4 & 3 & \underline{2} \end{bmatrix}$,

column maxima: $\begin{bmatrix} \underline{5} & -1 & 1 \\ -3 & \underline{5} & 1 \\ 4 & 3 & \underline{2} \end{bmatrix}$

The game is strictly determined, with a saddle point at row 3, column 3 and a value of 2.

2. Row minima: $\begin{bmatrix} \underline{1} & 2 & 3 \\ 3 & 2 & \underline{1} \end{bmatrix}$,

column maxima: $\begin{bmatrix} 1 & \underline{2} & \underline{3} \\ \underline{3} & \underline{2} & 1 \end{bmatrix}$

The game has no saddle point and so is not strictly determined.

3. Row minima: $\begin{bmatrix} \underline{0} & 1 \\ 1 & \underline{0} \\ 2 & \underline{-1} \end{bmatrix}$,

column maxima: $\begin{bmatrix} 0 & \underline{1} \\ 1 & 0 \\ \underline{2} & -1 \end{bmatrix}$

The game has no saddle point and so is not strictly determined.

4. Row minima: $\begin{bmatrix} 2 & \underline{1} & 2 \\ \underline{-1} & 0 & 3 \\ 4 & 1 & \underline{-4} \end{bmatrix}$,

column maxima: $\begin{bmatrix} 2 & \underline{1} & 2 \\ -1 & 0 & \underline{3} \\ \underline{4} & \underline{1} & -4 \end{bmatrix}$

The game is strictly determined, with a saddle point at row 1, column 2 and a value of 1.

5. $\begin{bmatrix} \frac{3}{4} & \frac{1}{4} \end{bmatrix}\begin{bmatrix} 0 & 24 \\ 12 & -36 \end{bmatrix}\begin{bmatrix} \frac{1}{3} \\ \frac{2}{3} \end{bmatrix} = [7]$

7

6. $\begin{bmatrix} \frac{1}{2} & \frac{1}{2} \end{bmatrix}\begin{bmatrix} -6 & 6 & 0 \\ 0 & -12 & 24 \end{bmatrix}\begin{bmatrix} \frac{1}{3} \\ \frac{1}{3} \\ \frac{1}{3} \end{bmatrix} = [2]$

2

7. $[0.2 \;\; 0.3 \;\; 0.5]\begin{bmatrix} 1 & 0 \\ -3 & 1 \\ 0 & 5 \end{bmatrix}\begin{bmatrix} 0.4 \\ 0.6 \end{bmatrix} = [1.4]$

1.4

8. $[0.1 \;\; 0.1 \;\; 0.8]\begin{bmatrix} 0 & 1 & 3 \\ -1 & 0 & 2 \\ -3 & -2 & 0 \end{bmatrix}\begin{bmatrix} 0.4 \\ 0.3 \\ 0.3 \end{bmatrix} = [-1.3]$

-1.3

Copyright © 2018 Pearson Education, Inc.

9. Add 4 to each entry to get $\begin{bmatrix} 1 & 8 \\ 6 & 2 \end{bmatrix}$. Then use the simplex method.

$$\begin{array}{c|ccccc|c} & z_1 & z_2 & t & u & M & \\ \hline t & 1 & 8 & 1 & 0 & 0 & 1 \\ u & \underline{6} & 2 & 0 & 1 & 0 & 1 \\ \hline M & -1 & -1 & 0 & 0 & 1 & 0 \end{array}$$

$$\begin{array}{c|ccccc|c} & z_1 & z_2 & t & u & M & \\ \hline t & 0 & \underline{\frac{23}{3}} & 1 & -\frac{1}{6} & 0 & \frac{5}{6} \\ z_1 & 1 & \frac{1}{3} & 0 & \frac{1}{6} & 0 & \frac{1}{6} \\ \hline M & 0 & -\frac{2}{3} & 0 & \frac{1}{6} & 1 & \frac{1}{6} \end{array}$$

$$\begin{array}{c|ccccc|c} & z_1 & z_2 & t & u & M & \\ \hline z_2 & 0 & 1 & \frac{3}{23} & -\frac{1}{46} & 0 & \frac{5}{46} \\ z_1 & 1 & 0 & -\frac{1}{23} & \frac{4}{23} & 0 & \frac{3}{23} \\ \hline M & 0 & 0 & \frac{2}{23} & \frac{7}{46} & 1 & \frac{11}{46} \end{array}$$

$v = \frac{1}{M} = \frac{46}{11}$

R's optimal strategy:

$[vt \;\; vu] = \left[\frac{46}{11} \cdot \frac{2}{23} \;\; \frac{46}{11} \cdot \frac{7}{46}\right] = \left[\frac{4}{11} \;\; \frac{7}{11}\right]$

C's optimal strategy:

$$\begin{bmatrix} vz_1 \\ vz_2 \end{bmatrix} = \begin{bmatrix} \frac{46}{11} \cdot \frac{3}{23} \\ \frac{46}{11} \cdot \frac{5}{46} \end{bmatrix} = \begin{bmatrix} \frac{6}{11} \\ \frac{5}{11} \end{bmatrix}$$

10. Add 7 to each entry to get $\begin{bmatrix} 10 & 1 \\ 3 & 11 \end{bmatrix}$. Then apply the simplex method.

$$\begin{array}{c|ccccc|c} & z_1 & z_2 & t & u & M & \\ \hline t & \underline{10} & 1 & 1 & 0 & 0 & 1 \\ u & 3 & 11 & 0 & 1 & 0 & 1 \\ \hline M & -1 & -1 & 0 & 0 & 1 & 0 \end{array}$$

$$\begin{array}{c|ccccc|c} & z_1 & z_2 & t & u & M & \\ \hline z_1 & 1 & \frac{1}{10} & \frac{1}{10} & 0 & 0 & \frac{1}{10} \\ u & 0 & \underline{\frac{107}{10}} & -\frac{3}{10} & 1 & 0 & \frac{7}{10} \\ \hline M & 0 & -\frac{9}{10} & \frac{1}{10} & 0 & 1 & \frac{1}{10} \end{array}$$

$$\begin{array}{c|ccccc|c} & z_1 & z_2 & t & u & M & \\ \hline z_1 & 1 & 0 & \frac{11}{107} & -\frac{1}{107} & 0 & \frac{10}{107} \\ z_2 & 0 & 1 & -\frac{3}{107} & \frac{10}{107} & 0 & \frac{7}{107} \\ \hline M & 0 & 0 & \frac{8}{107} & \frac{9}{107} & 1 & \frac{17}{107} \end{array}$$

$v = \frac{1}{M} = \frac{107}{17}$

R's optimal strategy:

$[vt \;\; vu] = \left[\frac{107}{17} \cdot \frac{8}{107} \;\; \frac{107}{17} \cdot \frac{9}{107}\right] = \left[\frac{8}{17} \;\; \frac{9}{17}\right]$

C's optimal strategy:

$$\begin{bmatrix} vz_1 \\ vz_2 \end{bmatrix} = \begin{bmatrix} \frac{107}{17} \cdot \frac{10}{107} \\ \frac{107}{17} \cdot \frac{7}{107} \end{bmatrix} = \begin{bmatrix} \frac{10}{17} \\ \frac{7}{17} \end{bmatrix}$$

11. Row 2, column 3 is a saddle point: [0 1]

12.

$$\begin{array}{c|cccccc|c} & z_1 & z_2 & s & t & u & M & \\ \hline s & 1 & 3 & 1 & 0 & 0 & 0 & 1 \\ t & 3 & 1 & 0 & 1 & 0 & 0 & 1 \\ u & \underline{4} & 2 & 0 & 0 & 1 & 0 & 1 \\ \hline M & -1 & -1 & 0 & 0 & 0 & 1 & 0 \end{array}$$

$$\begin{array}{c|cccccc|c} & z_1 & z_2 & s & t & u & M & \\ \hline s & 0 & \underline{\frac{5}{2}} & 1 & 0 & -\frac{1}{4} & 0 & \frac{3}{4} \\ t & 0 & -\frac{1}{2} & 0 & 1 & -\frac{3}{4} & 0 & \frac{1}{4} \\ z_1 & 1 & \frac{1}{2} & 0 & 0 & \frac{1}{4} & 0 & \frac{1}{4} \\ \hline M & 0 & -\frac{1}{2} & 0 & 0 & \frac{1}{4} & 1 & \frac{1}{4} \end{array}$$

$$\begin{array}{c|cccccc|c} & z_1 & z_2 & s & t & u & M & \\ \hline z_2 & 0 & 1 & \frac{2}{5} & 0 & -\frac{1}{10} & 0 & \frac{3}{10} \\ t & 0 & 0 & \frac{1}{5} & 1 & -\frac{4}{5} & 0 & \frac{2}{5} \\ z_1 & 1 & 0 & -\frac{1}{5} & 0 & \frac{3}{10} & 0 & \frac{1}{10} \\ \hline M & 0 & 0 & \frac{1}{5} & 0 & \frac{1}{5} & 1 & \frac{2}{5} \end{array}$$

$v = \frac{1}{M} = \frac{5}{2}$

$$\begin{bmatrix} vz_1 \\ vz_2 \end{bmatrix} = \begin{bmatrix} \frac{5}{2} \cdot \frac{1}{10} \\ \frac{5}{2} \cdot \frac{3}{10} \end{bmatrix} = \begin{bmatrix} \frac{1}{4} \\ \frac{3}{4} \end{bmatrix}$$

Copyright © 2018 Pearson Education, Inc.

13. The payoff matrix is $\begin{matrix} & 2 & 6 \\ 2 & -3 & 2 \\ 6 & 6 & -3 \end{matrix}$. Add 4 to each entry to get $\begin{bmatrix} 1 & 6 \\ 10 & 1 \end{bmatrix}$. Then apply the simplex method.

$$\begin{array}{c} \\ t \\ u \\ M \end{array}\begin{array}{c} \begin{matrix} z_1 & z_2 & t & u & M & \end{matrix} \\ \left[\begin{array}{ccccc|c} 1 & \underline{6} & 1 & 0 & 0 & 1 \\ 10 & 1 & 0 & 1 & 0 & 1 \\ \hline -1 & -1 & 0 & 0 & 1 & 0 \end{array}\right] \end{array}$$

$$\begin{array}{c} \\ z_2 \\ u \\ M \end{array}\begin{array}{c} \begin{matrix} z_1 & z_2 & t & u & M & \end{matrix} \\ \left[\begin{array}{ccccc|c} \frac{1}{6} & 1 & \frac{1}{6} & 0 & 0 & \frac{1}{6} \\ \frac{59}{6} & 0 & -\frac{1}{6} & 1 & 0 & \frac{5}{6} \\ \hline -\frac{5}{6} & 0 & \frac{1}{6} & 0 & 1 & \frac{1}{6} \end{array}\right] \end{array}$$

$$\begin{array}{c} \\ z_2 \\ z_1 \\ M \end{array}\begin{array}{c} \begin{matrix} z_1 & z_2 & t & u & M & \end{matrix} \\ \left[\begin{array}{ccccc|c} 0 & 1 & \frac{10}{59} & -\frac{1}{59} & 0 & \frac{9}{59} \\ 1 & 0 & -\frac{1}{59} & \frac{6}{59} & 0 & \frac{5}{59} \\ \hline 0 & 0 & \frac{9}{59} & \frac{5}{59} & 1 & \frac{14}{59} \end{array}\right] \end{array}$$

$$v = \frac{1}{M} = \frac{59}{14}$$

a. Carol's optimal strategy is $\begin{bmatrix} \frac{59}{14} \cdot \frac{5}{59} \\ \frac{59}{14} \cdot \frac{9}{59} \end{bmatrix} = \begin{bmatrix} \frac{5}{14} \\ \frac{9}{14} \end{bmatrix}$.

Ruth's optimal strategy is $\begin{bmatrix} \frac{59}{14} \cdot \frac{9}{59} & \frac{59}{14} \cdot \frac{5}{59} \end{bmatrix} = \begin{bmatrix} \frac{9}{14} & \frac{5}{14} \end{bmatrix}$.

b. Since the value is positive, the game favors Ruth (the row player). The value of the game is $\frac{59}{14} - 4 = \frac{3}{14}$.

14. a.

	Strong	Avg.	Weak
A	3000	2000	1000
B	6000	2000	–3000
C	15,000	1000	–10,000

b. Row 1, column 3 is a saddle point. The investors optimal strategy is to buy stock A.

15. For the given conditions, the matrices will not have a saddle point and therefore, will not be strictly determined.

16. The optimal strategy will remain the same as long as the row 1, column 1 value is less than the smallest value in the matrix or $2 + h < 6$ or $h < 4$.

17. a. Since both lines are straight lines, the intersection of the two lines would be the only solution to the system of equations. Moving from that intersection point would increase the value of one line while decreasing the value of the other.

b. You would use the equations $y = a_{11}r + a_{12}(1-r)$ and $y = a_{21}r + a_{22}(1-r)$ and find the point of intersection as in part a.

18.

$$\begin{aligned} a_{11}r + a_{21}(1-r) &= a_{12}r + a_{22}(1-r) \\ a_{11}r + a_{21} - a_{21}r &= a_{12}r + a_{22} - a_{22}r \\ a_{11}r - a_{21}r - a_{12}r + a_{22}r &= a_{22} - a_{21} \\ r(a_{11} - a_{21} - a_{12} + a_{22}) &= a_{22} - a_{21} \\ r &= \frac{a_{22} - a_{21}}{a_{11} - a_{21} - a_{12} + a_{22}} \end{aligned}$$

 Copyright © 2018 Pearson Education, Inc.

Chapter 10

Exercises 10.1

1. $i = \frac{0.03}{12} = 0.0025$

$n = (12)(2) = 24$

3. $i = \frac{0.022}{2} = 0.011$

$n = (2)(20) = 40$

5. $i = \frac{0.045}{12} = 0.00375$

$n = (12)(3.5) = 42$

7. $i = \frac{0.028}{1} = 0.028$

$n = (1)(4) = 4$

$P = \$500$

$F = \$558.40$

9. $i = \frac{0.024}{2} = 0.012$

$n = (2)(9.5) = 19$

$P = \$7174.85$

$F = \$9000$

11. $i = \frac{0.06}{12} = 0.005$

$n = (12)(30) = 360$

$P = \$3000$

$F = \$18,067.73$

13. $\left(1+\frac{0.021}{12}\right)^{12\times 2}(\$1000) = \$1042.86$

15. $F = \left(1+\frac{0.027}{12}\right)^{12\times 3}(\$6000) = \$6505.63$

$int. = F - P = \$6505.63 - \$6000 = \$505.63$

17. $\left(1+\frac{0.04}{4}\right)^{4\times 6}(\$10,000) = \$12,697.35$

19. $\left[\frac{1}{\left(1+\frac{0.024}{12}\right)^{12\times 25}}\right](\$100,000) = \$54,914.06$

21. $\left[\frac{1}{\left(1+\frac{0.034}{4}\right)^{4\times 3}}\right](\$10,000) = \$9034.19$

23. For $P = \$1400$,

$F = \left(1+\frac{0.025}{1}\right)^{9}(\$1400) = \$1748.41.$

\$1400 now is more profitable.

25. $r_{\text{eff}} = \left(1+\frac{0.622}{52}\right)^{52} - 1 \approx 0.8558$

This interest rate is better than 85%.

27. $F = \left(1+\frac{0.016}{4}\right)^{0.25\times 4}(\$1000) = \$1004$

$int. = F - P = \$1004 - \$1000 = \$4.00$

$F = \left(1+\frac{0.016}{4}\right)^{0.25\times 4}(\$1004) = \$1008.02$

$int. = F - P = \$1008.02 - \$1004 = \$4.02$

$F = \left(1+\frac{0.016}{4}\right)^{0.25\times 4}(\$1008.02) = \$1012.05$

$int. = F - P = \$1012.05 - \$1008.02 = \$4.03$

29. $F = \left(1+\frac{0.03}{4}\right)^{2\times 4}(\$2000) = \$2123.20$

$int. = F - P = \$2123.20 - \$2000 = \$123.20$

$F = \left(1+\frac{0.03}{4}\right)^{3\times 4}(\$2000) = \$2187.61$

$int. = F - P = \$2187.61 - \$2123.20 = \$64.41$

Copyright © 2018 Pearson Education, Inc.

31. $B_5 = \left(1+\frac{0.026}{4}\right)^{4\times5}(\$1000) = \$1138.35$

$B_4 = \left(1+\frac{0.026}{4}\right)^{4\times4}(\$1000) = \$1109.23$

$int. = B_5 - B_4 = \$1138.35 - \$1109.23 = \$29.12$

33. $P = \left[\frac{1}{\left(1+\frac{0.04}{4}\right)^{4\times1} - 1}\right](\$406.04) = \$10,000$

35. $1500(1+r)^7 = 2100$

$(1+r)^7 = 1.4$

$(1+r) = \sqrt[7]{1.4}$

$r \approx 4.92\%$

37. a. $\frac{r}{12}(\$10,000.00) = \$20.00$, so r $=0.024=2.4\%$ compounded monthly

b. $(1.002)^3(\$10,000.00) = \$10,060.12$
$\$10,060.12 - \$10,040.04 = \$20.08$

c. $(1.002)^{24}(\$10,000.00) = \$10,491.20$
$[(1.002)^{24} - (1.002)^{23}](\$10,000) = \$20.94$

39. a. $r = 0.015$
$n = \frac{6}{12} = \frac{1}{2}$
$P = \$500$
$F = \$503.75$

b. $r = 0.025$
$n = 2$
$P = \$500$
$F = \$525$

41. $F = (1+3\cdot0.012)(\$1000) = \$1036$

43. $P = \left[\frac{1}{(1+10\cdot0.02)}\right](\$3000) = \$2500$

45. $\left(1+\frac{6}{12}r\right)(\$980) = \$1000$
$r \approx 0.0408 = 4.08\%$

47. $(1+n\cdot0.015)(\$500) = \$800,\ n = 40$ years

49. $(1+n\cdot0.02)P = 2P,\ n = 50$ years

51. $F = (1 + nr)P;\ P = \frac{F}{1+nr}$

53. $\left(1+\frac{0.04}{4}\right)^{4\times1}(\$100) = \$104.06$

$\frac{\$4.06}{\$100} = 0.0406 = 4.06\%$

55. $r_{\text{eff}} = \left(1+\frac{0.04}{2}\right)^2 - 1 = 0.0404$; 4.04%

57. $r_{\text{eff}} = \left(1+\frac{0.044}{12}\right)^{12} - 1 \approx 0.0449$; 4.49%

59. $\frac{r}{4}(\$10,000) = \$100$, so $r = 0.04$ and $i = 0.01$

$\frac{1}{(1.01)^{12}}(\$10,000) = \$8874.49$

61. (a)

63. Since we start with $1000 and this amount doubles every six years, we have $2000 at the end of six years, $4000 at the end of twelve years and $8000 at the end of eighteen years. So, it will take 18 years for the investment to grow to $8000.

65. Assume an initial investment of $100.00. Then over a 10 year period, the amount of growth would be $100(1+0.04)^{10} = 148.02$ about a 48% increase; so answer d) is correct.

Copyright © 2018 Pearson Education, Inc.

67. Assume \$100 invested initially. Then 100 would increase to 102.5 after the first year. The total amount of the investment after three years would then be $100(1.025)(1.03)(1.084) = \$114.44$. If the same amount was invested at an interest rate of r % compounded annually, the amount would be the same , so;

$$100(1+r)^3 = 114.44$$
$$(1+r)^3 = 1.1444$$
$$(1+r) = \sqrt[3]{1.1444}$$
$$r \approx 4.6\%$$

69. $N = 24, I\% = 2.1, PV = -1000, PMT = 0,$ $P/Y = 12,$ find $FV.$
$= FV(0.175\%, 24, 0, -1000)$
compound interest $PV = 1000, i = 0.175\%,$ $n = 24,$ annual

71. $N = 300, I\% = 2.4, PMT = 0, FV = -100000,$ $P/Y = 12,$ find $PV.$
$= PV(0.2\%, 300, 0, -100000)$
compound interest $FV = 100000, i = 0.2\%,$ $n = 300,$ annual

73. $N = 7, PMT = 0, PV = 1500, FV = -2100,$ $P/Y = 1,$ find $I\%.$
$= RATE(7, 0, 1500, -2100)$
compound interest $PV = 1500, FV = 2100,$ $n = 7,$ annual

75. After 25 years: $(1.027)^{25}(\$1000) = \$1946.53$
After 35 years: $(1.027)^{35}(\$1000) = \$2540.77$
25; 35

77. $(1.019)^n(\$500,000)$ passes \$1,000,000 when $n = 37$ years.

79. After year 20, option A would have a value of $F = (1+20(0.04))(\$1000) = \$1800$ and option B would have a value of $F = (1+0.03)^{20}(\$1000) = \$1806.11.$

Exercises 10.2

1. $i = \dfrac{0.021}{12} = 0.00175$

$n = (12)(10) = 120$

$R = \$100$

$F = \$13,340.09$

3. $i = \dfrac{0.034}{4} = 0.0085$

$n = (4)(5) = 20$

$R = \$2000$

$P = \$36,642.08$

5. $\left[\dfrac{\left(1+\frac{0.026}{2}\right)^{2\times5} - 1}{\frac{0.026}{2}}\right](\$1500) = \$15,908.62$

7. $\left[\dfrac{\frac{0.018}{12}}{\left(1+\frac{0.018}{12}\right)^{12\times1} - 1}\right](\$1681.83) = \$139.00$

9. $\left[\dfrac{1-(1.00125)^{-24}}{0.00125}\right](\$3000) = \$70,887.09$

11. $\left[\dfrac{\frac{0.016}{4}}{1-\left(1+\frac{0.016}{4}\right)^{-4\times3}}\right](\$47,336.25) = \$4048.00$

13. a. $\left[\dfrac{\left(1+\frac{0.03}{12}\right)^{12\times4} - 1}{\frac{0.03}{12}}\right](\$500) = \$25,465.60$

Copyright © 2018 Pearson Education, Inc.

b. 48 ($500) = $24,000

c. $25,465.60 – 48($500) = $1465.60

d.

Month	Interest	Balance
1		$500.00
2	$0.0025\times 500 = \$1.25$	500 + 1.25 + 500 = $1001.25
3	$0.0025\times 1001.25 = \$2.50$	1001.25 + 2.50 + 500 = $1503.75

15.

Quarter	Interest	Balance
1		$10000.00
2	$0.0055\times 10000 = \$55.00$	$10000+55-1000=9055$
3	$0.0055\times 9055 = \$49.80$	$9055+49.80-1000=8104.80$
4	$0.0055\times 8104.80 = \$44.58$	

17. $13{,}340.09-(100)(120)=\$1340.09$

19. $(2000)(20)-36{,}642.08=\$3357.92$

21. $\left[\dfrac{\left(1+\frac{0.03}{12}\right)^{12\times 10}-1}{\frac{0.03}{12}}\right](\$1000)=\$139{,}741.42$

Lump sum is better.

23. a. $\left[\dfrac{\left(1+\frac{0.033}{12}\right)^{12\times 1}-1}{\frac{0.033}{12}}\right](\$200)=\$2436.63$

b. The extra $200 a month is better.

25. Jack withdraws for 12(.75) = 9 months.

$\left[\dfrac{1-\left(1+\frac{0.018}{12}\right)^{-12\times 0.75}}{\frac{0.018}{12}}\right](\$200)=\$1786.57$

27. $\left(1+\dfrac{0.08}{4}\right)^{4\times 11}(\$1000)+\left[\dfrac{\left(1+\frac{0.08}{4}\right)^{4\times 11}-1}{\frac{0.08}{4}}\right](\$100)=\$9340.32$

29. $\left[\dfrac{\left(1+\frac{0.021}{12}\right)^{12\times 10}-1}{\frac{0.021}{12}}\right](\$100)+\left(1+\dfrac{0.021}{12}\right)^{12\times 3}(\$1000)=\$14{,}405.06$

 Copyright © 2018 Pearson Education, Inc.

31. The future value of the annuity will be $10R + \$3400$, so:

$$10R+\$3400=\left[\frac{(1+0.02)^{10}-1}{0.02}\right]R$$

$$10R+\$3400=10.949721R$$
$$\$3400=0.949721R$$
$$\$3580=R$$

33. (a)

35. Present value:

$$\left[\frac{1-(1.015)^{-30}}{0.015}\right]\left(\frac{0.04}{2}\right)(\$5000)+\frac{\$5000}{(1.015)^{30}}$$
$$=\$5600.40$$

37. a. $(\$200,000)\left(\frac{0.09}{12}\right)=\$1500$

b. $$\left[\frac{\frac{0.09}{12}}{1-\left(1+\frac{0.09}{12}\right)^{-12\times 5}}\right](\$200,000)$$
$$=\$4151.67$$

39. $$\left[\frac{\frac{0.04}{2}}{\left(1+\frac{0.04}{2}\right)^{2\times 15}-1}\right](\$1,000,000)=\$24,649.92$$

41. $$\left[\frac{\left(1+\frac{0.048}{12}\right)^{12\times 15}-1}{\frac{0.048}{12}}\right](\$30)=\$7886.136075$$

Since the problem was in millions, the face value will be \$7886.136075 million or \$7,886,130,075.

43. $$\left[\frac{\frac{0.026}{4}}{\left(1+\frac{0.026}{4}\right)^{4\times 15}-1}\right](\$5,000,000)=\$68,404.06$$

45. $0.05P = \$12,000$, so $P = \$240,000$.

47. a. $P=(1.03)^7(\$20,000)=\$24,597.48$

b. $R=\left[\frac{0.03}{1-(1.03)^{-4}}\right](\$24,597.48)=\$6617.39$

49. $P=\frac{R}{i}=\frac{\$90,000}{0.03}=\$3,000,000$

$$\frac{\$3,000,000}{(1.03)^9}=\$2,299,250.20$$

51. $N=10$, $I\%=2.6$, $PV=0$, $PMT=-1500$, $P/Y=2$, find FV.

$=FV(1.3\%,10,-1500,0)$

annuity FV, $n=10$, $i=1.3\%$, $pmt=\$1500$

53. $N=12$, $I\%=1.8$, $PV=0$, $FV=-1681.83$, $P/Y=12$, find PMT.

$=PMT(0.15\%,12,0,-1681.83)$

annuity pmt, $n=12$, $i=.15\%$, $FV=\$1681.83$

55. $N=24$, $I\%=1.5$, $PMT=-3000$, $FV=0$, $P/Y=12$, find PV.

$=PV(0.125\%,24,-3000,0)$

annuity PV, $i=.125\%$, $n=24$, $pmt=\$3000$

57. $N=12$, $I\%=1.6$, $PV=-47336.25$, $FV=0$, $P/Y=4$, find PMT.

$=PMT(0.4\%,12,-47336.25,0)$

annuity pmt, $n=12$, $i=.4\%$, $PV=\$47336.25$

59. Set up a table with $Y_1=(1.05\,^\wedge X-1)/.05*1000$. After 2, 3, and 4 years, Y_1 equals \$2050, \$3152.50 and \$4310.13. Y_1 equals \$30,539 after 19 years, and exceeds \$50,000 after 26 years.

61. Use $Y_1=(1.001\,^\wedge X-1)/0.001*15$. $Y_1=503$ after 33 weeks.

Copyright © 2018 Pearson Education, Inc.

63. The original \$5000 will go through 10 years of interest and 10 years of payments. Therefore, the original money will be worth $\$5000(1.06)^{10}(0.997)^{10} = \$8689.21$. The \$5000 deposited in the second year will go through 9 years of interest and 9 years of payment, therefore it will be worth $\$5000(1.06)^{9}(0.997)^{9} = \$8222.03$. Continue this pattern for the 10 years and adding all the values together will give you a balance of \$68,617.21 at the 0.3% payment. Using the same method for the 1.5% payment will yield a balance of \$63,882.62 after the 10 years. The difference would then be \$4734.59.

Exercises 10.3

1. $$\left[\frac{\frac{0.064}{4}}{1-\left(1+\frac{0.064}{4}\right)^{-4\times 5}}\right](\$10,000) = \$588.22$$

3. $$\left[\frac{\frac{0.052}{2}}{1-\left(1+\frac{0.052}{2}\right)^{-2\times 3}}\right](\$4000) = \$728.63$$

5. $$\left[\frac{1-\left(1+\frac{0.078}{52}\right)^{-52\times 2}}{\frac{0.078}{52}}\right](\$23.59) = \$2270.00$$

7. $$\left[\frac{\frac{0.054}{12}}{1-\left(1+\frac{0.054}{12}\right)^{-12\times 25}}\right](\$100,000) = \$608.13$$

9. $$\left[\frac{1-\left(1+\frac{0.045}{12}\right)^{-12\times 30}}{\frac{0.045}{12}}\right](\$724.56) = \$143,000.00$$

11. $$\left(1+\frac{0.042}{12}\right)(\$10,000) - \$1019.35 = \$9015.65$$

13. $$\left(1+\frac{r}{52}\right)(\$2000) - \$100 = \$1903$$

$$1+\frac{r}{52} = 1.0015$$

$$\frac{r}{52} = 0.0015$$

$$r = 0.078 = 7.8\%$$

 Copyright © 2018 Pearson Education, Inc.

15. $\left[\dfrac{\frac{0.06}{12}}{1-\left(1+\frac{0.06}{12}\right)^{-4}}\right]830 = 210.10$

Payment	Amount	Interest	Applied to Principal	Unpaid balance
1	$210.10	$4.15	$205.95	$624.05
2	210.10	3.12	206.98	417.07
3	210.10	2.08	208.02	209.05
4	210.10	1.05	209.05	0.00

17. a. $\dfrac{0.048}{12}(\$204,700) = \$818.80$

b. $\$1073.99 - \$818.80 = \$255.19$

c. $\$204,700 - \$255.19 = \$204,444.81$

19. a. $\left[\dfrac{\frac{0.06}{12}}{1-\left(1+\frac{0.06}{12}\right)^{-12\times 3}}\right](\$9480) = \$288.40$

b. $(\$288.40)(3)(12) = \$10,382.40$

c. $\$10,382.40 - \$9480 = \$902.40$

d. $\left[\dfrac{1-\left(1+\frac{0.06}{12}\right)^{-12(2)}}{\frac{0.06}{12}}\right](\$288.40) = \$6507.13$

e. $\left[\dfrac{1-\left(1+\frac{0.06}{12}\right)^{-12(1)}}{\frac{0.06}{12}}\right](\$288.40) = \$3350.90$

f. $12(\$288.40) - (\$6507.13 - \$3350.90) = \$304.57$

g.

Payment	Amount	Interest	Applied to Principal	Unpaid balance
1	$288.40	$47.40	$241.00	$9239.00
2	288.40	46.20	242.20	8996.80
3	288.40	44.98	243.42	8753.38
4	288.40	43.77	244.63	8508.75

Copyright © 2018 Pearson Education, Inc.

21. **a.** Monthly payment $\left[\dfrac{\frac{0.009}{12}}{1-\left(1+\frac{0.009}{12}\right)^{-60}}\right](\$18,000) = \$306.91$

b. amount of loan = \$18,000 – \$500 = \$17,500

Monthly payment $\left[\dfrac{\frac{0.06}{12}}{1-\left(1+\frac{0.06}{12}\right)^{-12\times 5}}\right](\$17,500) = \$338.32$

c. Option a is more favorable.

23. Option 1: Pay 500 for first five months. Loan amount is 5000

Monthly payment $R = \left[\dfrac{\frac{0.09}{12}}{1-\left(1+\frac{0.09}{12}\right)^{-13}}\right](\$5000) = \$405.11$

Total amount of payments: $(\$405.11)(13) + \$500 = \$5766.43$

Option 2: r = 0.06. n = 18 and P = 5500

Monthly payment $R = \left[\dfrac{\frac{0.06}{12}}{1-\left(1+\frac{0.06}{12}\right)^{-18}}\right](\$5500) = \$320.27$

Total amount of payments: $(\$320.27)(18) = \$5764.86$, so Option 2 costs slightly less.

25. $\left[\dfrac{1-\left(1+\frac{0.064}{2}\right)^{-2\times 8}}{\frac{0.064}{2}}\right](\$905.33) + \left[\dfrac{1}{\left(1+\frac{0.064}{2}\right)^{2\times 8}}\right](\$5,000) = \$14,220.61$

27. $\left[\dfrac{1-\left(1+\frac{0.06}{12}\right)^{-12\times 3}}{\frac{0.06}{12}}\right](\$100) = \$3287.10$

\$6287.10 – (\$2000 + \$3287.10) = \$1000.00

$\left(1+\dfrac{0.06}{12}\right)^{12\times 3}(\$1000) = \$1196.68$

29. Assume you have a \$100,000 mortgage, find your payments, and your balance after 15 years. The payment will be

$\left[\dfrac{\frac{0.068}{12}}{1-\left(1+\frac{0.068}{12}\right)^{-12\times 30}}\right](\$100,000) = \$651.93.$

The balance after 15 years will be $\left[\dfrac{1-\left(1+\frac{0.068}{12}\right)^{-12\times 15}}{\frac{0.068}{12}}\right](\$651.93) = \$73,441.68.$

Therefore, the percent paid is $\dfrac{100,000-73,441.68}{100,000} = 0.2656 = 26.56\%.$

 Copyright © 2018 Pearson Education, Inc.

31. $\left[\dfrac{\frac{0.06}{12}}{1-\left(1+\frac{0.06}{12}\right)^{-12\times 25}}\right](\$50,000)=\$322.15$ per month

$\left[\dfrac{1-\left(1+\frac{0.06}{12}\right)^{-12(25-10)}}{\frac{0.06}{12}}\right](\$322.15)=\$38,175.91$

$\$150,000-\$38,175.91=\$111,824.09$

33. The loan amount (F) is 36 times the payment amount (R) minus the interest. The loan amount is also

$$F=\left[\frac{1-\left(1+\frac{0.06}{12}\right)^{-12\times 3}}{\frac{0.06}{12}}\right]R$$

$F=32.8710R$

Therefore,

$$\begin{aligned}36R-\$1085.16&=32.8710R\\ 3.1290R&=\$1085.16\\ R&=\$346.81\end{aligned}$$

35. (a)

37. **a.**

$$\begin{aligned}I_n+Q_n&=R \text{ and } I_n=iB_{n-1}\\ iB_{n-1}+Q_n&=R\\ B_{n-1}&=\frac{R-Q_n}{i}\end{aligned}$$

b.

$$\begin{aligned}(1+i)\frac{R-Q_n}{i}-R&=\frac{R-Q_{n+1}}{i}\\ (1+i)(R-Q_n)-iR&=R-Q_{n+1}\\ (1+i)R-(1+i)Q_n-iR&=R-Q_{n+1}\\ R+iR-(1+i)Q_n-iR&=R-Q_{n+1}\\ -(1+i)Q_n&=-Q_{n+1}\\ (1+i)Q_n&=Q_{n+1}\end{aligned}$$

c. The amount of the portion applied to the principal in the next month is equal to the amount of the portion applied to the principal in the previous month multiplied by 1 plus the interest rate.

d.

$$\begin{aligned}Q_{11}&=(1+i)Q_{10}\\ &=(1+0.01)(\$100)\\ &=(1.01)(\$100)\\ Q_{11}&=\$101.00\end{aligned}$$

and

$$\begin{aligned}Q_{12}&=(1+i)Q_{11}\\ &=(1+0.01)(\$101)\\ &=(1.01)(\$101)\\ Q_{11}&=\$102.01\end{aligned}$$

39. $N=20, I\%=6.4, PV=-10000,$
$FV=0,\ P/Y=4,$ find $PMT.$
$=PMT(1.6\%,20,-10000,0)$

annuity pmt, $PV=\$10000$, $i=1.6\%$,
$n=20$

Copyright © 2018 Pearson Education, Inc.

41. $N = 104, I\% = 7.8, PMT = -23.59,$
$FV = 0, \ P/Y = 52,$ find $PV.$

$= PV(.15\%, 104, -23.59, 0)$

annuity PV, $pmt = \$23.59, i = 0.15\%,$
$n = 104$

43. $N = 300, I\% = 5.4, PV = -100000,$
$FV = 0, \ P/Y = 12,$ find $PMT.$

$= PMT(.45\%, 300, -100000, 0)$

annuity pmt, $PV = \$100000, i = 0.45\%,$
$n = 300$

45. $N = 360, I\% = 4.5, PMT = -724.56,$
$FV = 0, \ P/Y = 12,$ find $PV.$

$= PV(0.375\%, 360, -724.56, 0)$

annuity PV, $pmt = \$724.56, i = 0.375\%,$
$n = 360$

47. After 1 month:
$$\left(1 + \frac{0.048}{12}\right)(\$4193.97) - \$100 = \$4110.75$$

After 2 months:
$$\left(1 + \frac{0.048}{12}\right)(\$4110.75) - \$100 = \$4027.19$$

After 3 months:
$$\left(1 + \frac{0.048}{12}\right)(\$4027.19) - \$100 = \$3943.30$$

The loan will be paid off after 46 months.

49. Enter 10000, then run
1.0075 * Ans – 166.68 repeatedly. The balance drops below \$7500 after 25 months, below \$5000 after 46 months, and below \$2500 after 65 months.

51. The balance B must drop to where $\left(\frac{0.085}{12}\right)B < \$250,$ or $B < \$35{,}294.12.$ Let
$Y_1 = Y_6(300 - X) * 1000$ where
$Y_6 = ((1 + I)^\wedge X - 1)/(I(1 + I)^\wedge X).$ Make a table. Then $B \le \$35{,}294.12$ after 260 payments which means that the next payment, after 261 months, is the first one where at least 75% goes toward debt reduction.

Exercises 10.4

1. deferred

3. $[\text{amount after taxes}] = (1 - 0.45)(300{,}000)$
$= \$165{,}000$

5. $$\left[\frac{\left(1 + \frac{0.06}{1}\right)^{1\times 52} - 1}{\frac{0.06}{1}}\right](\$5000) = \$1,641,407.11$$

7. If we assume a marginal tax bracket of 20%,
$$\left[\frac{\left(1 + \frac{.06}{1}\right)^{1\times 52} - 1}{\frac{0.06}{1}}\right](.8)(\$5000) = \$1,313,125.69$$

 Copyright © 2018 Pearson Education, Inc.

9. a. For Earl:

[earnings after income tax]
$= [1 - \text{tax bracket}] \cdot [\text{amount}]$
$= [.60] \cdot [5000]$
$= 3000$

$$\left[\frac{\left(1+\frac{0.06}{1}\right)^{1\times 12} - 1}{\frac{0.06}{1}}\right](\$3000) = \$50,609.82$$

This money then earns interest compounded annually for 36 years and grows to

$$\$50,609.82 \cdot (1.06)^{36} = \$50,609.82(8.147252) = \$412,330.96$$

b. For Larry:

$$\left[\frac{\left(1+\frac{0.06}{1}\right)^{1\times 36} - 1}{\frac{0.06}{1}}\right](\$3000) = \$357,362.60$$

c. Earl paid in $12 \times \$3000 = \$36,000$ while Larry paid in $36 \times \$3000 = \$108,000.$ Larry paid in more.

d. Earl has $54,968.36 more than Larry.

11. $R = \dfrac{P(1+rt)}{12t} = \dfrac{4000(1+0.10\cdot 1)}{12(1)}$
$= \$366.67$

13. $R = \dfrac{P(1+rt)}{12t} = \dfrac{3000(1+0.09\cdot 3)}{12(3)}$
$= \$105.83$

15. $r = \dfrac{12Rt - P}{Pt} = \dfrac{12(171.21)(1) - 2000}{2000(1)}$
≈ 0.0273 or about 2.73%

17. $r = \dfrac{12Rt - P}{Pt} = \dfrac{12(608.44)(3) - 20,000}{20,000(3)}$
≈ 0.0317 or about 3.17%

19. a. $[\text{total repayment}] = \dfrac{[\text{loan amount}]}{1 - rt} = \dfrac{880}{1 - 0.06(2)}$
$= 1000$

$[\text{monthly payment}] = \dfrac{[\text{total payment}]}{12t} = \dfrac{1000}{12(2)}$
$= \$41.67$

b. $R = \dfrac{P(1+rt)}{12t} = \dfrac{880(1+0.06\cdot 2)}{12(2)}$
$= \$41.07$

The monthly payment is less.

21. False. The effective rate differs from the APR only when discount points are involved.

23. False. The longer the mortgage will be held, the lower the effective rate.

Copyright © 2018 Pearson Education, Inc.

25. False. The up-front fees must change proportionally for there to be no effect on the APR.

27. $$\textit{Payment} = \left[\frac{\frac{0.09}{12}}{1-\left(1+\frac{0.09}{12}\right)^{-12\times 25}}\right](\$250,000) = \$2097.99$$

New P = 250,000 – 5000 = 245,000
Using the Excel function 12*RATE(300, –2097.99, 245000, 0) gives 0.09249 or 9.25%; (d).

29. $$\textit{Payment} = \left[\frac{\frac{0.055}{12}}{1-\left(1+\frac{0.055}{12}\right)^{-12\times 20}}\right](\$250,000) = \$1719.72$$

New P = 250,000 – 10,000 = 240,000
Using the Excel function 12*RATE(240, –1719.72, 240000, 0) gives 0.06002 or about 6%; (a).

31. [monthly payment] = $581.03
Using the Excel function 12*RATE(48, –581.03, 99000, –94341.50) gives 0.05999 or about 6%; (a).

33. $$\textit{Payment} = \left[\frac{\frac{0.06}{12}}{1-\left(1+\frac{0.06}{12}\right)^{-12\times 30}}\right](\$100,000) = \$599.55$$

New P = 100,000 – 3000 = 97,000
The mortgage has 360 – 84 = 276 months to go. Therefore,

$$\textit{balance} = \left[\frac{1-\left(1+\frac{0.06}{12}\right)^{-276}}{\frac{0.06}{12}}\right](\$599.55) = \$89,639.31$$

Using the Excel function 12*RATE(84, –599.55, 97000, –89639.31) gives 0.06560 or about 6.56%; (c).

35. APR: $n = 20 \cdot 12 = 240$

$$R = \frac{0.005(1.005)^{240}}{1.005^{240}-1} \cdot 80,000 = 573.14$$

P = 80,000 – .03(80,000) = 77,600
12*RATE(n, R, P, 0) = 12*RATE(240, –573.14, 77600, 0) = 6.38%
effective rate: $m = 10 \cdot 12 = 120$
R = 573.14, P = 77,600

$$B = \frac{1.005^{120}-1}{0.005(1.005)^{120}} \cdot 573.14 = 51,624.70$$

12*RATE(m, R, P, B) = 12*RATE(120, –573.14, 77600, –51624.70) = 6.47%

37. APR: $n = 15 \cdot 12 = 180$

$$R = \frac{0.0075(1.0075)^{180}}{1.0075^{180}-1} \cdot 120,000 = 1217.12$$

P = 120,000 – 0.01(120,000) = 118,800

$$\frac{(1+i)^{180}-1}{i(1+i)^{180}} \cdot 1217.12 = 118,800 \text{ yields}$$

9.17%.
effective rate: $m = 5 \cdot 12 = 60$
R = 1217.12, P = 118,800

$$B = \frac{1.0075^{120}-1}{0.0075(1.0075)^{120}} \cdot 1217.12 = 96,081.51$$

$$1217.12 - 96,081.51i = (1217.12 - 118,8000i) \cdot (1+i)^{60} \text{ yields}$$

9.27%.

 Copyright © 2018 Pearson Education, Inc.

39. The salesman is comparing the future value of the savings account to the sum of the present values of the loan payments at time of payment. A proper comparison would be the future value of the savings account, \$1083.14, to the future value of the series of payments (assuming 4% interest)

$$\left[\frac{\left(1+\frac{0.04}{12}\right)^{24}-1}{\frac{0.04}{12}}\right](\$43.87)=\$1094.24$$

41. a. $P=\$200{,}000$ and $i=\dfrac{0.069}{12}=0.00575$

$$Payment = iP = (.00575)(\$200,000) = \$1150$$

b. $$Payment=\left[\frac{\frac{0.069}{12}}{1-\left(1+\frac{0.069}{12}\right)^{-12\times 10}}\right](\$200,000)=\$2311.87$$

43. a. For the first 5 years; $Payment=\left[\dfrac{\frac{0.06}{12}}{1-\left(1+\frac{0.06}{12}\right)^{-12\times 25}}\right](\$250,000)=\$1610.75$

b. The balance after 5 years; $balance=\left[\dfrac{1-\left(1+\frac{0.06}{12}\right)^{-12\times 20}}{\frac{0.06}{12}}\right](\$1610.75)=\$224,829.73$

c. For the sixth year, $P=\$224{,}829.73$, $n=240$, and $i=0.044+0.025=0.069$.

$$Payment=\left[\frac{\frac{0.069}{12}}{1-\left(1+\frac{0.069}{12}\right)^{-240}}\right](\$224,829.73)=\$1729.63$$

45. a. The balance at the beginning of the 7th year;

$$balance=\left[\frac{1-\left(1+\frac{0.069}{12}\right)^{-228}}{\frac{0.069}{12}}\right](\$1729.63)=\$219,418.04$$

b. Without the cap, $P=\$219{,}418.04$, $n=228$, and $i=0.077+0.025=0.102$.

$$Payment=\left[\frac{\frac{0.102}{12}}{1-\left(1+\frac{0.102}{12}\right)^{-228}}\right](\$219,418.04)=\$2181.80$$

Copyright © 2018 Pearson Education, Inc.

c. Without the cap, the percentage increase from the sixth year to the seventh year would be $\frac{2181.80-1729.63}{1729.63}=0.2614=26.14\%$

Since this percentage is greater than the 7 % cap, the monthly payment will be $(1.07)(\$1729.63)=\$1850.70$.

d. The interest due in the 73rd month would be $(0.0085)(\$219,418.04)=\$1865.05$.

e. Since the interest owed is more than the payment made, the balance will increase by $\$1865.05-\$1850.70=\$14.35$. Therefore the new balance will be $\$219,418.04+\$14.35=\$219,432.39$.

47.

```
N=36
I%=11.08292218
PV=10000
PMT=-327.78
FV=0
P/Y=12
C/Y=12
PMT:END BEGIN
```

The APR is about 11.08%.

49.

	A	B	C	D
1	n =	300		
2	i =	0.00542		
3	loan amount =	300,000.00		
4	points:	3.00		
5				
6	1/a$_n$ =	0.0067520716		
7	monthly payment =	2,025.62		
8	APR =	0.06831867		
9				
10	formula in B6:	=(B2*(1+B2)^B1)/(-1+(1+B2)^B1)		
11	formula in B7:	=B6*B3		
12	formula in B8:	=12*RATE(B1,-B7,B3-0.01*B4*B3,0)		

The APR is about 6.83.%.

Copyright © 2018 Pearson Education, Inc.

51

	A	B	C	D
1	n =	300		
2	i =	0.00542		
3	loan amount =	140,000.00		
4	points:	3		
5				
6	$1/a_n$ =	0.0067520716		
7	monthly payment =	945.29		
8	m =	84		
9	a_m=	127.1356748173		
10	unpaid balance =	120180.0853		
11	APR =	0.070787924		
12				
13				
14	formula in B6:	=(B2*(1+B2)^B1)/(-1+(1+B2)^B1)		
15	formula in B7:	=B6*B3		
16	formula in B9:	=(-1+(1+B2)^(B1-B8))/(B2*(1+B2)^(B1-B8))		
17	formula in B10:	=B9*B7		
18	formula in B11:	=12*RATE(B8,-B7,B3-0.01*B4*B3,-B10)		

The APR is about 7.08%.

53. Mortgage *A* costs an extra $1750 up front.
[difference in monthly payments] = $1181.61 – $1159.84 = $21.77
Using the Excel function NPER(.002, 21.77, –1750) gives 87.72 months.

55. Mortgage *A* costs an extra $2000 up front.
[difference in monthly payments] = $1303.85 – $1283.93 = $19.92
Using the Excel function NPER(.004, 19.92, –2000) gives 128.63 months.

Exercises 10.5

1. $y_n = \left(1+\frac{0.04}{4}\right)y_{n-1}$

$y_n = 1.01y_{n-1}$

$y_0 = 1000$

3. $y_n = y_{n-1} + 0.025(2500)$

$y_n = y_{n-1} + 62.5n$

$y_0 = 2500$

5. $y_n = \left(1+\frac{0.48}{12}\right)y_{n-1} - 1073.99$

$y_n = 1.004y_{n-1} - 1073.99$

$y_0 = 204,700$

7. $y_n = \left(1+\frac{0.039}{52}\right)y_{n-1} - 100$

$y_n = 1.00075y_{n-1} - 100$

$y_0 = 25,000$

9. $y_n = \left(1+\frac{0.028}{2}\right)y_{n-1} + 1000$

$y_n = 1.014y_{n-1} + 1000$

$y_0 = 2500$

11. a. Based on the initial difference equation:
$y_0 = \$37,780$
$r = (0.008)(4) = 0.032 = 3.2\%$
deposit $= \$1196$

Copyright © 2018 Pearson Education, Inc.

b. $y_7 = -149000 + 187280(1.008)^7$
$y_7 = 84,591.85$

$y_8 = -149000 + 187280(1.008)^8$
$y_8 = 92,173.16$

$y_9 = -149000 + 187280(1.008)^9$
$y_9 = 100,000$

Yes, after 9 years.

13. a. $y_0 = 1$

$y_1 = y_0 + 5$; $y_2 = y_1 + 5$
$y_1 = 1 + 5$; $y_2 = 6 + 5$
$y_1 = 6$; $y_2 = 11$

$y_3 = y_2 + 5$; $y_4 = y_3 + 5$
$y_3 = 11 + 5$; $y_4 = 16 + 5$
$y_3 = 16$; $y_4 = 21$

b. $y_n = 1 + 5n$

15. a. $y_0 = 7$

$y_1 = 0.4y_0 + 3$
$y_1 = 0.4(7) + 3$
$y_1 = 5.8$

$y_2 = 0.4y_1 + 3$
$y_2 = 0.4(5.8) + 3$
$y_2 = 5.32$

$y_3 = 0.4y_2 + 3$
$y_3 = 0.4(5.32) + 3$
$y_3 = 5.128$

$y_4 = 0.4y_3 + 3$
$y_4 = 0.4(5.128) + 3$
$y_4 = 5.0512$

b. $y_n = 5 + 2(0.4)^n$

17. a. $y_0 = 2$

$y_1 = -5y_0$
$y_1 = -5(2)$
$y_1 = -10$

$y_2 = -5y_1$
$y_2 = -5(-10)$
$y_2 = 50$

$y_3 = -5y_2$
$y_3 = -5(50)$
$y_3 = -250$

$y_4 = -5y_3$
$y_4 = -5(-250)$
$y_4 = 1250$

b. $y_n = 2(-5)^n$

19. Based on the information given, $y_n = 1.00175y_{n-1}$ when $y_0 = 1000$. Therefore, the difference equation is $y_n = y_0(1.00175)^n$. Now substitute the known values and solve for the unknown:

$y_{24} = 1000(1.00175)^{24}$
$y_{24} = \$1042.86$

21. Based on the information given, $y_n = 1.011y_{n-1}$ when $y_{12} = 40,100$. Therefore, the difference equation is $y_n = y_0(1.011)^n$. Now substitute the known values and solve for the unknown:

$y_{12} = y_0(1.011)^{12}$
$40,100 = y_0(1.140286)$
$y_0 = \$35,166.61$

Copyright © 2018 Pearson Education, Inc.

23. Based on the information given, $y_n = y_{n-1} + 0.045(1000)$ when $y_0 = 1000$. Therefore, the difference equation is $y_n = y_0 + 45n$. Now substitute the known values and solve for the unknown:

$$y_3 = y_0 + 45n$$
$$y_3 = 1000 + 45(3)$$
$$y_3 = \$1135$$

25. Based on the information given, $y_n = y_{n-1} + 0.025y_0$ when $y_{10} = 2000$. Therefore, the difference equation is $y_n = y_0 + (0.025n)y_0$. Now substitute the known values and solve for the unknown:

$$y_{10} = y_0 + (0.025)(10)y_0$$
$$2000 = y_0 + 0.25y_0$$
$$2000 = 1.25y_0$$
$$y_0 = \$1600$$

27. Based on the information given, $y_n = 1.025y_{n-1} + 1500$ when $y_0 = 17000$. Therefore, the difference equation is

$y_n = -60,000 + 77,000(1.025)^n$. Now substitute the known values and solve for the unknown:

$$y_{10} = -60000 + 77000(1.025)^{10}$$
$$y_{10} = -60000 + 98566.51$$
$$y_{10} = \$38,566.51$$

29. Based on the information given, $y_n = 1.01125y_{n-1} + 1500$ when $y_{16} = 73000$. Therefore, the difference equation is $y_n = -133333.33 + (y_0 + 133333.33)(1.01125)^n$. Now substitute the known values and solve for the unknown:

$$y_{16} = -133333.33 + (P + 133333.33)(1.01125)^{16}$$
$$73000 = -133333.33 + (P + 133333.33)(1.01125)^{16}$$
$$206,333.33 = (P + 133333.33)(1.1960148)$$
$$172,517.37 = P + 133333.33$$
$$P = \$39,184.04$$

Copyright © 2018 Pearson Education, Inc.

31. Based on the information given, $y_n = 1.0375y_{n-1} + R$ when $y_0 = 10500$. Therefore, the difference equation is $y_n = -\frac{R}{0.0375} + \left(10500 + \frac{R}{0.0375}\right)(1.0375)^n$. Now substitute the known values and solve for the unknown:

$$\begin{aligned} y_7 &= -\frac{R}{0.0375} + \left(10500 + \frac{R}{0.0375}\right)(1.0375)^7 \\ 20000 &= -\frac{R}{0.0375} + \left(10500 + \frac{R}{0.0375}\right)(1.0375)^7 \\ 20000 &= -\frac{R}{0.0375} + 13586.45131 + \frac{1.2939R}{0.0375} \\ 6413.548689 &= \frac{0.2939R}{0.0375} \\ R &= \$818.33 \end{aligned}$$

33. Based on the information given, $y_n = 1.008y_{n-1} - 700$ when $y_0 = 43000$. Therefore, the difference equation is $y_n = 87{,}500 - 44{,}500(1.008)^n$. Now substitute the known values and solve for the unknown:

$$\begin{aligned} y_{20} &= 87{,}500 - 44{,}500(1.008)^{20} \\ y_{20} &= 87{,}500 - 52{,}188 \\ y_{20} &= \$35{,}312 \end{aligned}$$

35. Based on the information given, $y_n = 1.00425y_{n-1} - R$ when $y_0 = 300{,}080$. Therefore, the difference equation is $y_n = \frac{R}{0.00425} + \left(300{,}080 - \frac{R}{0.00425}\right)(1.00425)^n$. Now substitute the known values and solve for the unknown:

$$\begin{aligned} y_{300} &= \frac{R}{0.00425} + \left(300{,}080 - \frac{R}{0.00425}\right)(1.00425)^{300} \\ 0 &= \frac{R}{0.00425} + \left(300{,}080 - \frac{R}{0.00425}\right)(1.00425)^{300} \\ 0 &= \frac{R}{0.00425} + 1{,}070{,}999.26 - \frac{3.5690R}{0.00425} \\ -1{,}070{,}999.26 &= -\frac{2.5690R}{0.00425} \\ R &= \$1771.80 \end{aligned}$$

 Copyright © 2018 Pearson Education, Inc.

37. Based on the information given, $y_n = 1.0125y_{n-1} - R$ when $y_0 = 2710$. Therefore, the difference equation is $y_n = \frac{R}{0.0125} + \left(2710 - \frac{R}{0.0125}\right)(1.0125)^n$. Now substitute the known values and solve for the unknown:

$$y_{12} = \frac{R}{0.0125} + \left(2710 - \frac{R}{0.0125}\right)(1.0125)^{12}$$
$$0 = \frac{R}{0.0125} + \left(2710 - \frac{R}{0.0125}\right)(1.0125)^{12}$$
$$0 = \frac{R}{0.0125} + 3145.64 - \frac{1.1608R}{0.0125}$$
$$-3145.64 = -\frac{0.1608R}{0.0125}$$
$$R = \$244.60$$

39. Based on the information given, $y_n = (1+i)y_{n-1} - R$ when $y_0 = P$. Therefore, the difference equation is $y_n = \frac{R}{i} + \left(P - \frac{R}{i}\right)(1+i)^n$.

41. a. Based on the information given, the difference equation is $y_n = 1.01y_{n-1} - 600$ when $y_0 = 50,000$.

b. $y_n = \frac{600}{0.01} + \left(50,000 - \frac{600}{0.01}\right)(1.01)^n$

$y_n = 60,000 - 10,000(1.01)^n$

c. $y_{10} = 60,000 - 10,000(1.01)^{10}$

$y_{10} = 48,954$ people

43. a. Based on the information given, the difference equation is $y_n = 0.92y_{n-1} + 8$ when $y_0 = 0$.

b. $y_n = \frac{8}{0.08} + \left(0 - \frac{8}{0.08}\right)(0.92)^n$

$y_n = 100 - 100(0.92)^n$

$y_n = 100(1 - (0.92)^n)$

c. $y_{11} = 100(1 - (0.92)^{11})$

$y_{11} = 60$ doctors

45. a. Based on the information given, the difference equation is $y_n = 0.7y_{n-1} + 3.6$ when $y_0 = 0$.

b. $y_n = \frac{3.6}{0.3} + \left(0 - \frac{3.6}{0.3}\right)(0.7)^n$

$y_n = 12 - 12(0.7)^n$

$y_n = 12(1 - (0.7)^n)$

c. $y_6 = 12(1 - (0.7)^6)$

$y_6 = 10.6$ units

47. a. Based on the information given, the difference equation is $y_n = 0.8y_{n-1} + 14$ when $y_0 = 40$.

b. $y_n = \frac{14}{0.2} + \left(40 - \frac{14}{0.2}\right)(0.8)^n$

$y_n = 70 - 30(0.8)^n$

c. $y_5 = 70 - 30(0.8)^5$

$y_5 = 60.2^\circ F$

49. a. Based on the information given, the difference equation is $y_n = (1-2p)y_{n-1} + p$ when $y_0 = 1$.

Copyright © 2018 Pearson Education, Inc.

b. $y_n = \frac{p}{2p} + \left(1 - \frac{p}{2p}\right)(1-2p)^n$

$y_n = \frac{1}{2} + \frac{1}{2}(1-2p)^n$

51. Based on the information given, the difference equation is $y_n = 1.18y_{n-1}$ when $y_0 = 1000$. Therefore the difference equation will be:

$$y_n = \frac{0}{-0.18} + \left(1000 - \frac{0}{-0.18}\right)(1.18)^n$$
$$y_n = 1000(1.18)^n$$

Now solve the equation:

$$3000 = 1000(1.18)^n$$
$$3 = 1.18^n$$
$$n = \frac{\ln 3}{\ln 1.18}$$
$$n = 6.64$$

Therefore it will take approximately 7 years for the investment to triple.

53. a. Based on the information given, the difference equation is $y_n = 0.7y_{n-1} + 22.5$ when $y_0 = 600$ and the difference equation will be $y_n = 75 + 525(0.7)^n$.

b. $y_2 = 75 + 525(0.7)^2$

$y_2 = 332.25°$

c.
$$80 = 75 + 525(0.7)^n$$
$$5 = 525(0.7)^n$$
$$0.00952 = 0.7^n$$
$$n = \frac{\ln 0.00952}{\ln 0.7}$$
$$n \approx 13.05$$

Therefore it will take approximately 14 minutes for the temperature to drop below 80 degrees.

55. a. 32.65% will be using the drug after 1 year.

b. More than half the doctors will be using the drug after 15 months.

c. More than 99% of the doctors will be using the drug after 32 months.

57. a. Based on the information given, the difference equation is $y_n = 0.87y_{n-1} + 130$ when $y_0 = 130$ and the difference equation will be $y_n = 1000 - 870(0.87)^n$.

b. $y_{24} = 1000 - 870(0.87)^{24}$

$y_{24} = 969.24$ mg

Therefore, after 24 hours, there will be approximately 969.24 mg of caffeine remaining in the body

Copyright © 2018 Pearson Education, Inc.

Chapter 10 Review Exercises

1. $100(1.03)^{10} \approx \$134.39$

2. $\left[\dfrac{\frac{0.027}{12}}{\left(1+\frac{0.027}{12}\right)^{12\times10}-1}\right](\$240{,}000)=\$1744.37$

3. The monthly mortgage payment should not exceed $\left(\dfrac{39{,}216}{12}\right)(0.25)=\$817.00.$

$$\left[\dfrac{1-\left(1+\frac{0.042}{12}\right)^{-12\times30}}{\frac{0.042}{12}}\right](\$817)=\$167{,}069.80$$

4. $50\left(1+\dfrac{0.0219}{365}\right)^{365}=\$51.11$

5. 2.92% compounded daily yields $\left(1+\dfrac{0.0292}{365}\right)^{365}-1=0.0296=2.96\%$ annually. 3% compounded annually is better.

6. $\left[\dfrac{\left(1+\frac{0.03}{12}\right)^{12\times5}-1}{\frac{0.03}{12}}\right](\$200)=\$12{,}929.34$

7. a. $\left[\dfrac{\frac{0.045}{12}}{1-\left(1+\frac{0.045}{12}\right)^{-12\times15}}\right](\$200{,}000)=\$1529.99$

b. $\left[\dfrac{1-\left(1+\frac{0.045}{12}\right)^{-12(15-5)}}{\frac{0.045}{12}}\right](\$1529.99)=\$147{,}627.70$

8. $(\$35{,}000)(1.005)^{120}=\$63{,}678.89$

9. $\dfrac{\$50{,}000}{\left(1+\frac{0.03}{12}\right)^{12\times10}}=\$37{,}054.78$

10. $\dfrac{\$10{,}000}{\left(1+\frac{0.027}{12}\right)^{12\times2}}+\dfrac{\$5000}{\left(1+\frac{0.027}{12}\right)^{12\times3}}=\$14{,}086.28$

11. $\left[\dfrac{\frac{0.06}{12}}{1-\left(1+\frac{0.06}{12}\right)^{-12\times4}}\right](\$12{,}000-\$3{,}000)=\$211.37$

Copyright © 2018 Pearson Education, Inc.

12. $\left[\dfrac{\frac{0.04}{2}}{1-\left(1+\frac{0.04}{2}\right)^{-2\times 5}}\right]\left(1+\dfrac{0.04}{2}\right)^{2\times 2}(\$100,000)=\$12,050.34$

13. $\dfrac{\$30,000}{\left(1+\frac{0.024}{12}\right)^{12\times 15}}=\$20,937.82$

$\$105,003.50-\$20,937.82=\$84,065.68$

$\left[\dfrac{\frac{0.024}{12}}{1-\left(1+\frac{0.024}{12}\right)^{-12\times 15}}\right](\$84,065.68)=\$556.59$

14. $\dfrac{\$100,000}{\left(1+\frac{0.06}{12}\right)^{12\times 10}}=\$54,963.27$

$\left[\dfrac{\frac{0.06}{12}}{1-\left(1+\frac{0.06}{12}\right)^{-12\times 10}}\right](\$500,000-\$54,963.27)=\$4940.82$

15. $\left[\dfrac{\left(1+\frac{0.06}{12}\right)^{12\times 30}-1}{\frac{0.06}{12}}\right](\$100)=\$100,451.50$

16. $\left[\dfrac{1-\left(1+\frac{0.045}{12}\right)^{-12\times 10}}{\frac{0.045}{12}}\right](\$2000)=\$192,978.65$

17. Investment A: $\left[\dfrac{(1+0.025)^{10}-1}{0.025}\right]1000=\$11,203.38$

Investment B: $5000(1+0.025)^5+5000=\$10,657.04$
Thus Investment A is the better investment.

18. Present value of annuity is $\left[\dfrac{1-\left(1+\frac{0.048}{12}\right)^{-12\times 5}}{\frac{0.048}{12}}\right](\$5)=\$266.24$

The present value of \$1000 is $\dfrac{\$1000}{\left(1+\frac{0.048}{12}\right)^{12\times 5}}=\$787.00$

Yes, it is a bargain, since the present value is 266.24 + 787.00 = \$1053.24.

19. $\left(1+\dfrac{0.022}{2}\right)^2-1=0.022121=2.2121\%$

20. $\left(1+\dfrac{0.027}{12}\right)^{12}-1=0.027337=2.7337\%$

Copyright © 2018 Pearson Education, Inc.

21. $\left(1+\frac{0.022}{4}\right)^{4\times15}(\$10,000)+\left[\frac{\left(1+\frac{0.022}{4}\right)^{4\times15}-1}{\frac{0.022}{4}}\right](\$1000)=\$84,753.66$

22. $R=\left[\frac{\frac{0.06}{12}}{1-\left(1+\frac{0.06}{12}\right)^{-36}}\right](\$10,000)=\$304.22$

Paym	Amount	Interest	Applied to Principal	Unpaid balance
1	\$304.22	\$50.00	\$254.22	\$9745.78
2	304.22	48.73	255.49	9490.29
3	304.22	47.45	256.77	9233.52
4	304.22	46.17	258.05	8975.47
5	304.22	44.88	259.34	8716.13
6	304.22	43.58	260.64	8455.49

23. $\left[\frac{(1.0015)^{120}-1}{0.0015}\right](\$200)(1.0015)^{120}=\$31,451.59$

24. $\left[\frac{\frac{0.018}{12}}{1-\left(1+\frac{0.018}{12}\right)^{-12\times5}}\right](\$300,000)=\$5232.12$

25. $\left[\frac{\frac{0.048}{12}}{1-\left(1+\frac{0.048}{12}\right)^{-12\times30}}\right](\$150,000)=\$787.00$

26. a. $[\text{amount after taxes}]=(1-0.30)(30{,}000)$
$=\$21{,}000$

b. $30,000\cdot(1.06)^5=40,146.77$
$[\text{amount after taxes}]=(1-0.35)(40,146.77)=\$26,095.40$

27. a. $[\text{amount after taxes}]=(1-0)(30,000)=\$30,000$

b. $30{,}000\cdot(1.06)^5=\$40,146.77$

28. $r=\frac{12Rt-P}{Pt}=\frac{12(228.42)(2)-5000}{5000(2)}=0.048208=4.82\%$

29. Loan A is better, because the monthly payments for Loan B will be $\frac{3000(1.06)}{12}=\$265.$

30. $P = \$90{,}000 - 0.02(\$90{,}000) = \$88{,}200$
$n = 15 \cdot 12 = 180$
$$R = \frac{0.005}{1-(1.005)^{-80}} \cdot \$90,000 = \$759.47$$

31. $m = 6 \cdot 12 = 72$
$R = \$716.43$
$P = \$100{,}000 - 0.3(\$100{,}000) = \$97{,}000$
$B = \$81{,}298.32$

32. Mortgage A costs an extra \$2000 up front.
[difference in monthly payments] = \$1413.56 – \$1350.41 = \$63.15
$i = 0.035/12 = 0.00291667$, $R = \$63.15$, $P = \$2000$
Using the Excel function NPER(.00291667, 63.15, –2000) gives 33.3 months.

33. Through technology, you can find $x = 0.06$ on a graphing calculator by finding the intersection of $Y_1 = 245000$ and $Y_2 = ((1+X/12)\hat{}\ 240-1)/(X/12(1+X/12)\hat{}\ 240)*1755.21$.

34. Solve $567.79 - i(86{,}837.98) = [567.79 - i(97{,}000)](1+i)^{96}$, using technology, to find $i = 0.005$; the interest rate is $12i = .06$. On a graphing calculator, find the intersection of $Y_1 = 567.79 - X(86837.98)$ and $Y_2 = (567.79 - X*97000)(1+X)\hat{}\,96$.

35. **a.** $P = \$380{,}000$ and $i = \dfrac{0.069}{12} = 0.00575$
$Payment = iP = (0.00575)(\$380,000) = \$2185$

b. $$Payment = \left[\frac{\frac{0.069}{12}}{1-\left(1+\frac{0.069}{12}\right)^{-12\times 15}}\right](\$380,000) = \$3394.34$$

36. **a.** For the first 5 years; $$Payment = \left[\frac{\frac{0.063}{12}}{1-\left(1+\frac{0.063}{12}\right)^{-12\times 25}}\right](\$220,000) = \$1458.08$$

b. The balance after 5 years; $$balance = \left[\frac{1-\left(1+\frac{0.063}{12}\right)^{-12\times 20}}{\frac{0.063}{12}}\right](\$1458.08) = \$198,690.34$$

c. For the sixth year, $P = \$198{,}690.34$, $n = 240$, and $i = 0.0455 + 0.028 = 0.0735$.
$$Payment = \left[\frac{\frac{0.0735}{12}}{1-\left(1+\frac{0.0735}{12}\right)^{-240}}\right](\$198,690.34) = \$1582.46$$

37. **a.**

$y_1 = -3y_0 + 8 \quad y_2 = -3y_1 + 8 \quad y_3 = -3y_2 + 8$

$y_0 = 3;\ y_1 = -3(1) + 8;\ y_2 = -3(5) + 8;\ y_3 = -3(-7) + 8$

$y_1 = 5 \quad y_2 = -7 \quad y_3 = 29$

b. $y_n = 2 - (-3)^n$

 Copyright © 2018 Pearson Education, Inc.

c. $y_4 = 2-(-3)^4 = 2-81 = -79$

38. a.

$$\begin{array}{llll} & y_1 = y_0 - \frac{3}{2} & y_2 = y_1 - \frac{3}{2} & y_3 = y_2 - \frac{3}{2} \\ y_0 = 10; & y_1 = 10 - \frac{3}{2}; & y_2 = \frac{17}{2} - \frac{3}{2}; & y_3 = 7 - \frac{3}{2} \\ & y_1 = \frac{17}{2} & y_2 = 7 & y_3 = \frac{11}{2} \end{array}$$

b. $y_n = 10 - \frac{3}{2}n$

c. $y_6 = 10 - \frac{3}{2}(6) = 10 - 9 = 1$

39. Based on the information given, the difference equation is $y_n = 1.007625y_{n-1}$ when $y_0 = P$ and the difference equation will be $y_n = P(1.007625)^n$.

$$2474 = P(1.007625)^{28}$$
$$P = \frac{2474}{(1.007625)^{28}}$$
$$P = 2000$$

Therefore you would need to deposit \$2000.

40. Based on the information given, the difference equation is $y_n = 1.001y_{n-1}$ when $y_0 = 1000$ and the difference equation will be $y_n = 1000(1.001)^n$.

$$y_{104} = 1000(1.001)^{104}$$
$$y_{104} = 1109.54$$

Therefore your balance after 2 years will be \$1109.54.

41. Based on the information given, $y_n = 1.0005y_{n-1} + R$ when $y_0 = 0$. Therefore, the difference equation is $y_n = -\frac{R}{0.0005} + \left(0 + \frac{R}{0.0005}\right)(1.0005)^n$. Now substitute the known values and solve for the unknown:

$$y_{1092} = -\frac{R}{0.0005} + \left(0 + \frac{R}{0.0005}\right)(1.0005)^{1092}$$
$$36,000 = -\frac{R}{0.0005} + \frac{R}{0.0005}(1.0005)^{1092}$$
$$36,000 = -\frac{R}{0.0005} + \frac{1.7261R}{0.0005}$$
$$36,000 = \frac{0.7261R}{0.0005}$$
$$R = \$24.79$$

Copyright © 2018 Pearson Education, Inc.

42. Based on the information given, $y_n = 1.005y_{n-1} - R$ when $y_0 = 33{,}100$. Therefore, the difference equation is $y_n = \frac{R}{0.005} + \left(33{,}100 - \frac{R}{0.005}\right)(1.005)^n$. Now substitute the known values and solve for the unknown:

$$\begin{aligned} y_{240} &= \frac{R}{0.005} + \left(33{,}100 - \frac{R}{0.005}\right)(1.005)^{240} \\ 0 &= \frac{R}{0.005} + \left(33{,}100 - \frac{R}{0.005}\right)(1.005)^{240} \\ 0 &= \frac{R}{0.005} + 109{,}567.77 - \frac{3.3102R}{0.005} \\ -109{,}567.77 &= \frac{-2.3102R}{0.005} \\ R &= \$237.14 \end{aligned}$$

43. Based on the information given, $y_n = 0.92y_{n-1}$ when $y_0 = 100$.

Conceptual Exercises

44. No, not necessarily. It depends upon the number of compounding periods and the time.

45. The effective rate will be slightly higher than the nominal rate.

46. The payment will decrease because the amount being applied to the principle will decrease. The total amount of interest paid will increase due to the length of time being added to the loan.

47. No. Much more. For example, a house payment is a decreasing annuity. If you pay an additional 5% on the loan each month, the duration of the loan will decrease significantly more than just 5%.

48. When you have successive payments, the interest on the loan is re-calculated more frequently. The interest on the loan is always the interest on the unpaid balance. If more frequent payments are made, the interest will decrease faster.

Copyright © 2018 Pearson Education, Inc.

Chapter 11

Exercises 11.1

1. Statement

3. Statement

5. Not a statement—not a declarative sentence.

7. Not a statement—not a declarative sentence.

9. Statement

11. Not a statement—x is not specified.

13. Not a statement—not a declarative sentence.

15. Statement

17. $2+2=5$ and $4<7$

19. p: The Phelps library is in New York.
q: The Phelps library is in Dallas.
Then we have $p \vee q$

21. p: The Smithsonian Museum of Natural History has displays of rocks.
q: The Smithsonian Museum of Natural History has displays of bugs.
Then we have $p \wedge q$

23. p: Amtrak trains go to Chicago.
q: Amtrak trains go to Cincinnati.
Then we have $\sim p \wedge \sim q$ or $\sim(p \vee q)$

25. **a.** Ozone is opaque to ultraviolet light, and life on earth requires ozone.

b. Ozone is not opaque to ultraviolet light, or life on earth requires ozone.

c. Ozone is not opaque to ultraviolet light, or life on earth does not require ozone.

d. If life on Earth does not require ozone, then ozone is opaque to ultraviolet light.

Copyright © 2018 Pearson Education, Inc.

27. a. Florida borders Alabama or Florida borders Mississippi: $p \vee q$

b. Florida borders Alabama and Florida does not border Mississippi $p \wedge \sim q$

c. Florida borders Mississippi and Florida does not border Alabama $q \wedge \sim p$

d. Florida does not border Alabama and Florida does not border Mississippi $\sim p \wedge \sim q$

Exercises 11.2

1. Since r is a statement form, so is $\sim r$. Then $p \wedge \sim r$ is a statement form, and so is $\sim (p \wedge \sim r)$. Since q is a statement form, $\sim (p \wedge \sim r) \vee q$ is a statement form.

3. Since p is a statement form, so is $\sim p$. Since q and r are statement forms, so are $\sim p \vee r$ and $q \wedge r$, and hence also $(\sim p \vee r) \rightarrow (q \wedge r)$.

5.

p	q	p	$\wedge$	$\sim$	q
T	T	T	**F**	F	T
T	F	T	**T**	T	F
F	T	F	**F**	F	T
F	F	F	**F**	T	F
(1)	(2)		(4)	(3)	

7.

p	q	$(p$	$\vee$	$\sim q)$	$\wedge$	q
T	T	T	T	F	**T**	T
T	F	T	T	T	**F**	F
F	T	F	F	F	**F**	T
F	F	F	T	T	**F**	F
(1)	(2)		(5)	(3)	(6)	(4)

9.

p	q	$\sim$	$[(p \vee q)$	$\wedge$	$(p \wedge q)]$
T	T	**F**	T	T	T
T	F	**T**	T	F	F
F	T	**T**	T	F	F
F	F	**T**	F	F	F
(1)	(2)	(6)	(3)	(5)	(4)

11.

p	q	p	$\oplus$	$(\sim p$	$\vee$	$q)$
T	T	T	**F**	F	T	T
T	F	T	**T**	F	F	F
F	T	F	**T**	T	T	T
F	F	F	**T**	T	T	F
(1)	(2)		(5)	(3)	(4)	

Copyright © 2018 Pearson Education, Inc.

13.

p	q	r	(p	$\wedge$	$\sim$	r)	$\oplus$	q
T	T	T	T	F	F	T	**T**	T
T	T	F	T	T	T	F	**F**	T
T	F	T	T	F	F	T	**F**	F
T	F	F	T	T	T	F	**T**	F
F	T	T	F	F	F	T	**T**	T
F	T	F	F	F	T	F	**T**	T
F	F	T	F	F	F	T	**F**	F
F	F	F	F	F	T	F	**F**	F
(1)	(2)	(3)		(5)	(4)		(6)	

15.

p	q	r	$\sim$	[(p	$\wedge$	r)	$\vee$	q]
T	T	T	**F**	T	T	T	T	T
T	T	F	**F**	T	F	F	T	T
T	F	T	**F**	T	T	T	T	F
T	F	F	**T**	T	F	F	F	F
F	T	T	**F**	F	F	T	T	T
F	T	F	**F**	F	F	F	T	T
F	F	T	**T**	F	F	T	F	F
F	F	F	**T**	F	F	F	F	F
(1)	(2)	(3)	(6)		(4)		(5)	

17.

p	p	$\vee$	$\sim$	p
T	T	**T**	F	T
F	F	**T**	T	F
(1)		(3)	(2)	

19.

p	q	r	p	$\oplus$	(q	$\oplus$	r)
T	T	T	T	**T**	T	F	T
T	T	F	T	**F**	T	T	F
T	F	T	T	**F**	F	T	T
T	F	F	T	**T**	F	F	F
F	T	T	F	**F**	T	F	T
F	T	F	F	**T**	T	T	F
F	F	T	F	**T**	F	T	T
F	F	F	F	**F**	F	F	F
(1)	(2)	(3)		(5)		(4)	

Copyright © 2018 Pearson Education, Inc.

21.

p	q	r	$(p$	$\vee$	$q)$	$\wedge$	$(p$	$\vee$	$r)$
T	T	T	T	T	T	**T**	T	T	T
T	T	F	T	T	T	**T**	T	T	F
T	F	T	T	T	F	**T**	T	T	T
T	F	F	T	T	F	**T**	T	T	F
F	T	T	F	T	T	**T**	F	T	T
F	T	F	F	T	T	**F**	F	F	F
F	F	T	F	F	F	**F**	F	T	T
F	F	F	F	F	F	**F**	F	F	F
(1)	(2)	(3)		(4)		(6)		(5)	

23.

p	q	$(p$	$\vee$	$q)$	$\wedge$	$\sim$	$(p$	$\vee$	$q)$
T	T	T	T	T	**F**	F	T	T	T
T	F	T	T	F	**F**	F	T	T	F
F	T	F	T	T	**F**	F	F	T	T
F	F	F	F	F	**F**	T	F	F	F
(1)	(2)		(3)		(6)	(5)		(4)	

25.

p	q	r	$\sim$	$(p$	$\vee$	$q)$	$\wedge$	r
T	T	T	F	T	T	T	**F**	T
T	T	F	F	T	T	T	**F**	F
T	F	T	F	T	T	F	**F**	T
T	F	F	F	T	T	F	**F**	F
F	T	T	F	F	T	T	**F**	T
F	T	F	F	F	T	T	**F**	F
F	F	T	T	F	F	F	**T**	T
F	F	F	T	F	F	F	**F**	F
(1)	(2)	(3)	(5)		(4)		(6)	

27.

p	q	$\sim$	p	$\wedge$	$\sim$	q	$\sim$	$(p$	$\wedge$	$q)$
T	T	F	T	**F**	F	T	**F**	T	T	T
T	F	F	T	**F**	T	F	**T**	T	F	F
F	T	T	F	**F**	F	T	**T**	F	F	T
F	F	T	F	**T**	T	F	**T**	F	F	F
(1)	(2)	(3)		(5)	(4)		(4)		(3)	

Copyright © 2018 Pearson Education, Inc.

29.

p	q	$\sim$	$(p$	$\oplus$	$q)$	$(p$	$\wedge$	$q)$	$\vee$	$\sim$	$(p$	$\wedge$	$q)$
T	T	**T**	T	F	T	T	T	T	**T**	F	T	T	T
T	F	**F**	T	T	F	T	F	F	**T**	T	T	F	F
F	T	**F**	F	T	T	F	F	T	**T**	T	F	F	T
F	F	**T**	F	F	F	F	F	F	**T**	T	F	F	F
(1)	(2)	(4)		(			(4)		(6)	(5)		(3)	

31. For each of the four possible pairs of values for p and q, there are two possibilities, T or F. Thus there are $2^4 = 16$ possible truth tables.

33. a.

p
T
F
(1)

p	$\mid$	p
T	**F**	T
F	**T**	F
	(2)	

b.

p	q	$(p$	$\mid$	$p)$	$\mid$	$(q$	$\mid$	$q)$
T	T	T	F	T	**T**	T	F	T
T	F	T	F	T	**T**	F	T	F
F	T	F	T	F	**T**	T	F	T
F	F	F	T	F	**F**	F	T	F
(1)	(2)		(3)		(5)		(4)	

c.

p	q	$(p$	$\mid$	$q)$	$\mid$	$(p$	$\mid$	$q)$
T	T	T	F	T	**T**	T	F	T
T	F	T	T	F	**F**	T	T	F
F	T	F	T	T	**F**	F	T	T
F	F	F	T	F	**F**	F	T	F
(1)	(2)		(3)		(5)		(4)	

d.

p	q	p	$\mid$	$((p$	$\mid$	$q)$	$\mid$	$q)$
T	T	T	**F**	T	F	T	T	T
T	F	T	**F**	T	T	F	T	F
F	T	F	**T**	F	T	T	F	T
F	F	F	**T**	F	T	F	T	F
(1)	(2)		(5)		(3)		(4)	

35. p has truth value T and q has truth value F.

a.

p	$\vee$	$\sim$	q
T	**T**	T	F

b.

$\sim$	p	$\wedge$	q
F	T	**F**	F

Copyright © 2018 Pearson Education, Inc.

c.
$p \quad \oplus \quad q$
T **T** F

d.
$\sim \quad p \quad \oplus \quad q$
F T **F** F

e.
$\sim \quad (p \quad \oplus \quad q)$
F T T F

f.
$(p \quad \vee \quad q) \quad \oplus \quad \sim \quad q$
T T F **F** T F

37. The truth value of p must be true.

39. The truth value of p must be false.

41. $(p \oplus q) \wedge \sim q$
T T F **T** T F
1 will be displayed

43. a. The calculator will display F.

b. The calculator will display F.

45. a. The calculator will display F.

b. The calculator will display F.

47.

p	q	$p \wedge \sim q$
T	T	F
T	F	T
F	T	F
F	F	F

49.

p	q	r	$p \wedge (\sim q \vee r)$
T	T	T	T
T	T	F	F
T	F	T	T
T	F	F	T
F	T	T	F
F	T	F	F
F	F	T	F
F	F	F	F

51. (Not p) And q, or $\sim p \wedge q$

Copyright © 2018 Pearson Education, Inc.

Exercises 11.3

1.

p	q	~	p	→	~	q
T	T	F	T	**T**	F	T
T	F	F	T	**T**	T	F
F	T	T	F	**F**	F	T
F	F	T	F	**T**	T	F
(1)	(2)	(3)		(5)	(4)	

3.

p	q	$(p$	$\oplus$	$q)$	→	q
T	T	T	F	T	**T**	T
T	F	T	T	F	**F**	F
F	T	F	T	T	**T**	T
F	F	F	F	F	**T**	F
(1)	(2)		(3)		(4)	

5.

p	q	r	(~	p	∧	$q)$	→	r
T	T	T	F	T	F	T	**T**	T
T	T	F	F	T	F	T	**T**	F
T	F	T	F	T	F	F	**T**	T
T	F	F	F	T	F	F	**T**	F
F	T	T	T	F	T	T	**T**	T
F	T	F	T	F	T	T	**F**	F
F	F	T	T	F	F	F	**T**	T
F	F	F	T	F	F	F	**T**	F
(1)	(2)	(3)	(4)		(5)		(6)	

7.

p	q	$(p$	→	$q)$	↔	(~	p	∨	$q)$
T	T	T	T	T	**T**	F	T	T	T
T	F	T	F	F	**T**	F	T	F	F
F	T	F	T	T	**T**	T	F	T	T
F	F	F	T	F	**T**	T	F	T	F
(1)	(2)		(4)		(6)	(3)		(5)	

9.

p	q	r	$(p$	→	$q)$	→	r
T	T	T	T	T	T	**T**	T
T	T	F	T	T	T	**F**	F
T	F	T	T	F	F	**T**	T
T	F	F	T	F	F	**T**	F
F	T	T	F	T	T	**T**	T
F	T	F	F	T	T	**F**	F
F	F	T	F	T	F	**T**	T
F	F	F	F	T	F	**F**	F
(1)	(2)	(3)		(4)		(5)	

Copyright © 2018 Pearson Education, Inc.

11.

p	q	$(p$	$\vee$	$q)$	$\leftrightarrow$	$(p$	$\wedge$	$q)$
T	T	T	T	T	**T**	T	T	T
T	F	T	T	F	**F**	T	F	F
F	T	F	T	T	**F**	F	F	T
F	F	F	F	F	**T**	F	F	F
(1)	(2)		(3)		(5)		(4)	

13. $((\sim p) \wedge (\sim q)) \rightarrow ((\sim p) \wedge q)$

15. $(((\sim p) \wedge (\sim q)) \vee r) \rightarrow ((\sim q) \wedge r)$

17.

$\sim$	p	$\rightarrow$	q
F	T	**T**	F

19.

q	$\rightarrow$	p
F	**T**	T

21.

$(p$	$\oplus$	$q)$	$\rightarrow$	p
T	T	F	**T**	T

23.

$(p$	$\wedge$	$\sim$	$q)$	$\rightarrow$	$(\sim$	p	$\oplus$	$q)$
T	T	T	F	**F**	F	T	F	F

25.

p	$\rightarrow$	$[p$	$\wedge$	$(p$	$\oplus$	$q)]$
T	**T**	T	T	T	T	F

27. $p \leftrightarrow q$

29. $q \rightarrow p$; hypothesis: q; conclusion: p

31. $q \rightarrow p$; hypothesis: q; conclusion: p

33. $\sim p \rightarrow \sim q$; hypothesis: $\sim p$; conclusion: $\sim q$

35. $\sim p \rightarrow \sim q$; hypothesis: $\sim p$; conclusion: $\sim q$;
F T **T** F T TRUE

37. $\sim q \rightarrow \sim p$; hypothesis: $\sim q$; conclusion: $\sim p$;
TRUE

39. **a.** hyp: I will run a marathon.
con: Amy watches my dogs.

b. hyp: A student earns a B.
con: A student passes the course.

c. hyp: The football team is my favorite.
con: The football team wears green uniforms

d. hyp: Isaac has wrinkled fingers.
con: Isaac was in the bathtub.

41. **a.** If City Sanitation collects the garbage, then the mayor calls.

b. The price of beans goes down if there is no drought.

c. Goldfish swim in Lake Erie if Lake Erie is fresh water.

d. Tap water is not salted if it boils slowly.

43. **a.** $Z = 0 + 0 = 0$, so the condition fails and $A = 4$.

b. $Z = 8 + (-8) = 0$, so the condition fails and $A = 4$.

c. $Z = -3 + 3 = 0$, so the condition fails and $A = 4$.

d. $X = -3 \le 0$, so the condition fails and $A = 4$.

e. $X = 8 > 0$ and $Z = 8 + (-3) = 5 \neq 0$, so the condition is met and $A = 6$.

f. $X = 3 > 0$ and $Z = 3 + (-8) = -5 \neq 0$, so the condition is met and $A = 6$.

45. **a.** $B = -6 < 0$, so the condition is met and $Y = 7$.

b. $B = -6 < 0$, so the condition is met and $Y = 7$.

c. $C = (-2)(6) = -12 < 10$ and $B \ge 0$, so the condition fails and $Y = 0$.

d. $B = -1 < 0$, so the condition is met and $Y = 7$.

e. $A = 4 \ge 0$ and $B = 3 \ge 0$, so the condition fails and $Y = 0$.

f. $A = 3 \ge 0$ and $B = 1 \ge 0$, so the condition fails and $Y = 0$.

Copyright © 2018 Pearson Education, Inc.

47. **a.** $A = -1 < 0$ and $B = -2 < 0$, so the condition is met and $C = (-1)(-2) + 4 = 6$.

b. $B = 8 \geq 6$, so the condition is met and $C = (-2)(8) + 4 = -12$

c. $B = 3 \geq 0$ and $B = 3 < 6$, so the condition fails and $C = 0$.

d. $A = 3 \geq 0$ and $B = -2 < 6$, so the condition fails and $C = 0$.

e. $B = 8 \geq 6$, so the condition is met and $C = (3)(8) + 4 = 28$.

f. $A = 3 \geq 0$ and $B = -3 < 6$, so the condition fails and $C = 0$.

Exercises 11.4

1.

$[(p$	$\rightarrow$	$q)$	$\wedge$	$q]$	$\rightarrow$	p
F	T	T	T	T	**F**	F

When p is false and q is true, the statement is FALSE.

3. Show that the corresponding bi-conditional is a tautology.

p	q	$(p$	$\rightarrow$	$q)$	$\leftrightarrow$	$(\sim$	$(p$	$\wedge$	$\sim$	$q))$
T	T	T	T	T	**T**	T	T	F	F	T
T	F	T	F	F	**T**	F	T	T	T	F
F	T	F	T	T	**T**	T	F	F	F	T
F	F	F	T	F	**T**	T	F	F	T	F
(1)	(2)		(3)		(7)	(6)		(5)	(4)	

5.

p	q	c	$(p$	$\rightarrow$	$q)$	$\leftrightarrow$	$[(p$	$\wedge$	$\sim$	$q)$	$\rightarrow$	$c)]$
T	T	F	T	T	T	**T**	T	F	F	T	T	F
T	F	F	T	F	F	**T**	T	T	T	F	F	F
F	T	F	F	T	T	**T**	F	F	F	T	T	F
F	F	F	F	T	F	**T**	F	F	T	F	T	F
(1)	(2)	(3)		(6)		(8)		(5)	(4)		(7)	

7. False: consider p FALSE and q TRUE.

9. $\sim(p \vee q) \wedge \sim q$

$\Leftrightarrow \sim(p \wedge \sim q) \wedge \sim q$ — DeMorgan's law (8a)

$\Leftrightarrow \sim p \wedge \sim q$ — Idempotent law (5b)

$\Leftrightarrow \sim(p \vee q)$ — DeMorgan's law (8a)

$\Leftrightarrow \sim(\sim p \rightarrow q)$ — Implication (10a)

11. $p \oplus q \Leftrightarrow (p \vee q) \wedge \sim(p \wedge q) \Leftrightarrow \sim[\sim(p \vee q) \vee \sim(\sim p \vee \sim q)]$

13. $(p \vee q) \rightarrow (q \wedge \sim r)$

$\Leftrightarrow \sim(p \vee q) \vee (q \wedge \sim r)$ — Implication (10a)

15. $\sim(p \wedge \sim q) \rightarrow (p \vee \sim r)$

$\Leftrightarrow (p \wedge \sim q) \vee (p \vee \sim r)$ — Implication (10a) and double negation (1)

Copyright © 2018 Pearson Education, Inc.

17. a. $\sim(p \wedge q) \Leftrightarrow \sim p \vee \sim q$
Arizona does not border California, or Arizona does not border Nevada.

b. $\sim(p \vee q) \Leftrightarrow \sim p \wedge \sim q$
There are no tickets available, and the agency cannot get tickets.

c. $\sim(p \vee q) \Leftrightarrow \sim p \wedge \sim q$
The killer's hat was neither white nor gray.

19. $\sim(p \vee \sim q \vee r) \Leftrightarrow \sim p \wedge q \wedge \sim r$

21. a. Jeremy does not take 12 credits and Jeremy does not take 15 credits.

b. Sandra does not receive a gift from Sally or Sandra does not receive a gift from Sacha.

23. a. I have a ticket to the theater, and I did not spend a lot of money.

b. Basketball is played on an indoor court, and the players do not wear sneakers.

c. The stock market is going up, and interest rates are not going down.

d. Humans have enough water, and humans are not staying healthy.

25. a. If x is an even number, then the sum $2 + x$ is even.
True
For example: let $x = 4$ (even).
The sum $2 + x = 2 + 4 = 6$ (even)

b. If "S" is next to "K", then the computer keyboard does not have the QWERTY layout.
True.

27. a. Contrapositive: If a dog is not a Chihuahua, then it is not small (False).
Inverse: If a dog is not small, then it is not a Chihuahua. (True)
Converse: If a dog is a Chihuahua, then it is small (True).

b. Contrapositive: If you stop, then the traffic light is not green (True).
Inverse: If the traffic light is not green, then you stop. (False)
Converse: If you do not stop, then the traffic light is green (False).

c. Contrapositive: If we are not French citizens, then we do not live in France (False).
Inverse: If we do not live in France, Then we are not French citizens (False)
Converse: If we are French citizens, then we live in France (False).

29. a. $(q \vee r) \rightarrow \sim p$

b. $(q \vee r) \wedge p$; We wait on some other person or some other time, and change will come.

31. a. Inverse: If you wish to claim the premium tax credit for 2015, then you need the information in Part II.
Converse: If you do not need the information in Part II, then you do not wish to claim the premium tax credit for 2015.
Contrapositive: If you need the information in Part II, then you wish to claim the premium tax credit for 2015.
Negation: You do not wish to claim the premium tax credit for 2015 and you need the information in Part II.

Copyright © 2018 Pearson Education, Inc.

b. Inverse: If your spouse does not itemize deductions, then you can take the standard deduction.
Converse: If you can't take the standard deduction, then your spouse itemizes deductions.
Contrapositive: If you take the standard deduction, then your spouse does not itemize deductions.
Negation: Your spouse itemizes deductions and you can take the standard deduction.

c. Inverse: If you do not check a box, then your tax will change and your refund will change.
Converse: If your tax or refund don't change, then you check a box.
Contrapositive: If your tax or refund change, then you don't check a box.
Negation: You check a box and your tax will change and your refund will change.

33.

p	q	p	$\wedge$	$(p$	$\rightarrow$	$q)$	$\rightarrow$	q
T	T	T	T	T	T	T	**T**	T
T	F	T	F	T	F	F	**T**	F
F	T	F	F	F	T	T	**T**	T
F	F	F	F	F	T	F	**T**	F
(1)	(2)		(4)		(3)		**(5)**	

Exercises 11.5

1. e = "The auto manufacturer follows environmental regulations."
l = "The auto manufacturer faces legal action."
1. $e \vee l$ hyp.
2. $\sim e$ hyp.
3. l disj. syll. (1, 2)

3. a = "My allowance comes this week."
p = "I pay the rent."
b = "My bank account will be in the black."
e = "I will be evicted."
1. $(a \wedge p) \rightarrow b$ hyp.
2. $\sim p \rightarrow e$ hyp.
3. $\sim e \wedge a$ hyp.
4. $\sim e$ subtr. (3)
5. p mod. tollens (2, 4)
6. a subtr. (3)
7. b mod. ponens (5, 6, 1)

5. p = "The price of oil increases.
a = "The OPEC countries are in agreement."
d = "There is a U.N. debate."
1. $p \rightarrow a$ hyp.
2. $\sim d \rightarrow p$ hyp.
3. $\sim a$ hyp.
4. $\sim p$ mod. tollens (1, 3)
5. d mod. tollens (2, 4)

7. g = "The germ is present."
r = "The rash is present."
f = "The fever is present."
1. $g \rightarrow (r \wedge f)$ hyp.
2. f hyp.
3. $\sim r$ hyp.
4. $\sim r \vee \sim f$ addition (3)
5. $\sim(r \wedge f)$ DeMorgan (4)
6. $\sim g$ mod tollens (1, 5)

9. c = "The material is cotton."
r = "The material is rayon."
d = "The material can be made into a dress."
1. $(c \vee r) \rightarrow d$ hyp.
2. $\sim d$ hyp.
3. $\sim(c \vee r)$ mod. tollens (1, 2)
4. $\sim c \wedge \sim r$ DeMorgan (3)
5. $\sim r$ subtraction (4)

11. s = "Salaries go up."
m = "More people apply."
If s is false and m is true, then
$(s \rightarrow m) \wedge (m \vee s)$ is true but s is false.
$(s \rightarrow m) \wedge (m \vee s) \nRightarrow s$.
The argument is invalid.

Copyright © 2018 Pearson Education, Inc.

13. y = "The balloon is yellow."
p = "The ribbon is pink."
h = "The balloon is filled with helium."

1. $y \vee p$	hyp.
2. $h \rightarrow \sim y$	hyp.
3. h	hyp.
4. $\sim y$	mod. ponens (2, 3)
5. p	disj. syllogism (1, 4)

The argument is valid.

15. p = "The papa bear sits."
m = "The mama bear stands."
b = "The baby bear crawls on the floor."

1. $p \rightarrow m$	hyp.
2. $m \rightarrow b$	hyp.
3. $\sim b$	hyp.
4. $\sim m$	mod. tollens (2, 3)
5. $\sim p$	mod. tollens (1, 4)

The argument is valid.

17. w = "Wheat prices are steady."
e = "Exports will increase."
s = "The GNP will be steady."
If w and s are true and e is false, then
$[w \rightarrow (e \vee s)] \wedge (w \wedge s)$ is true but e is false.
$[w \rightarrow (e \vee s)] \wedge (w \wedge s) \not\Rightarrow e.$
The argument is invalid.

19. a = "The Red Sox won the ALC."
w = "The Red Sox play in the World Series."
c = "The Red Sox fire their coach."
If w and c are true and a is false, then
$(a \rightarrow w) \wedge (w \vee c)$ is true but $c \rightarrow a$ is false.
$(a \rightarrow w) \wedge (w \vee c) \not\Rightarrow c \rightarrow a$
The argument is invalid.

21. s = "Sam goes to the store."
m = "Sam needs milk."
$H_1 = s \rightarrow m$
$H_2 = \sim m$
$C = \sim s$

1. s	$\sim C$
2. $s \rightarrow m$	H_1
3. m	$\sim H_2$; mod. ponens (1, 2)

23. n = "The newspaper reports the crime."
t = "Television reports the crime."
s = "The crime is serious."
k = "A person is killed."
$H_1 = (n \wedge t) \rightarrow s$
$H_2 = k \rightarrow n$
$H_3 = k$
$H_4 = t$
$C = s$

1. $\sim s$	$\sim C$
2. $(n \wedge t) \rightarrow s$	H_1
3. $\sim(n \wedge t)$	mod. tollens (1, 2)
4. $\sim n \vee \sim t$	DeMorgan (3)
5. t	H_4
6. $\sim n$	disj. syllogism (4, 5)
7. $k \rightarrow n$	H_2
8. $\sim k$	$\sim H_3$; mod. tollens (6, 7)

25. j = "Jimmy finds his keys."
h = "He does his homework."
$H_1 = \sim j \rightarrow h$
$H_2 = \sim h$
$C = j$

Direct proof

1. $\sim j \rightarrow h$	H_1
2. $\sim h$	H_2
3. j	mod. tollens (1, 2)

Indirect proof

1. $\sim j$	$\sim C$	
2. $\sim j \rightarrow h$	H_1	
3. h	mod. ponens (1, 2)	contradiction
4. $\sim h$	H_2	

27. m = "Marissa goes to the movies."
k = "Marissa is in a knitting class."
i = "Marissa is idle."
$H_1 = \sim m \rightarrow \sim i$
$H_2 = \sim k \rightarrow i$
$H_3 = \sim k$
$C = m$

Copyright © 2018 Pearson Education, Inc.

Direct proof

1. $\sim m \to \sim i$	H_1
2. $\sim k \to i$	H_2
3. $\sim k$	H_3
4. i	mod. ponens (2, 3)
5. m	mod. tollens (1, 4)

Indirect proof

1. $\sim m$	$\sim C$	
2. $\sim m \to \sim i$	H_1	
3. $\sim i$	mod. ponens (1, 2)	
4. $\sim k \to i$	H_2	
5. k	mod. tollens (3, 4)	contradiction
6. $\sim k$	H_3	

Exercises 11.6

1. **a.** "1 is even or 1 is divisible by 3" is FALSE.

b. "4 is even or 4 is divisible by 3" is TRUE.

c. "3 is even or 3 is divisible by 3" is TRUE.

d. "6 is even or 6 is divisible by 3" is TRUE.

e. "5 is even or 5 is divisible by 3" is FALSE.

3. Abby's statement: $\forall x \sim p(x)$, or $\sim[\exists x\, p(x)]$. This is surely false. Abby meant to say, "not all men cheat on their wives" $\sim[\forall x\, p(x)]$, or $\exists x \sim p(x)$.

5. **a.** $\forall x\, p(x)$

b. $\exists x \sim p(x)$

c. $\exists x\, p(x)$

d. $\sim[\forall x\, p(x)]$

e. $\forall x \sim p(x)$

f. $\sim[\exists x\, p(x)]$

g. (b) and (d); (e) and (f) · (b) and (d) both say that there are some university professors who don't like poetry; (e) and (f) both say that no university professors like it.

7. **a.** TRUE;
$p(4) = (4 \text{ is prime}) \to (4^2 + 1 \text{ is even})$ is TRUE, because the hypothesis is FALSE.

b. FALSE;
$p(2) = (2 \text{ is prime}) \to (2^2 + 1 \text{ is even})$ is FALSE.

9. **a.** T; every x is either even or odd.

b. $[\forall x\, p(x)] \vee [\forall x\, q(x)]$
F **F** F

c. T; 5 is odd, hence even or odd, for instance.

d. F; no x is both even and odd.

e. F; 5 is not both even and odd, for instance.

f. $[\exists x\, p(x)] \wedge [\exists x\, q(x)]$
T **T** T

g. F; (4 is even) $\to$ (4 is odd) is false, for instance.

h. $[\forall x\, p(x) \to [\forall x\, g(x)]$
F **T** F

11. **a.** Not every dog has his day.

b. No men fight wars.

c. Some women are unmarried.

d. There exists a pot without a cover.

e. All children have pets.

f. Every month has 30 days.

13. **a.** "The sum of any two nonnegative integers is greater than 12." FALSE: consider $x = 1$, $y = 2$. "There exist two nonnegative integers whose sum is not greater than 12."

b. "For any nonnegative integer, there is a nonnegative integer that, added to the first, makes a sum greater than 12." TRUE

c. "There is a nonnegative integer that, added to any other nonnegative integer, makes a sum greater than 12." TRUE (Try $x = 13$.)

d. "There are two nonnegative integers, the sum of which is greater than 12." TRUE (Try $x = 6$, $y = 7$.)

Copyright © 2018 Pearson Education, Inc.

15. a. FALSE: let $x = 2$, $y = 3$.

b. TRUE: for any x, let $y = x$.

c. TRUE: let $x = 1$.

d. FALSE: no y is divisible by every x.

e. TRUE: for any y, let $x = y$.

f. TRUE: any x divides itself.

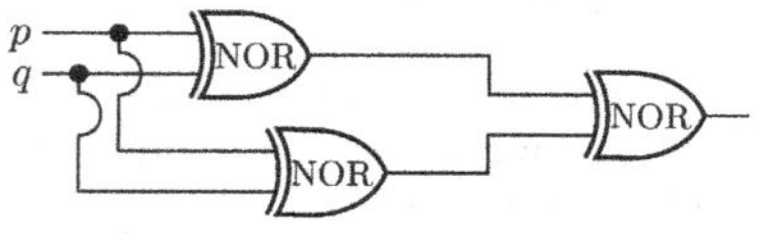

17. a. $S \subseteq T$ translates as $\forall x[x \geq 8 \rightarrow x \leq 10]$.

b. No; consider $x = 11$.

19. $S = \{2, 4, 6, 8\}$, $T = \{1, 2, 3, 4, 6, 8\}$.
So, $\forall x[x \in S \rightarrow x \in T)$.

Exercises 11.7

1. $(p \vee q) \wedge (p \vee \sim q)$

3. $((p \wedge q) \wedge \sim r) \wedge (\sim q \vee r)$

5.

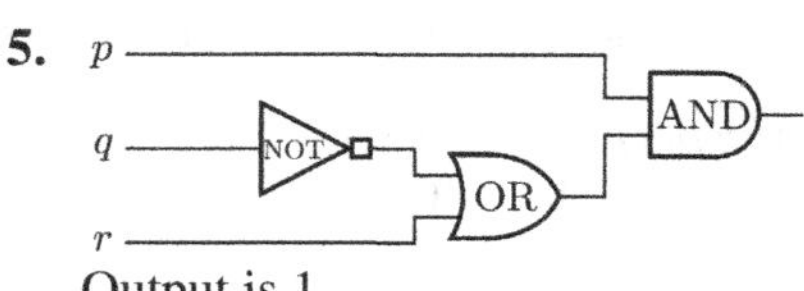

Output is 1

7.

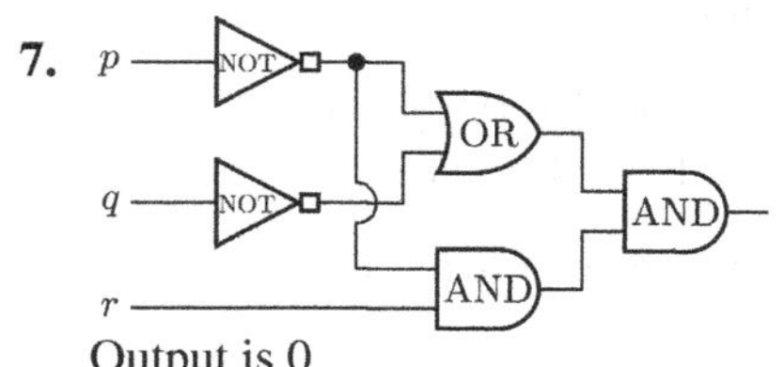

Output is 0

9. The circuit represents the logic statement
$(p \wedge q) \vee (p \wedge \sim q)$
$\Leftrightarrow p \wedge (q \vee \sim q)$ Distributive law (4b)
$\Leftrightarrow p \wedge t$ (7a)
$\Leftrightarrow p$ (6d)

11. The circuit represents the logic statement
$((p \wedge q) \wedge r) \vee \sim((p \vee q) \vee \sim r)$
$\Leftrightarrow (r \wedge (p \wedge q)) \vee (r \wedge \sim(p \vee q))$ Commutative law (2b) and DeMorgan's law (8a)
$\Leftrightarrow r \wedge [(p \wedge q) \vee \sim(p \vee q)]$ Distributive law (4b)

13.

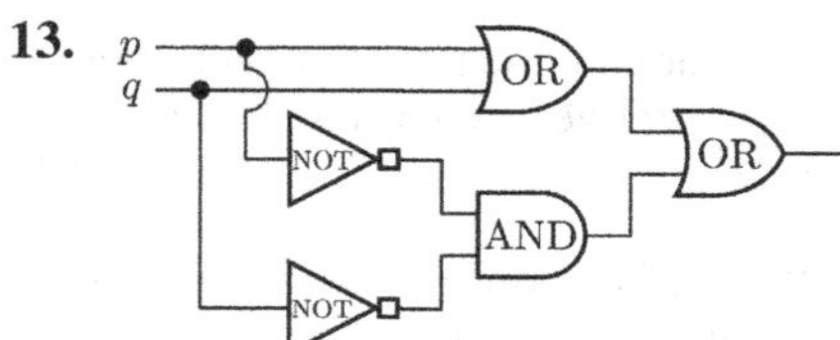

15.

Copyright © 2018 Pearson Education, Inc.

17. 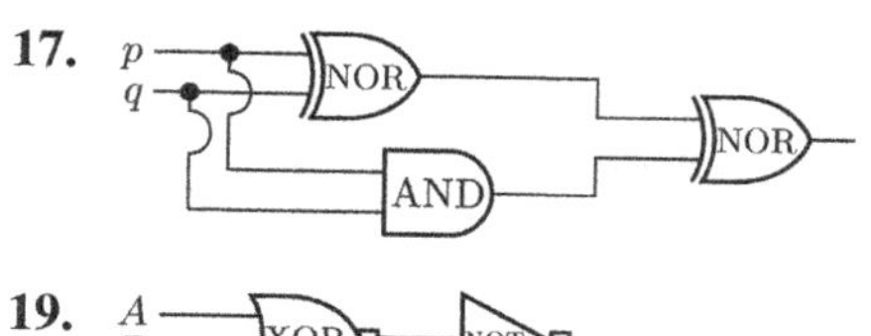

19. A B XOR NOT

21. (p And (Not q)) Or r

Chapter 11 Chapter Review Exercises

1. a. Statement

b. Not a statement—not a declarative sentence.

c. Statement

d. Not a statement—"he" is not specified.

e. Statement

2. a. If two lines are perpendicular, then their slopes are negative reciprocals of each other.

b. If goldfish can live in a fishbowl, then the water is aerated.

c. If it rains, then Jane uses her umbrella.

d. If Sally gives Morris a treat, then he ate all his food.

3. a. Contrapositive: If the Yankees are not playing in Yankee Stadium, then they are not in New York City; converse: If the Yankees are playing in Yankee Stadium, then they are in New York City.

b. Contrapositive: If the quake is not considered major, then the Richter scale does not indicate the earthquake is a 7; converse: If the quake is considered major, then the Richter scale indicates the earthquake is a 7.

c. Contrapositive: If a coat is not warm then it is not made of fur; converse: If a coat is warm then it is made of fur.

d. Conrapositive: If Jane is not in Moscow then she is not in Russia; converse: If Jane is in Moscow then she is in Russia.

4. a. $p =$ (two triangles are similar) and
$q =$ (their sides are equal).
$p \rightarrow q$ negated becomes $\sim(p \rightarrow q)$ or $p \wedge \sim q$, or "two triangles are similar but their sides are unequal."

b. $U =$ {real numbers} and $p(x) = (x^2 = 5)$.

$\exists x\, p(x)$ negated becomes $\forall x \sim q(x)$ or "For every real number x, $x^2 \neq 5$."

c. $U =$ {positive integers},
$p(n =$ (n is even), and $q(n) = (n^2$ is even).
$\forall n[p(n) \rightarrow q(n)]$ negated becomes $\exists n \sim [p(n) \rightarrow q(n)]$ or $\exists n[p(n) \wedge \sim q(n)]$, or "There exists a positive integer n such that n is even but n^2 is not even."

Copyright © 2018 Pearson Education, Inc.

d. $U = \{\text{real numbers}\}$ and $p(x) = (x^2 + 4 = 0)$.
$\exists x\, p(x)$ negated becomes $\forall x \sim p(x)$ or "For every real number x, $x^2 + 4 \neq 0$."

5. a.

p		p	$\vee$	$\sim$	p
T		T	**T**	F	T
F		F	**T**	T	F
(1)			(3)	(2)	

Tautology

b.

p	q		$(p$	$\to$	$q)$	$\leftrightarrow$	$(\sim$	p	$\vee$	$q)$
T	T		T	T	T	**T**	F	T	T	T
T	F		T	F	F	**T**	F	T	F	F
F	T		F	T	T	**T**	T	F	T	T
F	F		F	T	F	**T**	T	F	T	F
(1)	(2)			(4)		(6)	(3)		(5)	

Tautology

c. Let p and q be TRUE.
$(p \wedge \sim q) \leftrightarrow \sim(\sim p \wedge q)$
T F F T **F** T F T F T
Not a tautology

d. Let p, q, and r be FALSE.
$[p \to (q \to r)] \leftrightarrow [(p \to q) \to r]$
F T F T F **F** F T F F F
Not a tautology

6. a.

p	q	r		p	$\to$	$(\sim$	q	$\vee$	$r)$
T	T	T		T	**T**	F	T	T	T
T	T	F		T	**F**	F	T	F	F
T	F	T		T	**T**	T	F	T	T
T	F	F		T	**T**	T	F	T	F
F	T	T		F	**T**	F	T	T	T
F	T	F		F	**T**	F	T	F	F
F	F	T		F	**T**	T	F	T	T
F	F	F		F	**T**	T	F	T	F
(1)	(2)	(3)			(6)	(4)		(5)	

Copyright © 2018 Pearson Education, Inc.

b.

p	q	r	p	$\wedge$	$(q$	$\leftrightarrow$	$(r$	$\wedge$	$p))$
T	T	T	T	**T**	T	T	T	T	T
T	T	F	T	**F**	T	F	F	F	T
T	F	T	T	**F**	F	F	T	T	T
T	F	F	T	**T**	F	T	F	F	T
F	T	T	F	**F**	T	F	T	F	F
F	T	F	F	**F**	T	F	F	F	F
F	F	T	F	**F**	F	T	T	F	F
F	F	F	F	**F**	F	T	F	F	F
(1)	(2)	(3)		(6)		(5)		(4)	

7. a. True—a version of disjunctive syllogism

b. False—consider p false and q true.

8. a. True—contrapositive

b. False—consider p, q, and r false.

9. a. False—consider p false and q true.

b. True—modus tollens

10. a. $C = 3(4) + 5 = 17 > 0$ and $B = 5 > 0$, so the condition is met and $Z = 17$.

b. $B = 2 \not> 3$, so the condition fails and $Z = 100$.

c. $C = 3(-4) + 5 = -7 \not>$, so the condition fails and $Z = 100$.

d. $B = -2 \not> 3$, so the condition fails and $Z = 100$.

11. a. $C = 10$, so the condition is met and $Z = 5 \times 10 = 50$.

b. $C = 10$, so the condition is met and $Z = (5)(-5) = -25$.

c. $X = -10 \not> 0$, $C = 2 \not\geq 10$, so the condition fails and $Z = (-10) + (-5) = -15$.

d. $X = 2 > 0$, $Y = 5 > 0$, so the condition is met and $Z = 2 \times 5 = 10$.

12. a. Cannot be determined

b. TRUE, by the contrapositive

c. TRUE

d. Cannot be determined

e. Cannot be determined

13. a. Cannot be determined

b. Cannot be determined

c. TRUE, by contraposition and DeMorgan

Copyright © 2018 Pearson Education, Inc.

14. a. TRUE (given the additional assumption that at least two mathematicians exist)

b. FALSE; let $p(x)$ = "like rap music" and U = {mathematicians}.
$\forall x\, p(x) \quad \sim\exists x\, p(x)$

c. TRUE

15. a. Cannot be determined

b. Cannot be determined

c. True; $\exists x[\sim r(x)] \Leftrightarrow \sim\forall x[r(x)]$

16. t = "Taxes go up."
s = "I sell the house."
m = "I move to India."

1. $t \to (s \wedge m)$	hyp.
2. $\sim m$	hyp.
3. $\sim s \vee \sim m$	addition (2)
4. $\sim(s \wedge m)$	DeMorgan (3)
5. $\sim t$	mod. tollens (1, 4)

17. m = "I study mathematics."
b = "I study business."
p = "I can write poetry."

1. $m \wedge b$	hyp.
2. $b \to (\sim p \vee \sim m)$	hyp.
3. b	subtraction (1)
4. $\sim p \vee \sim m$	mod. ponens (2, 3)
5. m	subtraction (1)
6. $\sim p$	disj. syllogism (4, 5)

18. d = "I shop for a dress."
h = "I wear high heels."
s = "I have a sore foot."

1. $d \to h$	hyp.
2. $s \to \sim h$	hyp.
3. d	hyp.
4. h	mod. ponens (1, 3)
5. $\sim s$	mod. tollens (2, 4)

19. a = "Asters grow in the garden."
d = "Dahlias grow in the garden."
s = "It is spring."

1. $a \vee d$	hyp.
2. $s \to \sim a$	hyp.
3. s	hyp.
4. $\sim a$	mod. ponens (2, 3)
5. d	disj. syllogism (1, 4)

Copyright © 2018 Pearson Education, Inc.

20. t = "The professor gives a test."
h = "Nancy studies hard."
d = "Nancy has a date."
s = "Nancy takes a shower."
$H_1 = t \to h$
$H_2 = d \to s$
$H_3 = \sim t \to \sim s$
$H_4 = d$
$C = h$

1. $\sim h$	$\sim C$
2. $t \to h$	H_1
3. $\sim t$	mod. tollens (1, 2)
4. $\sim t \to \sim s$	H_3
5. $\sim s$	mod. ponens (3, 4)
6. $d \to s$	H_2
7. $\sim d$	$\sim H_4$; mod. tollens (5, 6)

Contradiction

21.

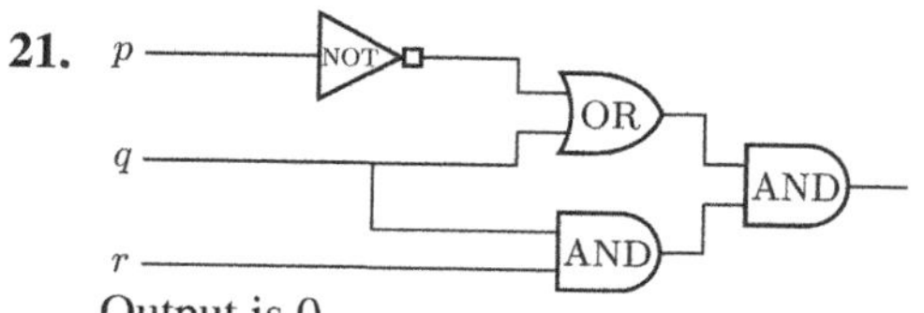

Output is 0

22. $((p \wedge \sim q) \wedge \sim(q \vee \sim r)) \wedge r$

$\Leftrightarrow ((p \wedge \sim q) \wedge (\sim q \wedge r)) \wedge r$	DeMorgan's law (8a)
$\Leftrightarrow ((p \wedge r) \wedge \sim q) \wedge r$	Distributive law (4b) and commutative law (2b)
$\Leftrightarrow (p \wedge (r \wedge r)) \wedge (\sim q \wedge r)$	Distributive law (4b)
$\Leftrightarrow p \wedge (\sim q \wedge r)$	Idempotent law (5b)

Conceptual Exercises

23.

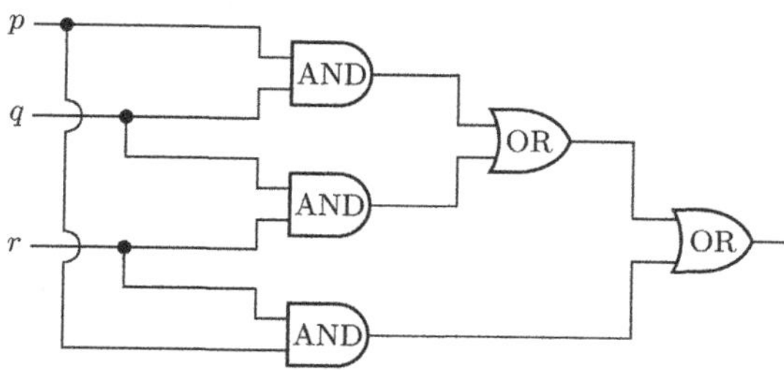

24.

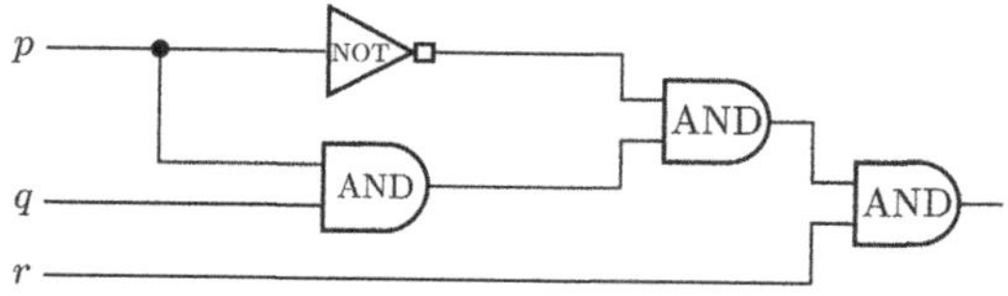

Copyright © 2018 Pearson Education, Inc.

25. (p NOR q) NOR (p AND q)

p	q	(p	NOR	q)	NOR	(p	AND	q)	$\leftrightarrow$	(p	XOR	q)
T	T	T	F	T	F	T	T	T	**T**	T	F	T
T	F	T	F	F	T	T	F	F	**T**	T	T	F
F	T	F	F	T	T	F	F	T	**T**	F	T	T
F	F	F	T	F	F	F	F	F	**T**	F	F	F
(1)	(2)		(3)		(6)		(4)		(7)		(5)	

Copyright © 2018 Pearson Education, Inc.

Chapter 12

Exercises 12.1

1. $a = 4$, $b = -6$, $\dfrac{b}{1-a} = \dfrac{-6}{1-4} = 2$

3. $a = -\dfrac{1}{2}$, $b = 0$, $\dfrac{b}{1-a} = \dfrac{0}{1+\frac{1}{2}} = 0$

5. $a = -\dfrac{2}{3}$, $b = 15$, $\dfrac{b}{1-a} = \dfrac{15}{1+\frac{2}{3}} = 9$

7. a. $y_0 = 10$, $y_1 = \dfrac{1}{2}(10) - 1 = 4$,

$y_2 = \dfrac{1}{2}(4) - 1 = 1$, $y_3 = \dfrac{1}{2}(1) - 1 = -\dfrac{1}{2}$,

$y_4 = \dfrac{1}{2}\left(-\dfrac{1}{2}\right) - 1 = -\dfrac{5}{4}$

b.

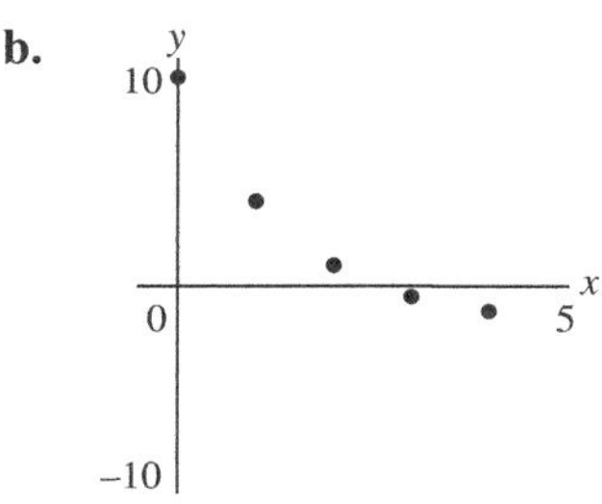

c. $y_n = \dfrac{-1}{1-\frac{1}{2}} + \left(10 - \dfrac{-1}{1-\frac{1}{2}}\right)\left(\dfrac{1}{2}\right)^n$

$= -2 + 12\left(\dfrac{1}{2}\right)^n$

9. a. $y_0 = 3.5$, $y_1 = 2(3.5) - 3 = 4$,
$y_2 = 2(4) - 3 = 5$, $y_3 = 2(5) - 3 = 7$,
$y_4 = 2(7) - 3 = 11$

b.

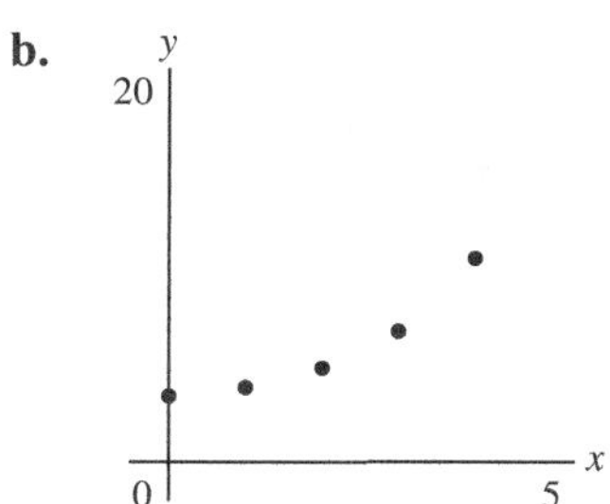

c. $y_n = \dfrac{-3}{1-2} + \left(3.5 - \dfrac{-3}{1-2}\right)(2)^n$

$= 3 + (.5)2^n$

11. a. $y_0 = 17.5$, $y_1 = -.4(17.5) + 7 = 0$,
$y_2 = -.4(0) + 7 = 7$,
$y_3 = -.4(7) + 7 = 4.2$,
$y_4 = -.4(4.2) + 7 = 5.32$

b.

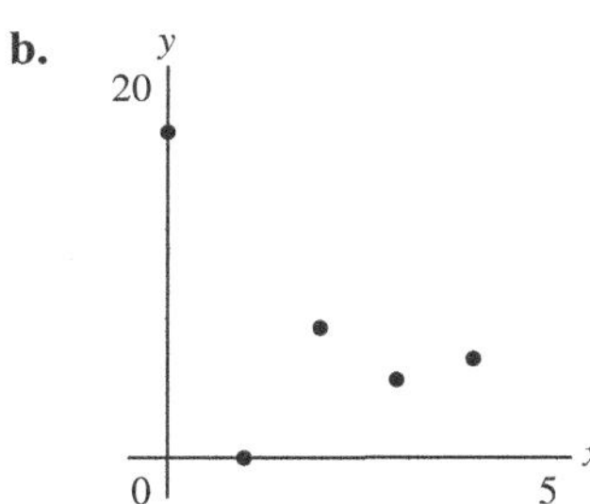

c. $y_n = \dfrac{7}{1+.4} + \left(17.5 - \dfrac{7}{1+.4}\right)(-.4)^n$

$= 5 + 12.5(-.4)^n$

Copyright © 2018 Pearson Education, Inc.

13. **a.** $y_0 = 15,\ y_1 = 2(15) - 16 = 14,$
$y_2 = 2(14) - 16 = 12,$
$y_3 = 2(12) - 16 = 8,$
$y_4 = 2(8) - 16 = 0$

b.

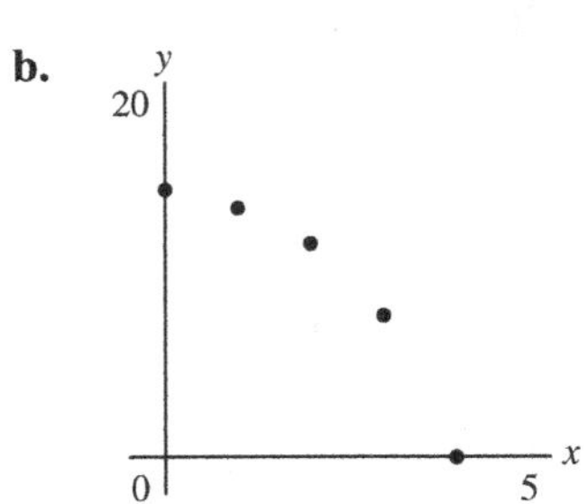

c. $y_n = \dfrac{-16}{1-2} + \left(15 - \dfrac{-16}{1-2}\right)(2)^n$
$= 16 - 2^n$

15. $y_0 = 6 - 5(.2)^0 = 1$
$y_1 = 6 - 5(.2)^1 = 5$
$y_2 = 6 - 5(.2)^2 = 5.8$
$y_3 = 6 - 5(.2)^3 = 5.96$
$y_4 = 6 - 5(.2)^4 = 5.992$

17. $y_n = 1.03y_{n-1},\ y_0 = 1000$

19. $y_n = .99y_{n-1} - 1{,}000{,}000,\ \ y_0 = 70{,}000{,}000$

 Copyright © 2018 Pearson Education, Inc.

21. a. $y_0 = 1,\ y_1 = 1+2 = 3,\ y_2 = 3+2 = 5,$
$y_3 = 5+2 = 7,\ y_4 = 7+2 = 9$

b.

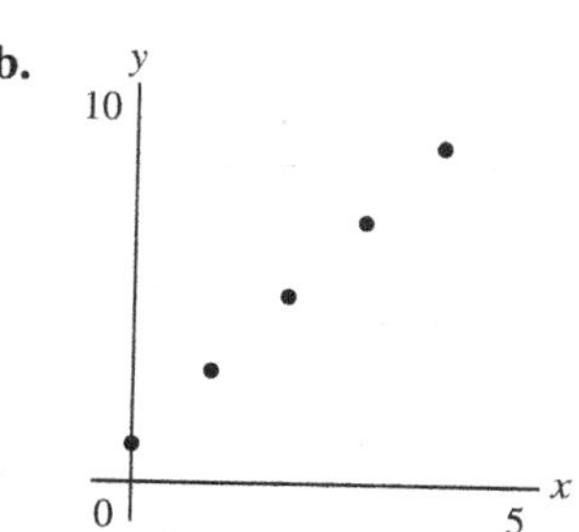

The points lie on a straight line.

c. $a = 1$, so the denominator of $\frac{b}{1-a}$ is zero.

23. $1.07(631) - 349 = \$326.17$

25. a. $y_n = 1.04y_{n-1} + 250,\ y_0 = 800$

b. $y_n = \frac{250}{1-1.04} + \left(800 - \frac{250}{1-1.04}\right)(1.04)^n$

$y_n = -6250 + 7050(1.04)^n$

c. $y_7 = -6250 + 7050(1.04)^7 \approx 3027.32$
about \$3027.32

27. $y_n = .978y_{n-1} - 10,\ y_0 = 1000$

Copyright © 2018 Pearson Education, Inc.

29. a.–b.

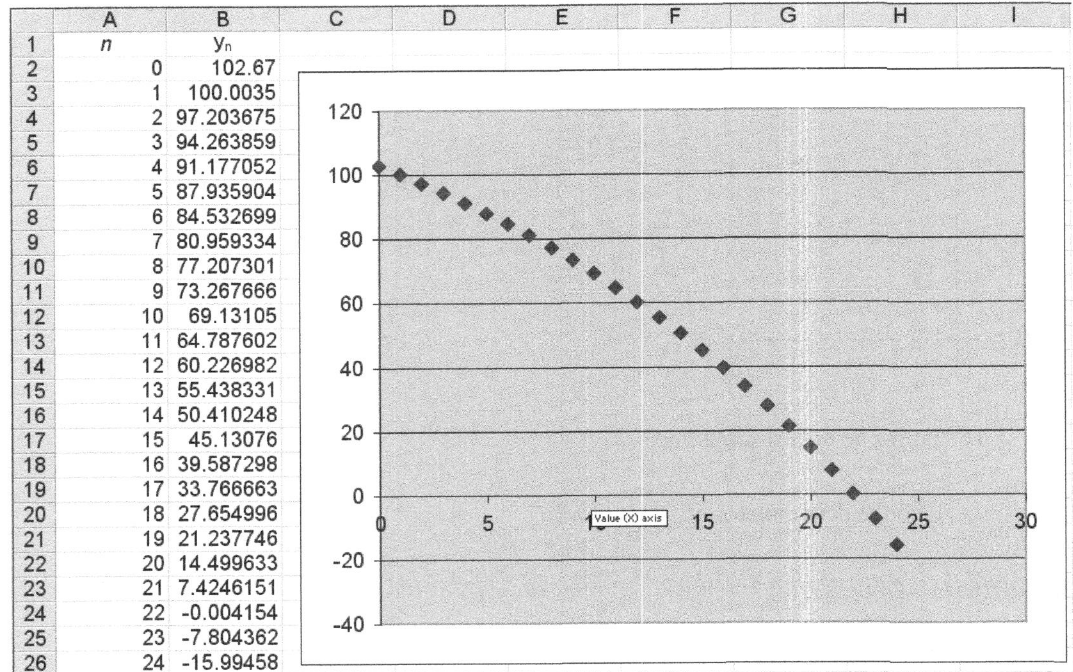

	A	B
1	*n*	y_n
2	0	102.67
3	1	100.0035
4	2	97.203675
5	3	94.263859
6	4	91.177052
7	5	87.935904
8	6	84.532699
9	7	80.959334
10	8	77.207301
11	9	73.267666
12	10	69.13105
13	11	64.787602
14	12	60.226982
15	13	55.438331
16	14	50.410248
17	15	45.13076
18	16	39.587298
19	17	33.766663
20	18	27.654996
21	19	21.237746
22	20	14.499633
23	21	7.4246151
24	22	-0.004154
25	23	-7.804362
26	24	-15.99458

c. $y_7 = 80.96$; $y_n \approx 0$ for $n = 22$

31. a.–b.

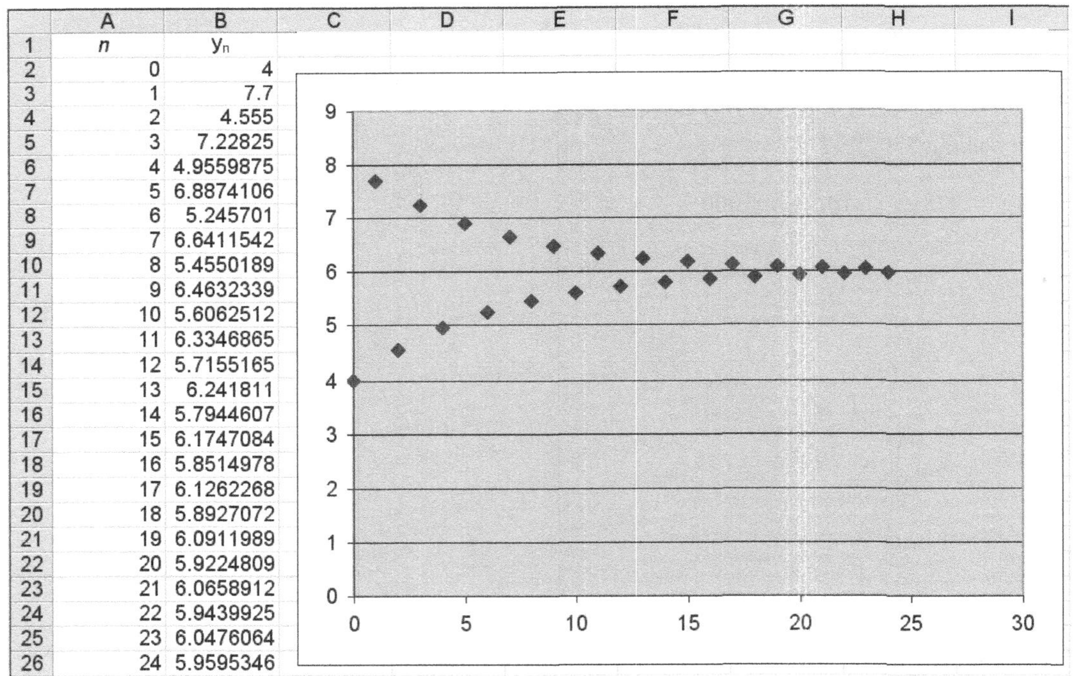

	A	B
1	*n*	y_n
2	0	4
3	1	7.7
4	2	4.555
5	3	7.22825
6	4	4.9559875
7	5	6.8874106
8	6	5.245701
9	7	6.6411542
10	8	5.4550189
11	9	6.4632339
12	10	5.6062512
13	11	6.3346865
14	12	5.7155165
15	13	6.241811
16	14	5.7944607
17	15	6.1747084
18	16	5.8514978
19	17	6.1262268
20	18	5.8927072
21	19	6.0911989
22	20	5.9224809
23	21	6.0658912
24	22	5.9439925
25	23	6.0476064
26	24	5.9595346

c. $y_{19} = 6.09120$; $|6 - y_n| < .2$ for $n \geq 15$

c. $y_4 = 5.8672$; $y_n < 4$ for $n \geq 9$

Copyright © 2018 Pearson Education, Inc.

33. a.–b.

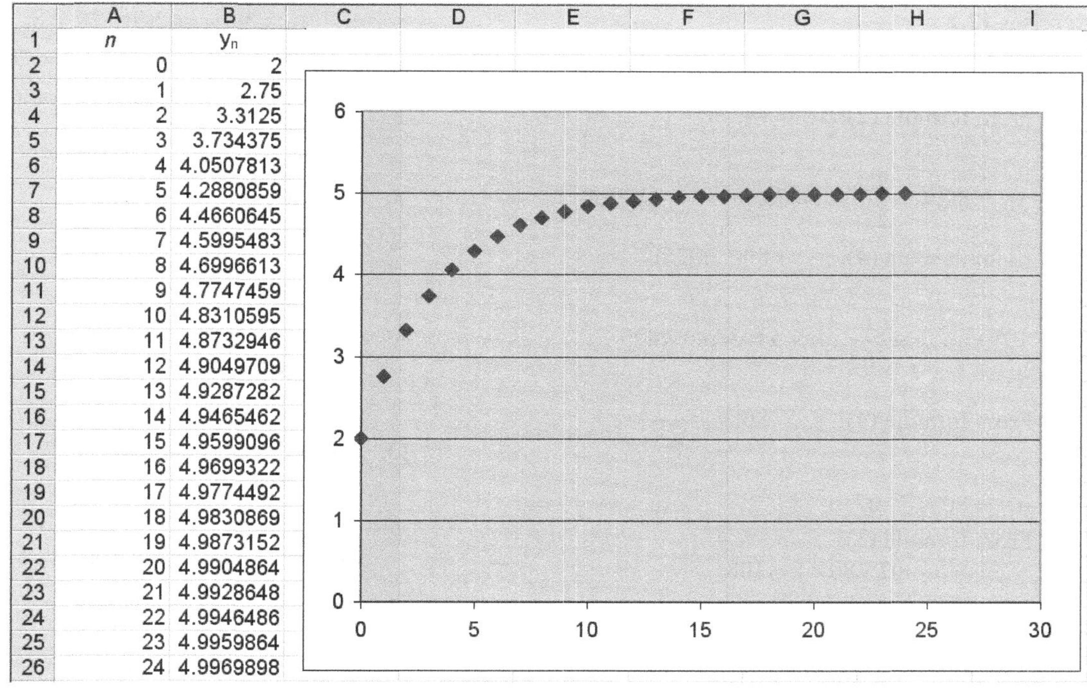

	A	B
1	*n*	y_n
2	0	2
3	1	2.75
4	2	3.3125
5	3	3.734375
6	4	4.0507813
7	5	4.2880859
8	6	4.4660645
9	7	4.5995483
10	8	4.6996613
11	9	4.7747459
12	10	4.8310595
13	11	4.8732946
14	12	4.9049709
15	13	4.9287282
16	14	4.9465462
17	15	4.9599096
18	16	4.9699322
19	17	4.9774492
20	18	4.9830869
21	19	4.9873152
22	20	4.9904864
23	21	4.9928648
24	22	4.9946486
25	23	4.9959864
26	24	4.9969898

c. $y_9 = 4.77475; |5 - y_n| < .5$ for $n \geq 7$

35. a.–b.

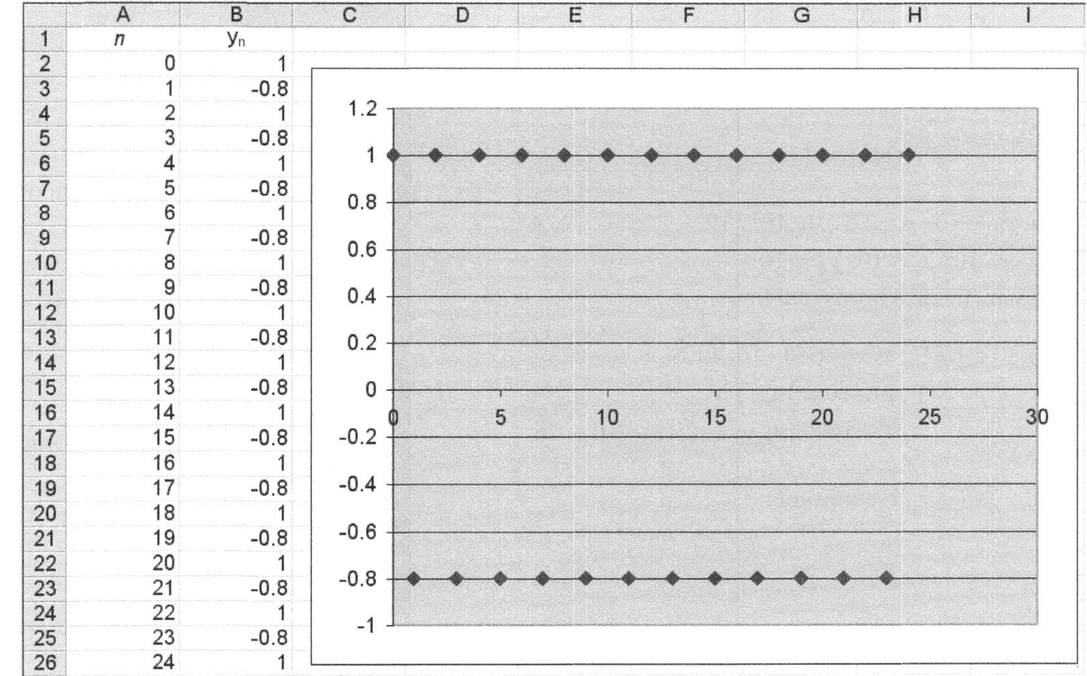

	A	B
1	*n*	y_n
2	0	1
3	1	-0.8
4	2	1
5	3	-0.8
6	4	1
7	5	-0.8
8	6	1
9	7	-0.8
10	8	1
11	9	-0.8
12	10	1
13	11	-0.8
14	12	1
15	13	-0.8
16	14	1
17	15	-0.8
18	16	1
19	17	-0.8
20	18	1
21	19	-0.8
22	20	1
23	21	-0.8
24	22	1
25	23	-0.8
26	24	1

c. The terms alternate between 1 and –.8.

Exercises 12.2

1. $b = 5,\ y_0 = 1$
From formula (2), $y_n = 1 + 5n$.

3. $y_0 = 80,\ i = \frac{.09}{12} = .0075,\ n = 5 \times 12 = 60$
From formula (4), $y_n = 80(1.0075)^{60}$

5. $y_0 = 80,\ i = \frac{1}{365},\ n = 5 \times 365 = 1825$
From formula (4), $y_n = 80\left(1 + \frac{1}{365}\right)^{1825}$.

7. $y_0 = 80,\ i = .07,\ n = 5$
From formula (3),
$y_n = 80 + .07 \times 80 \times 5 = 108$

9. $y_0 = 1$

a. $i = .40,\ n = 1,\ y_n = 1(1.40)^1 = \$1.40$

b. $i = \frac{.40}{2} = .20,\ n = 2 \times 1 = 2,$
$y_n = 1(1.20)^2 = \$1.44$

c. $i = \frac{.40}{4} = .10,\ n = 4 \times 1 = 4,$
$y_n = 1(1.10)^4 \approx \$1.46$

11. **a.** $y_0 = 10;\ y_1 = 2(10) - 10 = 10;$
$y_2 = 10;\ y_3 = 10;\ y_4 = 10$
points lie on a horizontal line

b. $y_0 = 11;\ y_1 = 2(11) - 10 = 12;$
$y_2 = 2(12) - 10 = 14;$
$y_3 = 2(14) - 10 = 18;\ y_4 = 2(18) - 10 = 26;$

points curve upward

c. $y_0 = 9;\ y_1 = 2(9) - 10 = 8;$
$y_2 = 2(8) - 10 = 6;\ \ y_3 = 2(6) - 10 = 2;$
$y_4 = 2(2) - 10 = -6;$ points curve downward

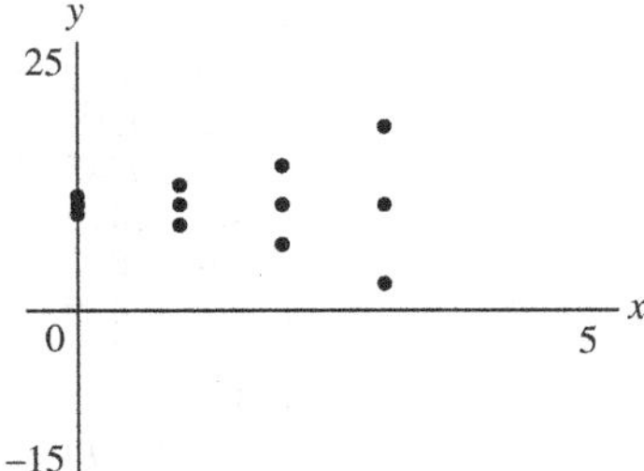

13. $a = .4,\ b = 3;\ y_n = \frac{3}{1 - .4} + \left(7 - \frac{3}{1 - .4}\right)(.4)^n$
$= 5 + 2(.4)^n;$
as n gets large, y_n approaches 5.

15. $a = -5,\ b = 0;$
$y_n = \frac{0}{1+5} + \left(2 - \frac{0}{1+5}\right)(-5)^n = 2(-5)^n;$
as n gets large, y_n gets arbitrarily large, alternating between being positive and negative.

17. **a.** $a = 1 + \frac{.066}{12} = 1.0055,\ b = -1600;$
$y_n = 1.0055y_{n-1} - 1600,\ \ y_0 = 250{,}525$

b. $n = 30 \times 12 = 360;\ y_{360} = 0$

19. **a.** $y_n = 1000(1.03)^n;\ y_0 = 1000$

b. $y_{20} = 1000(1.03)^{20}$
$y_{20} = \$1806.11$

21. $a = 1,\ b = \frac{50{,}000}{25} = 2000;$
$y_n = y_{n-1} - 2000,\ y_0 = 50{,}000;$
$y_n = 50{,}000 - 2000n$

 Copyright © 2018 Pearson Education, Inc.

23. $y_0 = 1000,\ i = \dfrac{.06}{4} = .015;\ \ y_n = 1.015y_{n-1}$

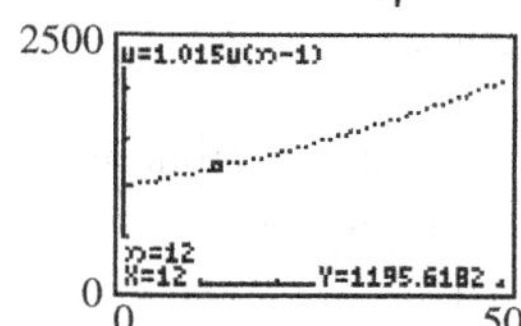

For $n = 3 \times 4 = 12,\ y_{12} = \$1195.62;$

$y_n = 1659$ for $n = 34$, or $8\frac{1}{2}$ years;

$y_n \geq 2000$ for $n \geq 47$, so it will double in 47 quarters.

25. $y_0 = 200,\ i = .045;\ y_n = y_{n-1} + 9$

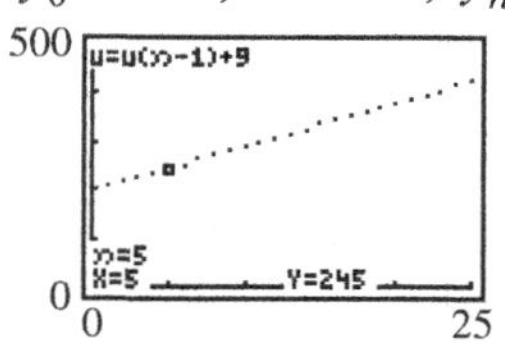

For $n = 5,\ y_5 = \$245;$

$y_n \geq 284$ for $n \geq 10$ years;

$y_n \geq 400$ for $n \geq 23$ years

27. 100 u=.85u(n-1)+9

n=100
X=100 Y=59.999995

0 0 100

y_n approaches 60.

Exercises 12.3

1. (a), (b), (d), (f), (h)

3. (b), (d), (e), (f)

5. (b), (d), (e), (f)

7. (a), (c), (h), and possibly (g)

9. Possible answer:

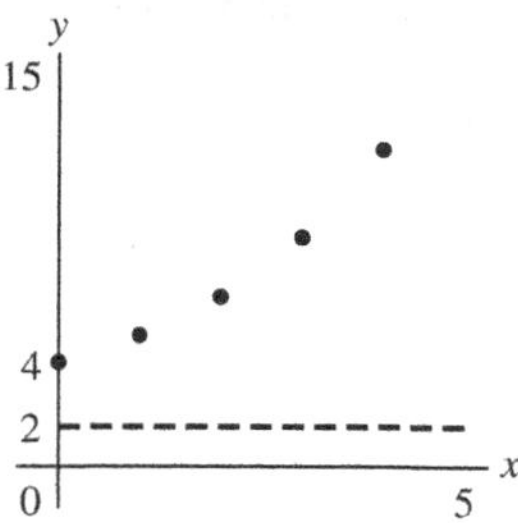

11. Possible answer:

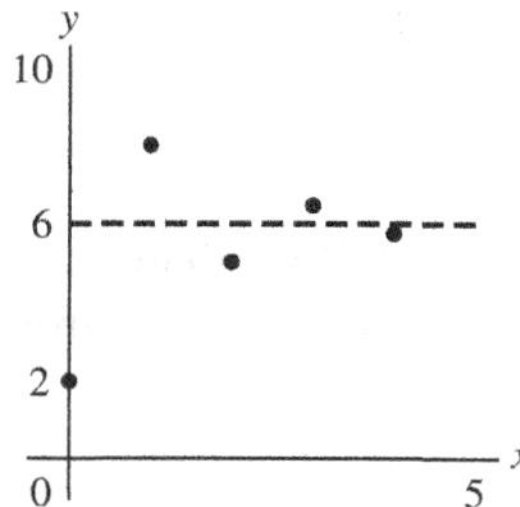

13. Possible answer:

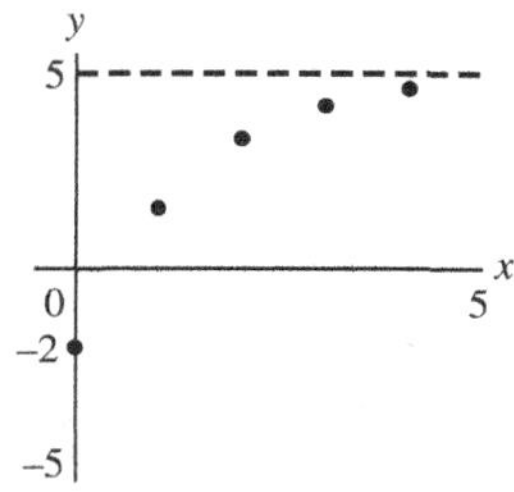

15. Draw $y = \dfrac{b}{1-a} = -2$ as a dashed line.

$a = 3 > 0$, so the graph is monotonic.

$|a| = 3 > 1$, so the graph is repelled from $y = -2$.

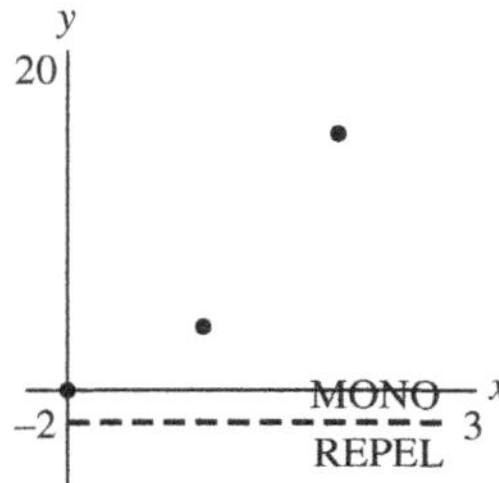

Copyright © 2018 Pearson Education, Inc.

17. Draw $y = \frac{b}{1-a} = 6$ as a dashed line.

$y_0 = 6$, so the graph is constant.

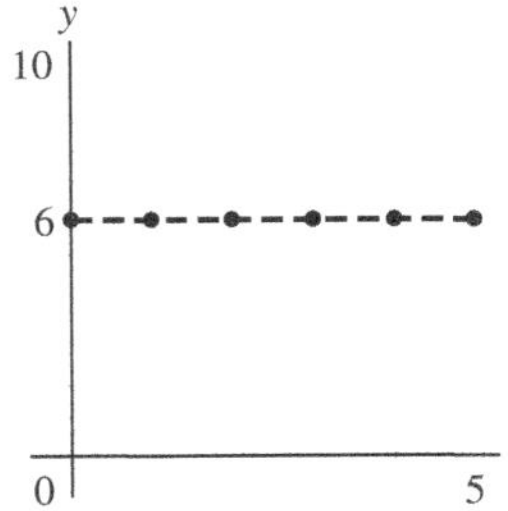

19. Draw $y = \frac{b}{1-a} = 4$ as a dashed line.

$a = -2 < 0$, so the graph is oscillating.

$|a| = 2 > 1$, so the graph is repelled from $y = 4$.

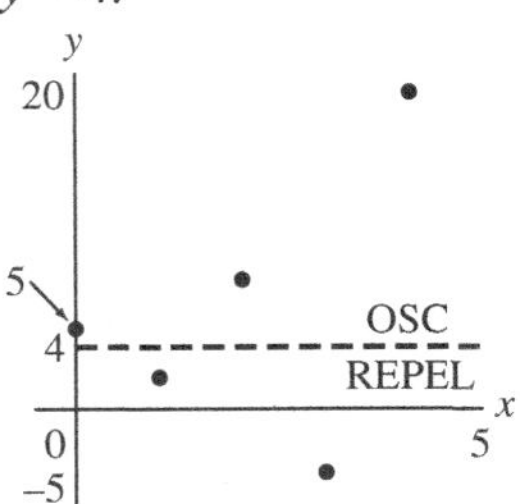

21. Draw $y = \frac{b}{1-a} = 10,000$ as a dashed line.

$a = .7 > 0$, so the graph is monotonic.

$|a| = .7 < 1$, so the graph is attracted to $y = 10,000$.

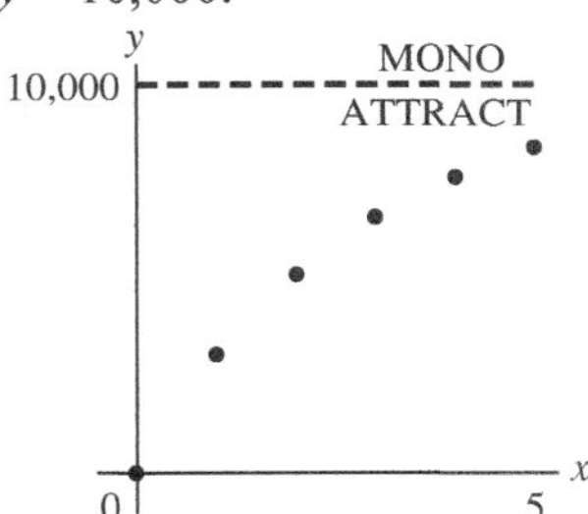

23. Draw $y = \frac{b}{1-a} = 1$ as a dashed line.

$a = -.6 < 0$, so the graph is oscillating.

$|a| = .6 < 1$, so the graph is attracted to $y = 1$.

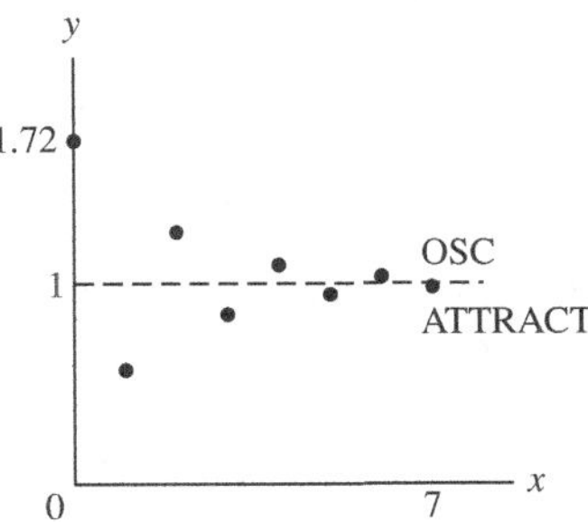

25. $a = 1 + i = 1.0035$, $b = -455$

The loan, y_0, must be less than $\frac{b}{1-a} = \$130,000$.

27. $a = 1 + i = 1.015$, $b = -120$

a. $y_n = 1.015y_{n-1} - 120$

b. The deposit, y_0, must be at least $\frac{b}{1-a} = \$8000$.

29. $a = 1 + i = 1.028$, $b = -1400$

$y_n = 1.028y_{n-1} - 1400$

The loan, y_0, must be less than $\frac{b}{1-a} = \$50,000$. If it is greater than or equal to \$50,000, the loan will never be paid off.

For Exercises 31–36, choose y_0 as the beginning value; choose $a > 0$ or < 0 depending on whether the graph is monotonic or oscillating, also $|a| > 1$ or < 1 depending on whether the graph is unbounded or approaches a value; and if it approaches or is repelled from a value, choose b so that $\frac{b}{1-a}$ equals that value.

31. Possible answer: $y_n = .5y_{n-1} + 4$, $y_0 = 1$

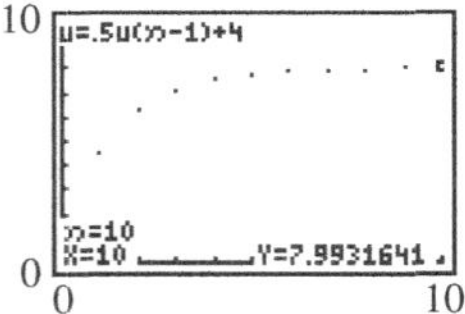

 Copyright © 2018 Pearson Education, Inc.

33. Possible answer: $y_n = 2y_{n-1},\ y_0 = 1$

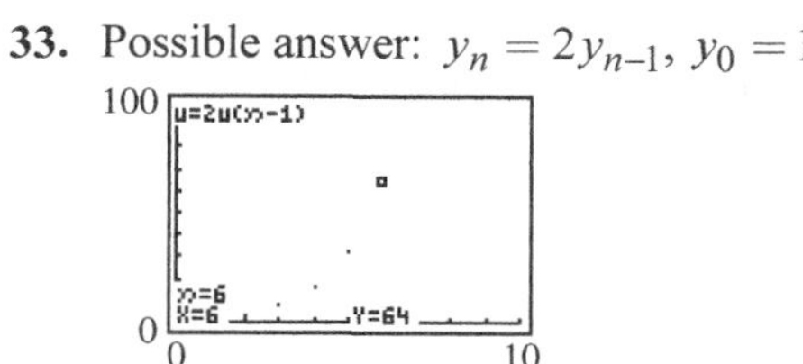

35. Possible answer: $y_n = -2y_{n-1},\ y_0 = 5$

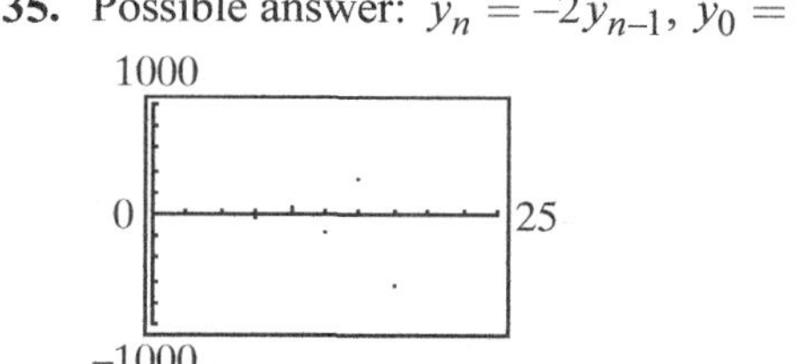

Exercises 12.4

1. $a = 1 + i = 1.0075,\ b = -261.50;$

$y_n = 1.0075y_{n-1} - 261.50,\quad y_0 = 32,500$

3. $a = 1 + i = 1.015,\ b = 200;$

$y_n = 1.015y_{n-1} + 200,\quad y_0 = 4000$

5. $a = 1 + i = 1.01,\ b = -660,\ \dfrac{b}{1-a} = 66,000$

$$y_{120} = 0 = 66,000 + (y_0 - 66,000)(1.01)^{120}$$

$$y_0 = \frac{-66,000}{(1.01)^{120}} + 66,000 \approx \$46,002.34$$

7. $a = 1 + i = 1.06,\ b = 300,$

$\dfrac{b}{1-a} = -5000,\ y_0 = 0$

$$y_{20} = -5000 + [0 - (-5000)](1.06)^{20} \approx \$11,035.68$$

9. $a = 1 + i = 1.005,\quad y_0 = 0$

$$y_{144} = 6000 = \frac{b}{-.005} + \left(0 - \frac{b}{-.005}\right)(1.005)^{144} = \frac{(1.005)^{144} - 1}{.005}b$$

$$b = \frac{6000 \times .005}{(1.005)^{144} - 1} \approx \$28.55$$

11. $a = 1 + i = 1.005,\quad y_0 = 4000$

$$y_{36} = 0 = \frac{b}{-.005} + \left(4000 - \frac{b}{-.005}\right)(1.005)^{36} = 4000(1.005)^{36} + 200b[(1.005)^{36} - 1]$$

$$b = \frac{-4000(1.005)^{36}}{200[(1.005)^{36} - 1]} \approx -121.69$$

\$121.69

13. $a = 1 + i = 1.08,\ b = -4$
Use nMin = 0, u(n) = 1.08u(n–1)–4,
u(nMin) = {45}
u(10) ≈ \$39.205 million
u(n) ≤ 0 for n = 30 years

15. $a = 1 + i = 1.00225,\ b = 100$
Use nMin = 0, u(n) = 1.00225u(n – 1) + 100,
u(nMin) = {0}
u(5) ≈ \$502.26
u(10) ≈ \$1010.19
u(15) ≈ \$1523.86
u(n) ≥ \$1938.97 for n = 19
u(n) > \$3000 for n = 30

Exercises 12.5

1. $a = 1 + .03 - .01 = 1.02,\ b = 0,\ \dfrac{b}{1-a} = 0$

$y_n = 1.02y_{n-1},\ y_0 = 100$ million

$a > 0$: monotonic; $|a| > 1$: repelled from $y = 0$

y
200 million
MONO
REPEL
0
x

Copyright © 2018 Pearson Education, Inc.

3. $a = 1 - .25 = .75,\ b = 0,\ \dfrac{b}{1-a} = 0$

$y_n = .75y_{n-1}$

$a > 0$: monotonic; $|a| < 1$: attracted to $y = 0$

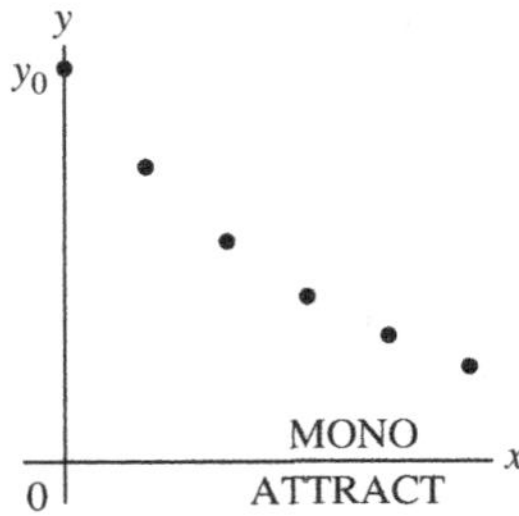

5. $y_n = y_{n-1} + .08(100 - y_{n-1}) = .92y_{n-1} + 8,$
$y_0 = 0$

$a = .92,\ b = 8,\ \dfrac{b}{1-a} = 100$

$a > 0$: monotonic; $|a| < 1$: attracted to
$y = 100$

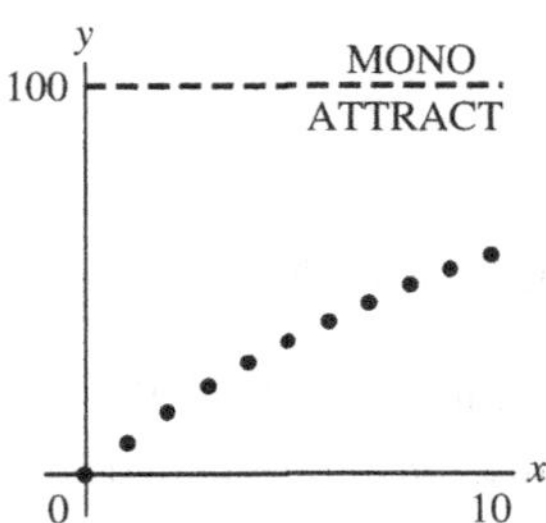

7. $y_n = y_{n-1} + .30(12 - y_{n-1})$
$= .7y_{n-1} + 3.6,\ y_0 = 0$

$a = .7,\ b = 3.6,\ \dfrac{b}{1-a} = 12$

$a > 0$: monotonic; $|a| < 1$: attracted to $y = 12$

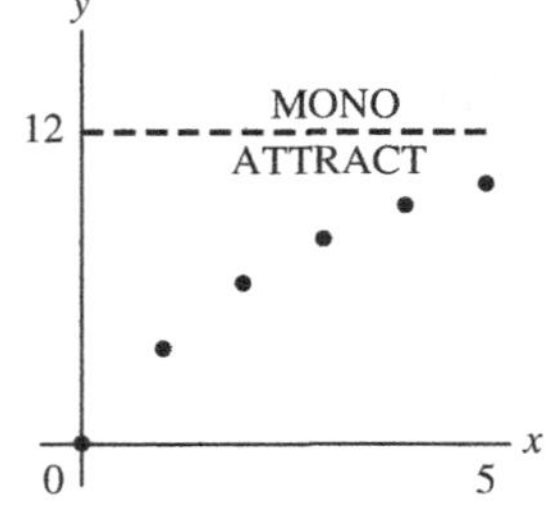

9. $a = 1 + i = 1.05,\ b = -1000,\ \dfrac{b}{1-a} = 20{,}000$

$y_n = 1.05y_{n-1} - 1000,\ y_0 = 30{,}000$

$a > 0$: monotonic; $|a| > 1$: repelled from
$y = 20{,}000$

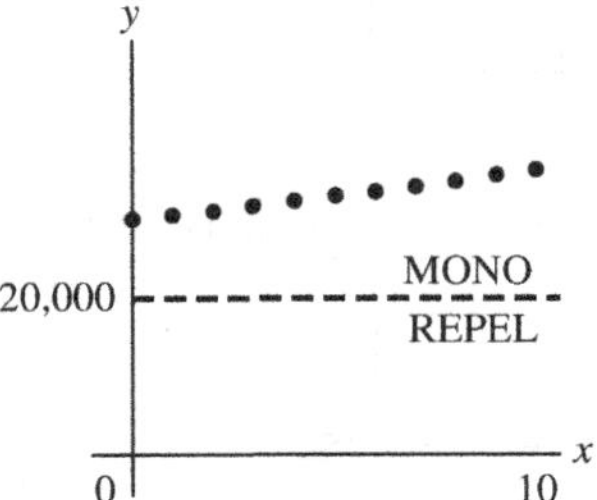

11. $y_n = y_{n-1} + .20(70 - y_{n-1}) = .8y_{n-1} + 14,$
$y_0 = 40$

$a = .8,\ b = 14,\ \dfrac{b}{1-a} = 70$

$a > 0$: monotonic; $|a| < 1$: attracted to
$y = 70$

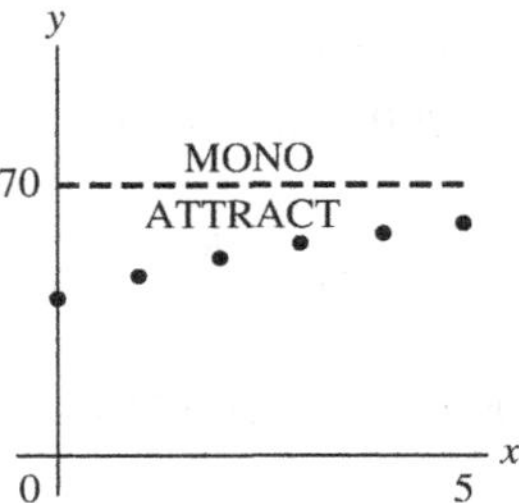

13. $p_n = 20 - .1q_n = 20 - .1(5p_{n-1} - 10)$
$= -.5p_{n-1} + 21,\ p_0 = 4.54$

$a = -.5,\ b = 21,\ \dfrac{b}{1-a} = 14$

$a < 0$: oscillating, $|a| < 1$: attracted to
$y = 14$

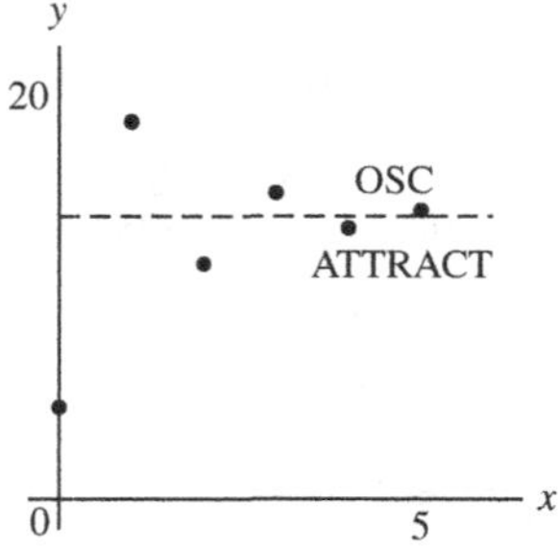

 Copyright © 2018 Pearson Education, Inc.

15. After 2 hours, .5 is left. After four hours, (.5)(.5) = .25 is left. Answer is (b).

17. Use nMin = 0,
u(n) = (1+.035–.02)u(n–1)–.0003,
u(nMin) = {5}
u(5) ≈ 5.38 million
u(n) > 6 for n = 13, or in the year 2022
u(n) ≥ 10 for n = 47, or in the year 2056

19. $y_n = y_{n-1} - .13y_{n-1}$

$y_n = (.87)y_{n-1},\ y_0 = 130$

$a = .87,\ b = 0,\ \frac{b}{1-a} = 0$

$y_n = 130(.87)^n$

After 5 hours, $y_n = 130(0.87)^5 = 64.79$ (This problem can be solved through trial and error or by the use of logarithms)

After 24 hours, $y_n = 130(0.87)^{24} = 4.6$ milligrams will remain.

Chapter 12 Review Exercises

1. a. $y_1 = -3(1)+8 = 5;\ y_2 = -3(5)+8 = -7;$
$y_3 = -3(-7)+8 = 29$

b. $a = -3,\ b = 8$

$$y_n = \frac{b}{1-a} + \left(y_0 - \frac{b}{1-a}\right)a^n$$
$$= 2 + (1-2)(-3)^n$$
$$= 2 - (-3)^n$$

c. $2 - (-3)^4 = -79$

2. a. $y_1 = 10 - \frac{3}{2} = \frac{17}{2};\ y_2 = \frac{17}{2} - \frac{3}{2} = 7;$

$y_3 = 7 - \frac{3}{2} = \frac{11}{2}$

b. $a = 1,\ b = -\frac{3}{2}$

$y_n = 10 - \frac{3}{2}n$

c. $10 - \frac{3}{2}(6) = 1$

3. $a = 1 + i = 1 + \frac{.0305}{4} = 1.007625,\ b = 0,\ \frac{b}{1-a} = 0$

$y_n = y_0(1.007625)^n$

Solve.

$2474 = y_0(1.007625)^{28}$

$y_0 = \frac{2474}{1.2370} \approx \$2000$

4. $a = 1 + i = 1.001,\ b = 0,\ \frac{b}{1-a} = 0,\ \ y_0 = 1000$

$y_{104} = 0 + (1000 - 0)(1.001)^{104}$
$= \$1109.54$

5. Draw $y = \frac{b}{1-a} = 6$ as a dashed line.

$a = -\frac{1}{3} < 0$, so the graph is oscillating.

$|a| = \frac{1}{3} < 1$, so the graph is attracted to $y = 6$.

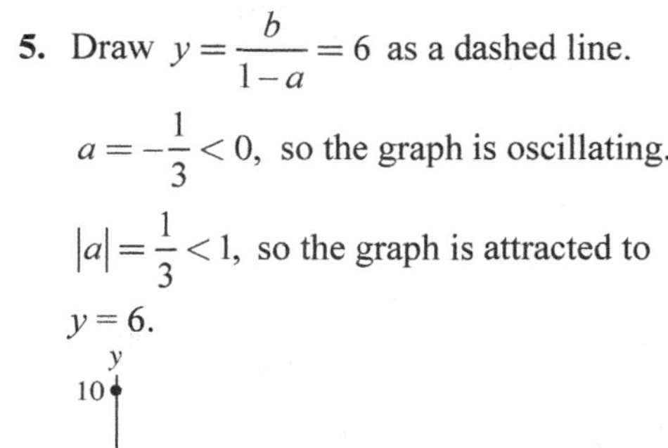

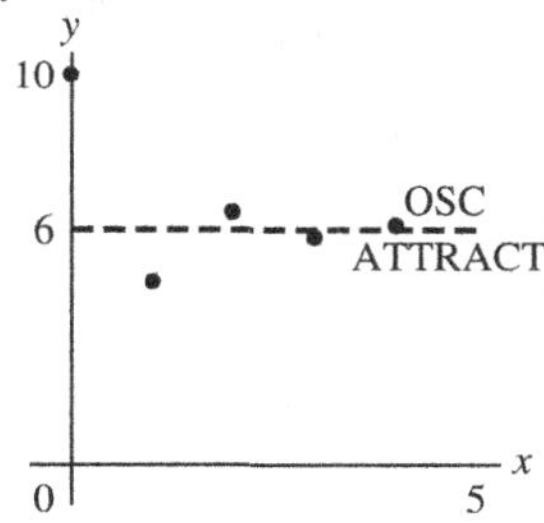

6. Draw $y = \frac{b}{1-a} = 4$ as a dashed line.

$a = 1.5 > 0$, so the graph is monotonic.

$|a| = 1.5 > 1$, so the graph is repelled from $y = 4$.

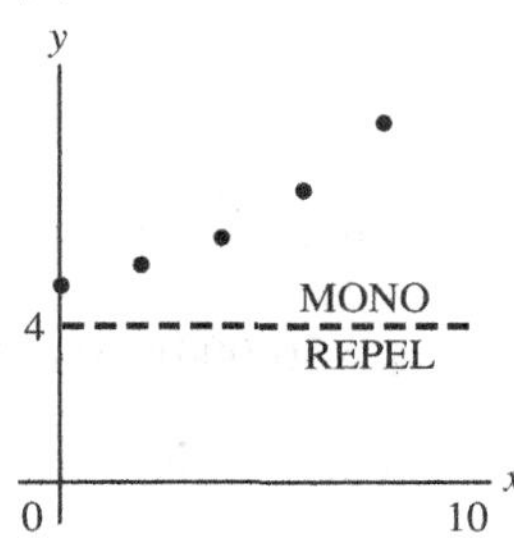

Copyright © 2018 Pearson Education, Inc.

7. a. $y_n = 1.03y_{n-1} - 600,\ y_0 = 120{,}000$

b. $\dfrac{b}{1-a} = 20{,}000;$

$y_{20} = 20{,}000 + (120{,}000 - 20{,}000)(1.03)^{20}$
$\approx 200{,}611$

8. a. $y_n = 1.0035y_{n-1} - 978.03,\ y_0 = 200{,}000$

b. $\dfrac{b}{1-a} = 279{,}437.14;$

$y_{84} = 279{,}437.14$
$+ (200{,}000 - 279{,}437.14)(1.0035)^{84}$
$\approx \$172{,}904.36$

9. $a = 1 + i = 1.0005,\ y_0 = 0,\ 21 \times 52 = 1092$

$y_{1092} = 36{,}000$

$= \dfrac{b}{-.0005} + \left(0 - \dfrac{b}{-.0005}\right)(1.0005)^{1092}$

$= 2000[(1.0005)^{1092} - 1]b$

$b = \dfrac{36{,}000}{2000[(1.0005)^{1092} - 1]} \approx \$24.79$

10. $a = 1 + i = 1.005,\ y_0 = 33{,}100$

$y_{240} = 0$

$= \dfrac{b}{-.005} + \left(33{,}100 - \dfrac{b}{-.005}\right)(1.005)^{240}$

$= 33{,}100(1.005)^{240} + 200[(1.005)^{240} - 1]b$

$b = \dfrac{-33{,}100(1.005)^{240}}{200[(1.005)^{240} - 1]} \approx -237.14$

\$237.14

11. $a = 1 + i = 1.08,\ b = -2400,\ \dfrac{b}{1-a} = \$30{,}000$

$y_{18} = 0 = 30{,}000 + (y_0 - 30{,}000)(1.08)^{18}$

$y_0 = \dfrac{-30{,}000}{(1.08)^{18}} + 30{,}000 \approx \$22{,}492.53$

12. $y_n = 1.00175y_{n-1} + 50,\ y_0 = 0$

$a = 1.00175,\ b = 50,\ \dfrac{b}{1-a} = -28{,}571.43$

$y_{48} = -28{,}571.43$
$+ [0 - (-28{,}571.43)](1.00175)^{48}$
$\approx \$2501.40$

13. $y_n = y_{n-1} + .10(1{,}000{,}000 - y_{n-1})$
$= .9y_{n-1} + 100{,}000,\ y_0 = 0$

$a = .9,\ b = 100{,}000,\ \dfrac{b}{1-a} = 1{,}000{,}000$

$a > 0$: monotonic, $|a| < 1$: attracted to $y = 1{,}000{,}000$

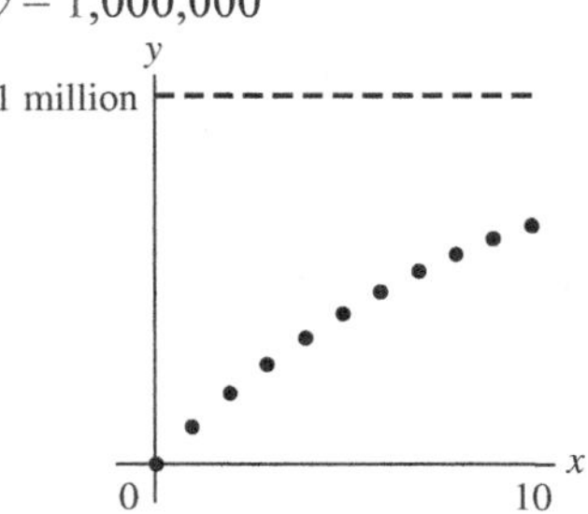

14. $y_n = y_{n-1} - .08y_{n-1} = .92y_{n-1},\ y_0 = 100$

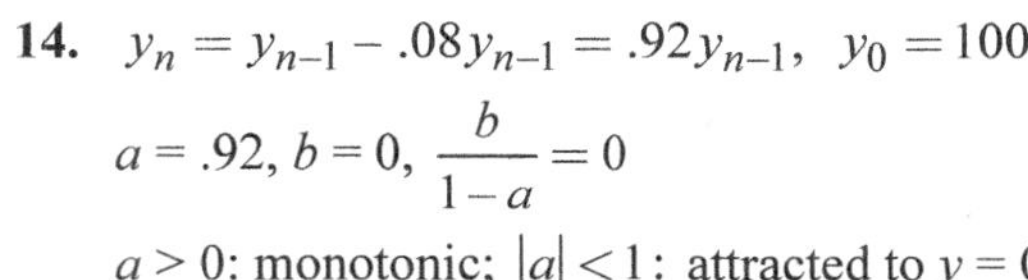

$a = .92,\ b = 0,\ \dfrac{b}{1-a} = 0$

$a > 0$: monotonic; $|a| < 1$: attracted to $y = 0$

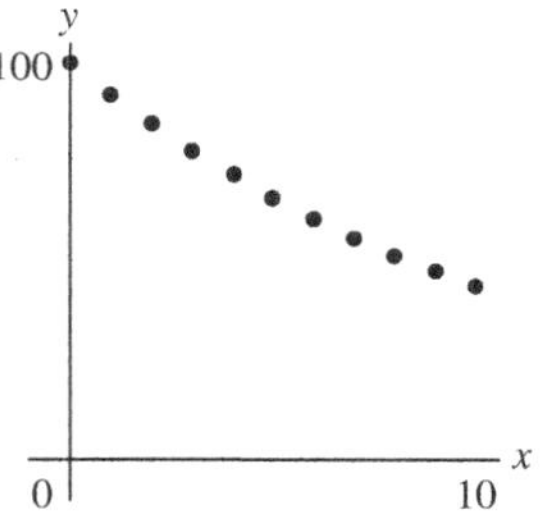

 Copyright © 2018 Pearson Education, Inc.